Adaptive Signal Processing for Radar

For a complete listing of the *Artech House Radar Library*,
turn to the back of this book . . .

Adaptive Signal Processing for Radar

Ramon Nitzberg

Artech House
Boston • London

Library of Congress Cataloging-in-Publication Data

Nitzberg, Ramon
Adaptive signal processing for radar / Ramon Nitzberg.
p. cm.
Includes bibliographical references and index.
ISBN 0-89006-586-1
1. Radar. 2. Adaptive signal processing. I. Title.
TK6575.N58 1991 91-29576
621.3848–dc20 CIP

685 Canton Street
Norwood, MA 02062

International Standard Book Number: 0-89006-586-1
Library of Congress Catalog Card Number: 91-29576

10 9 8 7 6 5 4 3 2 1

Contents

Preface

Several books on the general subject of adaptive processing have been published recently. This book differs from the others because, as indicated by the title, it emphasizes the implementation and analysis of various types of adaptive processing as applied to the radar field. Though many of the techniques discussed are applicable to fields such as communications, active and passive sonar, and IR, the special features of the radar can cause a significant difference in implementation and analysis of adaptive processing. For example, the doppler processing used for radar differs totally from that used in active sonar or communications. The adaptive antenna implementations also differ because the characteristics of the signal of interest are known exactly for radar but not for communications. That is, for radar the desired signal is, ideally, a reflection from a point reflector. Thus the desired signal is a delayed-scaled replica of the transmitted waveform and is known exactly. However, for the communications or passive sensor problem (sonar, IR, *et cetera*) the desired signal is not known exactly. Additionally, with radars it is often possible to obtain a time window such that an observed data block contains only interference. Therefore the pertinent characteristics of the interference can be estimated without being contaminated by the desired signal. It is often impossible for passive sensors or communications to isolate the desired signal and interference; thus, their estimation procedures must be different from those of radar estimation.

The material in this book emphasizes three techniques of adaptive processing. One technique is adaptive antennas, forms of which are commonly used in current radar implementations. Specifically, the Multiple Side Lobe Canceler is used to adapt the radar's receive-antenna pattern so to generate an approximate null response in the direction of interferers that are in the sidelobe region of the main antenna. This technique has been successful, but its performance tends to fall short of the theoretical upper-bound because of the limitations of practical implementation. However, as components improve so does the performance, and the more exotic techniques with superior performance discussed later in the book become of greater interest.

Another adaptive processing technique is adaptive thresholding (also known as Constant False Alarm Rate (CFAR) processing). This technique is also used in current radars. The basic CFAR technique is designed for an interference environment that is constant in power level over the test- and reference-cell window. Often the actual environment does not fit this design environment. Because the performance of CFAR techniques is environmentally sensitive, several other CFAR versions are synthesized and the performance analyzed for various interference environments.

The third adaptive technique discussed is adaptive doppler processing. This technique uses adaptive weights rather than fixed weights such as Taylor or Tchebychev to implement a doppler filter bank. The improved performance that is achieved using adaptive doppler processing is analyzed and the results presented in several graphs. For current component technology this technique is expensive to implement and is only used experimentally. The material can be used to perform a cost-performance trade to determine when the increased implementation cost is warranted.

In general, this book examines the detailed theory of the current techniques of adaptive radar processing, the effects that limit their performance (such as channel-to-channel mismatch), and the more exotic techniques and their performance. As this field is still evolving, and books of necessity are years out of date upon publication, some importand material is not included. Please accept my apologies.

Acknowledgments

Most of the material in this book was developed while I was employed by the General Electric Company in Syracuse, New York, and I appreciate the company's cooperation in allowing its publication. The many discussions I had with my co-workers were also helpful in developing this material. In particular, discussions with the late Sid Applebaum were of fundamental importance in clarifying many of the concepts and in suggesting analytical methods. Though not explicitly referenced, the ongoing work of many other researchers helped in generating the material in this book.

Ramon Nitzberg, Ph.D.

Chapter 1
Radar Fundamentals

1.1 INTRODUCTION

A radar or an active sonar operates by transmitting energy into the environment and obtaining information concerning the location of objects by detecting reflections of the transmitted energy. This process is identical to the familiar one of shouting and hearing an acoustic echo from a nearby object.

A passive sonar or IR sensor does not transmit energy; it detects energy generated by objects in the environment. Active sensors determine the range of the object by measuring the elapsed time between transmission and the echo's arrival, and combining the time delay with the speed of propagation. Range cannot be directly measured when using a passive sensor. Active or passive sensors can use directive antennas to determine the angular location of an object. Ideally, a directive antenna has zero response everywhere except over a small angular region. Thus, if energy is detected by a directive antenna, the angular location of the object is known to be within the nonzero response region. Actual antennas cannot have zero response over regions, but can be designed to have high response over a particular angular region and relatively low response over all other angles. Therefore, there is the potential for error when determining the true angle of the object.

1.2 MAXIMUM DETECTION RANGE

Detection of an object requires that the received, reflected energy be larger than the background energy. For any sensor system, the energy received from a reflector decreases as the object range increases. Thus, the difficulty of reflector detection increases with range. Because of the many similarities of active and passive sensors numerous aspects of nonadaptive and adaptive processing are applicable to any active or passive sensor. However, the emphasis here is on radar application.

There are many contributors to the background energy. The most fundamental are the random energy fluctuations that result from the random motion of electrons.

The presence of other reflectors (such as ground, weather, birds, insects), the reflected energy of which tends to obscure that of the reflector of interest, can also contribute to background energy. The reflector of interest is usually denoted as a *target,* whereas the reflectors that are not of interest are denoted as *clutter*. The energy due to the random electron effects is denoted as *noise*.

To illustrate the form of dependence on sensor parameters, an approximation for the maximum detection range of a conceptual radar will be determined. This radar generates a single pulse of energy, transmits it via a directional transmitting antenna, receives reflected energy via a directional receiving antenna, and detects a target if the observed energy at any time delay is large compared to the background noise energy. A pulse transmission is used to resolve reflections from targets at different ranges. The directional characteristic of the transmitting antenna causes the energy transmitted in the direction of the target to be greater than the value of the energy generated. The ratio of the two energies is the transmitted antenna gain, G_t. Similarly, the directional characteristics of the receive antenna cause a received antenna gain, G_r. If the transmitting antenna is omnidirectional (i.e., radiates equally in all directions), the transmitted power density at a range R equals $P/4\pi R^2$, where P is the peak-pulse transmitted power. Because of the transmitted antenna gain, the transmitted density in the target direction is $PG_t/4\pi R^2$. The product PG_t is termed the *effective radiated power*. The transmitted pulse is a gated portion of a sinusoidal source. The sinusoid (or carrier) frequency, denoted as f_0 ranges between a few megahertz and tens of thousands of megahertz, depending on the particular application.

When the wave impinges on a target, energy is reflected back toward the radar. The reflection can be viewed as being caused by an equivalent pulse source and transmitting antenna in the target. The equivalent effective radiated power of the reflector equals the product of the power density impinging on the target and the target scattering cross section, σ. Thus, the target's effective radiated power, P_e, is

$$P_e = PG_t\,\sigma/4\pi R^2 \tag{1.1}$$

The reflected power density at range R equals $P_e/4\pi R^2$. The power induced in the receiving antenna, P_r, equals the product of the power density impinging on the antenna and the antenna's effective capture area, A_r. Thus,

$$P_r = PG_t\sigma A_r/(4\pi R^2)^2 \tag{1.2}$$

The effective area is approximately equal to the physical area of the antenna when the antenna is steered in the direction of the target.

As an example of the magnitude of various parameters, typical radar system values are a peak transmitted power of 1 MW, transmitted antenna gain equal to one thousand, target cross section of 1 m^2, an effective receive antenna area of 10 m^2 and a target range of 200 km. Using equation 1.2, the peak receive power is 3.96

$\times 10^{-14}$ W. The average power of the background noise depends upon the bandwidth, B, of the amplifiers used in the receiver subsystem. The average noise power (in watts) is

$$P_n = kT_sB \quad (1.3)$$

where B is in hertz, and kT_s is a constant equal to 4×10^{-21} and is the product of Boltzmann's constant, k, and the effective system noise temperature in kelvins, T_s. For the assumed transmitted waveform of a gated sinusoid, the bandwidth is approximately equal to the inverse of the transmitted pulse duration, τ. For a typical duration of 1 μs, the bandwidth is approximately 1 MHz. By using (1.3), the average noise power is 4×10^{-15}. In this example, the peak power of the target reflection is larger by approximately a factor of 10 than the average noise power, and thus the probability of target detection is high. If the target range increases by a factor of 2, the reflected power decreases by a factor of 16, and thus the probability of target detection is very low.

A common method of increasing the maximum detection range is to transmit and process a train of N pulses rather than the single pulse transmission previously considered. For optimum signal processing, the performance improvement is equivalent to increasing the peak transmitted power by a factor of N.

1.3 CONCEPTUAL SYSTEM IMPLEMENTATION

A conceptual implementation of a basic radar system is shown in Figure 1.1. A single antenna is shown, as most radar systems use the same antenna for both transmitting and receiving. A duplexer is used to switch between the transmitting and receiving modes. The preferred transmitting technique, shown in Figure 1.1, generates a waveform at a low power level (usually in milliwatts) and amplifies the generated waveform to the megawatt level for transmission. Different transmitting waveforms can be used, depending on performance requirements and environmental conditions.

The received waveform is at a relatively low power level and requires amplification before processing can be completed. The waveform transmission is often at a relatively high carrier frequency and translation to a lower frequency is usually necessary to simplify the implementation of the amplifiers. The frequency translation is performed by a mixer. A common method of performing the frequency translation for both transmission and reception is illustrated in Figure 1.1, where typical center frequency values are indicated at several points.

The waveform generator creates a video waveform. This is mixed to the 3060 MHz transmitted carrier frequency. This carrier frequency is created by mixing the *stable local oscillator* (STALO) frequency of 3000 MHz with the *coherent local oscillator* (COHO) frequency of 60 MHz. The STALO is used to translate the received waveform to an *intermediate frequency* (IF) that is convenient for amplifier

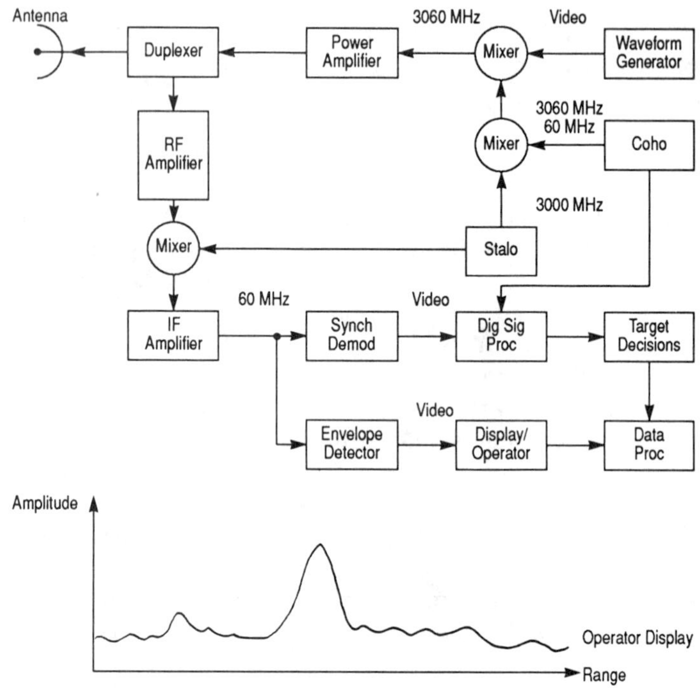

Figure 1.1 Conceptual implementation of basic radar system.

implementation. After amplification, the signal is presented to an operator or a signal processor. The operator or signal processor enters information to the data processor concerning the presence or absence of targets. A typical operator display of amplitude of the signal envelope *versus* range is shown at the bottom Figure 1.1. Large amplitudes, relative to the random noise fluctuation, indicate the presence of targets.

1.4 SYSTEM PERFORMANCE EQUATIONS

The antenna gain characteristic is sketched in Figure 1.2. The gain is shown *versus* a single angular dimension. In general, the gain varies similarly with both azimuth and elevation angles, although the two beamwidths are not necessarily identical. The region between the angles where the gain is reduced to one-half of its peak value is defined as the *beamwidth*. The beamwidth of an antenna of dimension D is approximately equal to λ/D, where λ is the wavelength and wavelength is related to carrier frequency f_0 by $\lambda = c/f_0$. Thus, for antenna width D_a and height D_e, the azimuth beamwidth, θ_a, and elevation beamwidth, θ_e, are approximated by

$$\begin{aligned} \theta_a &\approx \lambda/D_a \\ \theta_e &\approx \lambda/D_e \end{aligned} \tag{1.4}$$

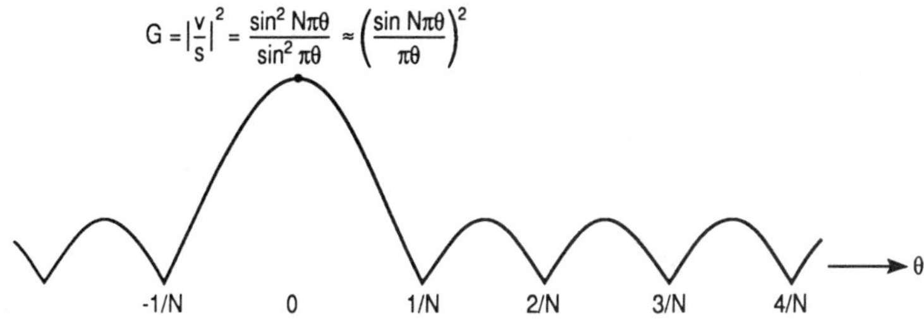

$\theta = 0 \rightarrow \text{PEAK}, G_0 = N^2$

$\theta = 1/N; 2/N; 3/N; \ldots, G = 0$

$\theta = \frac{1.5}{N}, G = \left(\frac{1}{\pi\, 1.5/N}\right), G_{SL} = G/G_0 = \left(\frac{1}{1.5\,\pi}\right)^2 \rightarrow -13.5 \text{ dB}$

- PEAK SIDELOBE = -13.2dB
- HIGH RESPONSE SIDELOBES: SIDELOBE JAMMING

Figure 1.2 Conceptual receiving antenna gain characteristic.

When the transmitted or received energy is concentrated in the solid angle region θ_a, θ_e due to antenna directivity, rather than over 4π steradians for an omnidirectional antenna, the antenna gain relative to the unit-gain omnidirection antenna is approximated by

$$G_r \approx \frac{4\pi}{\theta_a \theta_e} \tag{1.5}$$

Combining (1.4) and (1.5) gives

$$G_r = \frac{4\pi A}{\lambda^2} \tag{1.6}$$

where the effective antenna area has been approximated by the physical area, A. The operational system concept is that the antenna is pointed at a specified azimuth-elevation region and the waveform of N pulses is transmitted. After allowing sufficient time for energy to be received from a maximum range target (the dwell time), the antenna is steered to a different azimuth-elevation region. This action is continued until all angles have been interrogated, and then the cycle is repeated. As we shall discuss later, the mechanism of antenna steering depends on the particular antenna.

By combining equations (1.3), (1.4), and (1.6) for an N-pulse train, we obtain the maximum detection range as

$$R^4 = P_t\tau N G_t G_r \lambda^2 \sigma / \gamma k T_s (4\pi)^3 \qquad (1.7)$$

where γ is the ratio of target power to average noise power required for detection and is termed the *signal-to-noise ratio* (SNR).

From (1.7) we can see that detection range may be increased by increasing peak transmitted power or antenna gain, *et cetera*. However, there are performance constraints due to practical component limits and interrelationships among the parameters of (1.7). First, note that target detection is only an initial goal of the radar system. The ultimate goal is initiating and maintaining target tracks. For this purpose, an important system parameter is the data processor update period, T_u. This time strongly affects the tracking error and is especially important in establishing the maximum target maneuver rate that can be tracked. Short update periods improve target maneuver tracking, but degrade target detectability. The latter effect is because the update period is related to other terms in the maximum range equation.

The update period depends on the *pulse repetition interval* (PRI). Usually, the time between pulses is constant. The PRI is denoted as T and the dwell time, T_D, equals NT. As discussed in section 1.7, to avoid range ambiguity, the value of T depends on the maximum target range and the two-way propagation velocity. Thus,

$$T = 2R/c \qquad (1.8)$$

where c is the velocity of propagation, which equals the velocity of light.

The update period equals the number of resolvable angle-dwell positions multiplied by the dwell period. Assuming 4π-steradian coverage,

$$T_u = 4\pi NT / \theta_a \theta_e \qquad (1.9)$$

Combining (1.5), (1.7), and (1.9) gives

$$R^4 = P_a A T_u \sigma / \gamma k T_s (4\pi)^2 \qquad (1.10)$$

where the average transmitter power, P_a, is

$$P_a = P\tau / T \qquad (1.11)$$

An interesting interpretation of (1.10) is that the system's maximum detection range (in a thermal-noise environment) directly depends on the radar's parameters by the average transmitter power only (rather than the peak power, and does not depend on the pulse duration or the number of pulses), antenna area, and data processor update period. The antenna area is denoted as A, without subscript, to emphasize that the equation is applicable to a monostatic radar, which, by definition, uses the same antenna for transmitting and receiving.

The maximum radar range calculated by using (1.10) assumes that the required angular coverage equals 4π steradians. Often, the required angular coverage is less. Decreasing the required angular coverage increases detection range because the dwell time can be increased. Thus, for azimuth and elevation surveillance angles of θ_{sa} and θ_{se} respectively (θ_{se} is assumed small), from (1.10):

$$R^4 = P_a A T_u \theta_{sa} \theta_{se} \sigma / \gamma k T_s (4\pi)^3$$

For a specified angular surveillance region, a single transmit/receive beam has relatively short dwell time. The preceding derivation is based on a single transmitting-receiving beam architecture. A common alternative architecture uses a more complicated and expensive multiple-receiving-beam architecture, incorporating parallel receiving and processing channels. The advantage of the alternative architecture is that the maximum detection range is increased in a clutter environment and does not affect the detection range in a noise environment.

The multiple-receiving-beam architecture uses "beam-spoiling" techniques to generate a wide transmitting antenna beamwidth and multiple-offset-receiving antenna patterns to cover the surveillance angle. As the surveillance angle per dwell is increased, this allows increased dwell time, which, in turn, improves the performance of doppler processing techniques. However, the increased dwell time is precisely compensated by the decrease in transmitted antenna gain. Thus, the detection range in thermal-noise environments is unaffected. The decrease in transmitted antenna gain in clutter environments is usually unimportant, as it does not affect the signal-to-clutter ratio.

1.5 PROBABILITY OF TARGET DETECTION

Equation (1.10) describes the maximum target detection range as a dichotomy. For any range less than R, the target is detected; for any range greater than R (regardless of how small the differential), the target is not detected. This oversimplification is only approximately correct. An improved statement of reality depends on the concepts of *probability of detection* and *probability of false alarm.*

A target-present decision is made at each resolvable range position by comparing the envelope detector's voltage to a threshold value. When the envelope exceeds the threshold, a target-present decision is made; when the envelope is less than the threshold, a target-absent decision is made. The target present or absent decision is in error if no target is present when a target-present decision is made because the noise value happens to be large. Similarly, the decision is in error if a target is present when a target-absent decision is made because the sum of target plus noise happens to be small. The probability of the first type of error, a false alarm, is decreased by increasing the threshold. However, this, in turn, increases the probability of missing the target. In general, the probability of target detection depends on the probability of false alarm and the target SNR value. As discussed in more detail later, for many systems, the equation for the design (or allowable) probability of false alarm, P_{FD} is

$$p_{FD} = \exp(-T/p_n) \tag{1.12}$$

where T is the threshold value* and p_n is the noise power at the decisional device.

The target cross section varies considerably as a function of radar carrier frequency and relative angle. Thus, the cross section and the resulting SNR can be reasonably approximated as random variables. For aircraft targets, the probability of target detection, p_D, depends on the average SNR, γ_a, and is given by

$$p_D = \exp[-T/p_n(1 + \gamma_a)] \tag{1.13}$$

Combining (1.12) and (1.13) gives

$$p_D = p_{FD}^{1/(1 + \gamma_a)} \tag{1.14}$$

Figure 1.3 graphs the target detection probability *versus* SNR for false-alarm probabilities of 1×10^{-4}, 1×10^{-6}, and 1×10^{-8}. As developed below, these values are typically required to avoid overloading the data processor.

A surveillance radar needs to make a target present or absent decision once per resolvable volume element per radar scan. The number of resolvable range elements is approximately the ratio of the pulse repetition interval, T (proportional to the maximum target range), and the pulse duration. The example radar of Section 1.2 had a maximum range of 200 km. For a velocity of propagation equal to 3×10^8 m/s, the minimum allowable pulse repetition interval equals 1333 μs. Thus, the number of resolvable range cells equals 1333.

For a ground-based radar, the resolvable number of angle elements equals $2\pi/\theta_a\theta_e$. For a square 10 m^2 antenna and a carrier frequency of 3 GHz, from (1.4), each beamwidth equals 0.0316 radians (1.81°). The number of resolvable angle elements equals 6283 so that the number of resolvable volume elements per scan equals 8.4×10^6. Thus, if the false-alarm probability per decision equals 1×10^{-6}, the average number of false-track data points delivered to the data processor per scan equals 8.4, and this false alarm probability is about as high as can be tolerated. In

*Although the same symbol is used for threshold and pulse repetition interval, the meaning is always clear from the context.

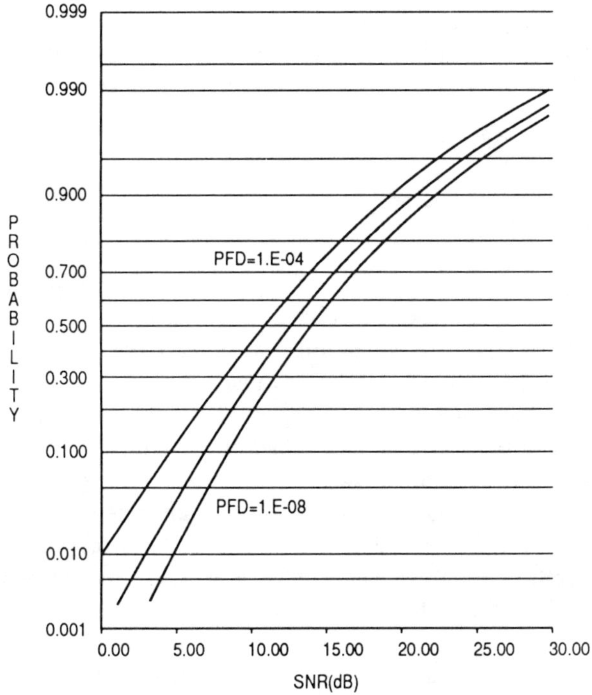

Figure 1.3 Detection probability *versus* SNR for several false-alarm probabilities.

addition, to avoid being inundated with false data, the data processor must be designed for eventually rejecting false-track data while maintaining track on true targets.

1.6 DATA PROCESSOR TRACKING CONSTRAINTS

The data processor uses the target data passed from the signal processor to establish target tracks. The new data is combined with prior data to estimate current target position and predict future position. The procedure is to compare the measured target position of a new data point with all predicted target positions. If a new data point is "close" to one of the predicted positions, the new measurement is combined with the history. The definition of "close" must take into account measurement errors due to noise effects as well as effects due to an unknown, newly initiated target maneuver. If the new data point is not close to any prior track, it is identified as a potential new track. Signal processor false alarms may give rise to an ever-increasing number of false tracks. Thus, provision for dropping tracks must be made if new data do not correlate. Also, a track usually is not initiated on the basis of a detection from a single scan. However, the need to prevent false tracks from inundating the data processor also decreases the probability of correctly initiating a track on a true target. As a consequence, analysis of track-initiation logic indicates that track performance is marginal unless the per-scan probability of detection is greater than approximately 0.7.

1.7 RANGE AMBIGUITY

As indicated previously, a benefit of transmission of an N-pulse train is an increase of the detectability of far-range, small cross section targets. However, this advantage is coupled with the disadvantage of target range ambiguity. As a specific example, let us assume that the pulse repetition interval is 2000 μs. This equals the time required for two-way transmission to a range of 300 km. If a target is detected at a measurement range of 200 km, in theory, the target may actually be at a range of $100 + 300 = 400$ km, or, generally, 100 km + any integral multiple of 300 km. The reason for this ambiguity is depicted in the sketches of Figure 1.4.

The top sketch of Figure 1.4 indicates successive pulse transmissions at intervals of T. The next sketch indicates the returns from two targets at ranges such that the time delays of reception are τ_{R1} and τ_{R2}, respectively. Note that the time delays are such that τ_{R1} is less than T whereas τ_{R2} is greater than T. The transmitted waveform is shown by the dashes. The next sketch indicates a conceptual implementation of the signal processor used to combine the returns from a four-pulse transmission waveform. A significant feature of the signal processor is the tapped delay line with tap spacing equal to the pulse repetition interval and the summation. This allows the returns from each range sample due to each of the transmitted pulses to be added. The gain values at each tap are discussed in Section 1.8. The last sketch shows the waveform at the summer output *versus* time. During the first period of duration T, the output consists of only the return from the near-range target due to transmitted pulse 1. The output during the second period of duration T is the return from the far-range target due to transmitted pulse 1 and the sum of the returns of the close-range target from transmitted pulses 1 and 2. Note that, referenced to the time of the second pulse transmission, the return due to the far-range target appears at an apparently closer range than the actual near-range target. The summed returns in the third and fourth intervals are similar, and they are decreased for the near-range target, but continue to increase for the far range target. The summations continue to decrease for successive time intervals. The output from the summer during the fourth period are used to make target present or absent decisions. The intrinsic assumption is that the returns due to transmission of the four pulses have been properly summed. We can see that this allows a far-range target to be ambiguously detected at the wrong range.

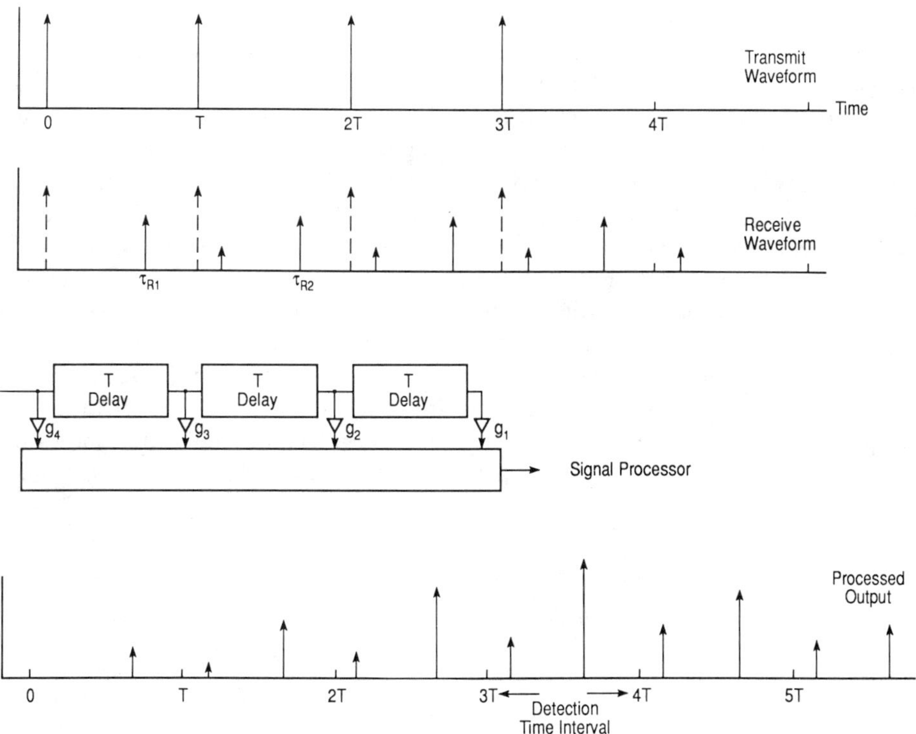

Figure 1.4 Repetitive pulse train waveforms and range ambiguities.

1.8 DETECTION OF TARGETS IN CLUTTER BY DOPPLER PROCESSING

The degree of detectability of targets in a noise environment is controlled by the ratio of the reflected signal energy and the thermal noise power. Often, however, even when the target SNR value is very large, the presence of reflectors of interest is difficult to detect because the reflected energy is less than that from other reflectors that are not of interest. Reflectors of interest are commonly termed *targets* and the others are called *clutter*. Note, however, that any reflector can be a target for some applications and clutter for others.

To be able to detect a "target" in the presence of "clutter," there needs to be some discriminant between the waveforms reflected from the two classes of reflectors. A useful discriminant is that often targets have high velocity, and therefore high doppler shifts, whereas clutter has zero or low velocity, and therefore zero or low doppler shifts.

The fundamental concept of doppler processing is that the range to a stationary reflector is the same for two consecutive pulse transmissions, whereas the range changes for a nonstationary reflector. The sketch of Figure 1.5(a) shows a transmitter waveform that can be used to exploit this differentiating characteristic of targets and clutter. The waveform is a gated sinusoid. The envelope of the gating waveform is a repetitive rectangle. The frequency of the sinusoid and the *pulse repetition frequency* (PRF) of the gating waveform are harmonically related so that the pulsed sinusoid has the same starting phase at each pulse. The differentiating phenomenon is depicted in the sketch of Figure 1.5(b), where the effects of noise are neglected. The uppermost sketch shows the return from a reflector due to transmission of a single pulse. The leftmost portion of each sketch indicates the transmission time of each return. The middle sketch shows the return from a stationary reflection due to transmission of the second pulse. This return is identical to the first. If these two returns are subtracted, the circuit output equals zero. The bottom sketch shows the return from a nonstationary reflector that is moving toward the radar. As the range

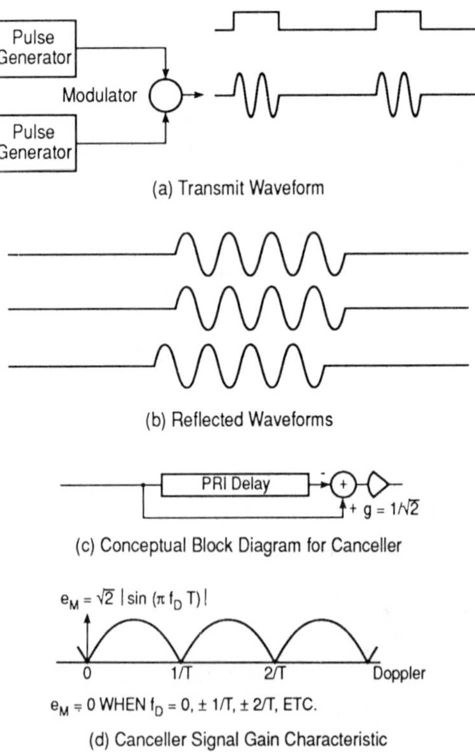

Figure 1.5 Doppler processing using a delay line canceller.

decreases, the reception delay decreases relative to that of the first pulse. If the returns are subtracted, the circuit output generally is nonzero.

Careful examination of the sketch indicates potential problems with the cancellation of nonstationary reflectors. We can see that almost exact cancellation of a nonstationary reflector will occur when the target velocity is such that the target range decreases by the equivalent of the time duration of one cycle of the transmitted pulsed sinusoid. Also note that the envelope of the second pulse return of a nonstationary reflector never occurs at the exact relative time as that of the first pulse. The envelope mismatch is usually negligible, but is significant for very high-speed targets.

The remainder of this section discusses delay-line canceller circuits and nonadaptive doppler filter banks. Later, we shall discuss adaptive doppler filtering.

1.8.1 Delay Line Canceller

The equation of a single pulse of a sinusoid is

$$e(t) = p(t)\cos\omega_0 t \tag{1.15}$$

where $p(t)$ is the rectangular function equal to unity for a time duration equal to the pulsewidth τ and zero elsewhere. The equation for an infinite pulse train is

$$e(t) = \sum_n p(t - nT)\cos\omega_0 t \tag{1.16}$$

with values of n ranging from minus to plus infinity.

The reflected waveform from a target is delayed by a time τ_R, which depends on the radial range of the target. For a target at initial range of R_0 and radial velocity v, the time delay is

$$\tau_R = 2(R_0 - vt)/c \tag{1.17}$$

The factor of 2 is due to the two-way transit to and from the target. The reflected sinusoid then is

$$\begin{aligned} \cos\omega_0(t - \tau_R) &= \cos(\omega_0 t + \omega_D t - \theta) \\ \omega_D &= 2\omega_0 v/c \\ \theta &= 2\omega_0 R_0/c \end{aligned} \tag{1.18}$$

Thus, the frequency is shifted by an amount that depends on the target's radial velocity and the phase is shifted by an amount that depends on the initial range.

There are two types of signal processors used to exploit the doppler characteristics of targets and clutter. The delay-line canceller is relatively simple to implement, but its performance is inferior to that of the doppler filter bank. A conceptual block diagram of a delay-line canceller is shown in Figure 1.5(c). It consists of a delay line with a delay equal to the pulse repetition interval, T, a subtracter, and an amplifier of gain $1/\sqrt{2}$. The amplifier is included to simplify the interpretation of the performance equations, but is not actually used in an implementation. In addition, the circuit is discussed as if its input is at the radial carrier frequency ω_0. In practice, the implementation is not done in this manner. Thus, this design and all others that are shown here and in later chapters are intended as conceptual, rather than actual, implementations.

A stationary target causes identical sinusoidal pulses at the two inputs to the subtracter so that, neglecting noise, the subtracter output is zero. The magnitude of the subtracter output for other target velocities, and therefore target detectability, depends on the target velocity. The amplifier gain is inserted so that the circuit noise gain is unity. Because target detectability depends on the signal-to-noise factor, changes in signal amplitude are easily related to SNR, and thus detectability. From (1.18), the canceller circuit output due to a reflector is

$$\begin{aligned} e_o &= \{\cos[(\omega_0 + \omega_D)t - \theta] - \cos[(\omega_0 + \omega_D)(t - T)] - \theta\}/\sqrt{2} \\ &= -2^{1/2} \sin[\pi(f_0 + f_D)T] \sin[(\omega_0 + \omega_D)t - \theta - (\omega_0 + \omega_D)T/2] \end{aligned} \quad (1.19)$$

Because the carrier frequency and PRF are harmonically related, $f_0 T$ equals an integer, and the output is zero when the target doppler frequency is zero. More generally, the magnitude of the sinusoidal output of the canceller circuit is

$$e_M = 2^{1/2}|\sin \pi f_D T| \quad (1.20)$$

and the magntiude is sketched *versus* doppler frequency in Figure 1.5(d). The peak value of $e_M = \sqrt{2}$. Thus, the energy increases by a factor of 2, and the output SNR value is twice the input SNR value. For many other frequencies, the value of e_M is less than unity, and thus the SNR value at the output is less than that at the input. Also, note that the zero output response also occurs at nonzero doppler frequencies, and therefore nonzero target velocities. These velocities are defined as *blind speeds*. They occur because the repetitive pulsed waveform has a doppler ambiguity at multiples of $1/T$, as it has a range ambiguity at range time delays of multiples of T. As a numerical example of the ambiguity, for a radar operating at a carrier frequency of 3 GHz and a PRF of 1000 Hz, the blind speeds are at multiples of 50 m/s (97.2 knots). Increasing the PRF to 10 kHz (or decreasing the carrier frequency to 300 MHz) causes the blind speeds to be located at multiples of 500 m/s (972 knots).

1.8.2 Doppler Filter Bank

The preceding delay-line canceller processes two pulses and achieves an SNR gain of 2 at a single doppler frequency of the unambiguous doppler range. For one-half of this range, there is no SNR gain, as the ratio of output to input SNR equals 1 or less. Improved performance is obtained by use of a doppler filter bank. This circuit sums N pulses with N generally ranging between 6 and 32. As for the delay-line canceller, this circuit has low response to clutter but has high response to targets at nonblind speed doppler frequencies. The ratio of output to input SNR is approximately N.

The conceptual implementation of a single filter is shown in Figure 1.4 for the special case of a 4-pulse processor. In general, the processor delays the returns due to the first transmitted pulse by a time of $(N - 1)T$, delays the returns due to the second transmitted pulse by a time of $(N - 2)T$, *et cetera*. When all the gain values equal $1/\sqrt{N}$, the sinusoidal output during the period when returns due to all N pulses are available is

$$e_0 = \sum_{m=1}^{N} \cos[(\omega_0 + \omega_D)(t - mT) - \theta]/\sqrt{N} \tag{1.21}$$

The common gain value of $1/\sqrt{N}$ is used so that the power gain to thermal noise equals 1.

The analysis is simplified by the use of a complex-valued signal representation. Because

$$\exp(j\phi) = \cos\phi + j\sin\phi$$

(1.21) can be expressed as the real part of

$$e_0' = \sum_{m=1}^{N} \exp j[(\omega_0 + \omega_D)(t - mT) - \theta]/\sqrt{N} \tag{1.22}$$

The preceding is a geometric series so that the magnitude of the sinusoidal output is

$$e_M = |\sin\pi f_D NT/N^{1/2} \sin\pi f_D T| \tag{1.23}$$

A sketch of a filter response is shown in Figure 1.6. As for the delay-line canceller, the response is periodic with the periodicity equal to the PRF. The response is maximum at zero doppler frequency. The maximum voltage gain equals $\sqrt{N}$. The maximum energy gain equals N so that, in turn, the maximum SNR gain equals N. The

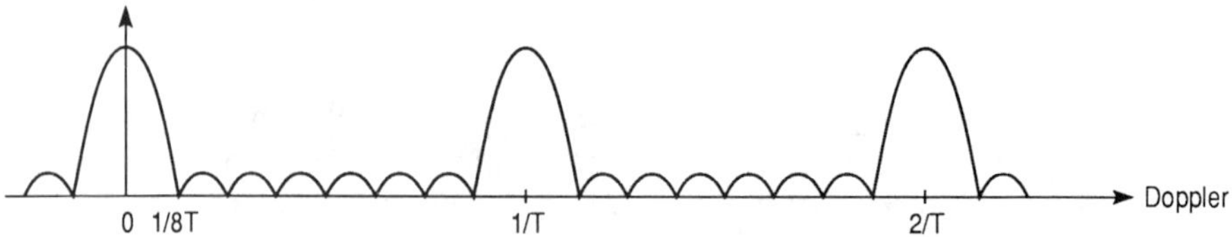

Figure 1.6 Eight pulse doppler filter response.

sketch is drawn for an eight-pulse processor and shows the first null of the response at a frequency of $1/8T$ and six sidelobe regions. In general, the first null is at $1/NT$. The filter energy gain is approximately equal to N over a frequency band of approximately $1/NT$. To achieve large gain over the total unambiguous doppler band, the use of a bank of filters is necessary with each filter tuned to a different doppler frequency. Usually, a bank of N filters is used to process N pulses.

A filter is tuned to a different doppler frequency by using delay-line tap gains that include phase shifts. For a steering phase shift at the mth tap equal to $m\phi_s$, (1.22) and (1.23) become

$$e_0' = \exp j[(\omega_0 + \omega_D)t - \theta] \sum \exp jm[\phi_s - (\omega_0 + \omega_D)T]/\sqrt{N}$$
$$e_M = |\sin\pi(f_D - f_s)NT/N^{1/2} \sin\pi(f_D - f_s)T|$$

where the steered frequency f_s is related to the phase shift as

$$\phi_s = 2\pi f_s T \tag{1.24}$$

Note that the delay-line canceller uses a single filter to cover the doppler band, whereas the bank uses N filters. The latter implementation is significantly more complex, but has an SNR gain of close to N, whereas the SNR gain of the canceller is usually much less. The average SNR gain of the canceller over the unambiguous doppler region equals 1.

1.9 NONCOHERENT SUMMING

Implementation of the doppler filter bank uses coherent integration. Thus, the summation of the returns from different pulse transmissions is made at a point prior to envelope detection. Therefore, the summation output depends on both the amplitudes and phases (or doppler frequency) of successive pulses. Coherent summation causes the output to be zero for certain target doppler frequencies. For some applications, clutter interference is not present, and the ability to reject the clutter response is not necessary. Then, the SNR loss of the delay-line canceller and the equipment complexity of the doppler filter bank can be avoided by summing after the envelope detector. This is termed *noncoherent integration,* as the output does not depend on phase (doppler frequency). However, as discussed later in detail, the doppler filter bank, in addition to its usefulness for clutter rejection, also is quite close to being the optimum processor for detection of targets in thermal-noise environments.

The conceptual block diagram of the noncoherent integrator is virtually identical to that of Figure 1.4. The only difference is that the circuit input is now obtained from a point after the envelope detector. Note that, unlike the doppler filter bank, only a single circuit is required because noncoherent integration is not sensitive to doppler frequency. The penalty for the equipment simplification is that the increase of SNR is less than that achieved by the doppler filter bank. The loss of noncoherent integration relative to the performance of the optimum coherent signal processor depends on the number of pulses processed. The loss in dB is approximately equal to

$$\begin{aligned} L_{\mathrm{dB}} &\approx 2 \log N \quad 1 \le N \le 10 \\ L_{\mathrm{dB}} &\approx 2 + 4 \log(N/10) \quad 10 \le N \le 1000 \end{aligned} \tag{1.25}$$

Chapter 2
Operational Environments

2.1 INTRODUCTION

A fundamental limitation to the ability of a sensor system to detect targets is thermal noise. Noise is always present. However, for many environmental conditions, the limit on detectability is not thermal noise, but other forms of masking waveforms. Often, signal processing specifically designed for such environments can improve target detectability. This chapter discusses the general characteristics of several interference environments. Later chapters will discuss the special signal processing techniques that are required.

2.2 THERMAL NOISE

Thermal noise is due to the random motion of atomic particles and, as the name implies, it is a function of temperature. The noise is present in components, such as resistors, and therefore amplifiers, mixers, *et cetera*. Noise is induced in the antenna due to random motion of electric charges in the atmosphere, radiation from the sun, distant galaxies, *et cetera*.

For our purposes, an important aspect of thermal noise is that its power density is constant over the radar band of the electromagnetic spectrum. The constancy leads to the nomenclature "white noise." The effective received noise power due to the combined effects of the antenna and receiver is directly proportional to bandwidth. Thus,

$$p_n = kT_sB \tag{2.1}$$

where k is Boltzmann's constant, B is the receiver bandwidth, and T_s is the effective temperature of the receiving system, which includes the antenna and amplifier noise. For a noiseless receiver connected to an antenna steered to an absorbing ground, T_s is nominally 290 K so that $kT_s = 4 \times 10^{-21}$, where B is measured in Hz. For practical receivers, the effective temperature is higher due to the receiver noise figure.

The receiver bandwidth is approximately equal to the transmitted signal bandwidth. This choice is obtained by a trade-off of noise power and peak target (signal) power at the output of the amplifier. For much smaller receiver bandwidths, the output noise power from the amplifier is considerably less, but the peak signal power is also much less as the amplifier transient response time is too long to respond fully to a short pulse, which is used to allow range resolution of multiple targets. The condition of receiver bandwidth being approximately equal to the transmitted signal bandwidth maximizes the output SNR. As discussed later, somewhat smaller bandwidths are often used to decrease the range sidelobes associated with pulse compression waveforms. However, the decreased bandwidth incurs an SNR loss relative to an optimum bandwidth matching condition. The general requirement to trade performance in one parameter to improve performance in another will be frequently seen in different contexts.

2.3 CLUTTER

A sensor can detect radiation (either reflected for active sensors or self-generated for passive sensors) from many different types of objects. The needs of the sensor user make some of these objects significant (targets), while other objects are annoyances (clutter, reverberation). For radars, the clutter reflections are often due to natural objects, such as the ground, clouds, and rain. Also, there is clutter due to artificial reflectors deliberately dispersed by an adversary to create echoes that obscure the reflections from targets. These dispersed reflectors are termed *chaff*.

Doppler-based signal processors are often used to discriminate between targets and clutter. Thus, the spectral characteristics of the clutter are important. Aircraft targets and ground clutter are relatively easy to distinguish, as the average velocity of ground clutter is zero and that of aircraft is usually different from zero. However, as indicated in the previous chapter, the blind speeds of doppler processors can cause difficulty in detecting aircraft at particular velocities.

The weather and chaff clutter tend to move with the prevailing winds so that their average velocity is not zero. Therefore, standard doppler-based signal processors often cannot easily distinguish between the targets and such clutter. Other discriminants, such as range extent or electromagnetic polarization characteristics, can sometimes be used.

2.4 RECEIVING ANTENNA CHARACTERISTICS

Artificial or man-made radiation (jamming or radio frequency interference, RFI), like thermal noise or clutter, can interfere with the detection of targets. The receiving antenna's angle-dependent gain characteristics have a major effect on the magnitude of the received interference.

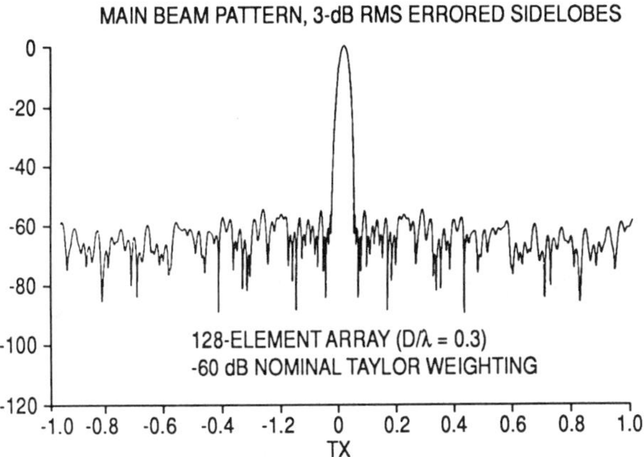

Figure 2.1 Typical antenna gain.

The idealization of the antenna pattern used in the previous chapter is that the antenna has constant nonzero gain over a small angular region and zero gain elsewhere. In practice, antennas have a gain *versus* angle characteristic like that sketched in Figure 2.1. If a constant power source is moved along a circle at constant range centered on the radar, the gain characteristic indicates the power received at the antenna as a function of the angle of the source. Note that if the only datum available to the radar is the antenna output voltage, there is a total angle ambiguity. Thus, the output voltage could be due to a source having a physical angle that is within the main-lobe response region, or the source, possibly of higher power or nearer range, could be at any other spatial angle. The inherent assumption of active and passive sensors is that the angular location of received radiation is that of the main-beam peak. The nonzero sidelobe pattern of the antenna causes two important effects pertinent to jamming vulnerability. One is that a sufficiently high-power jamming source can induce large voltages in the antenna. The other is that the angular location of the source will be misinterpreted.

2.5 JAMMING NOISE

In military radar applications, an adversary can decrease the maximum range capability of a radar by the use of noise generation. Jamming noise decreases the radar's range capability because the total received noise is the sum of the jamming noise and thermal noise. The increased noise power results in a decreased SNR, and thus a decreased maximum detection range.

The jamming noise generator can be carried on a penetrating aircraft or a nonpenetrating support aircraft as a *standoff jammer* (SOJ). A major difference between jamming noise and thermal noise is that jamming noise emanates from a single spatial angle, whereas thermal noise has essentially uniform density over spatial angles. Thus, modifying the antenna pattern to ensure a low response in the direction of the jammer (or jammers) will decrease the jamming effectiveness. The use of adaptive antennas, the response of which adapts to the environment, is an *electronic counter-countermeasure* (ECCM) technique.

2.6 FALSE TARGET JAMMING (PULSE JAMMING)

As discussed in Chapter 1, the false-alarm rate due to noise cannot be too large; otherwise, the false alarms will overwhelm the data processor. The effective false-alarm rate can be increased by enemy jamming that transmits replicas of the radar's pulses. The received pulses will appear to be targets. Often, the replicas are received through the radar's antenna sidelobes, and thus the angular location of the false target appears to be very different from that of the jamming source. Sidelobe false-target jamming can create many false targets, with each at apparently different angular locations. A large number of false targets will overwhelm the data processor and prevent the successful tracking of true targets. There are two radar ECCM techniques to reduce the effectiveness of false target jamming. One is to use a transmitted waveform that is difficult to repeat so that only a small number of false targets can be generated. The second technique is to use an auxiliary antenna. By comparing the energies received by the two antennas, the system can test whether the source is within or outside the antenna's main-lobe response region.

2.7 COMBINED ENVIRONMENTS

Note that a system should not be designed solely to cope with any single environmental condition, as the usual environment is the superposition of one or more conditions. Thus, as an example, techniques that can cope with a combined noise-jamming and clutter environment are of interest. Techniques that can successfully cope with a clutter environment, but which substantially degrade when presented with both environments, are of less interest.

2.8 QUANTITATIVE ENVIRONMENTAL CONSIDERATIONS

The example radar synthesized in Section 1.2 developed a per-pulse SNR of 10 at a range of 200 km for a 1 m^2 target. Thus, in a thermal-noise-only environment, the radar can easily detect the target at this and all smaller ranges. However, a combined environment of clutter plus jamming and thermal noise (the combined environment will be designated as *interference*) will often result in a per-pulse *signal-to-interference ratio* (SIR) much less than that required for detection. Signal processing techniques based on adaptive antennas and multiple-pulse adaptive doppler processing, can be used to increase the output SIR. This section develops an example of a hostile environment to indicate quantitatively the necessary improvement. Coupling the needed improvement with the attainable improvement indicates whether a particular technique is sufficient or whether an even better (and usually more expensive) technique should be implemented.

2.8.1 Ground Clutter

Ground clutter is due to the backscattering of radiation from the ground. If the ground were a perfect flat conducting sheet, all energy would be specularly reflected in the forward direction. Backscattering is thus a second-order effect caused by departures from specular reflection. As expected with second-order effects, the degree of backscattering varies greatly with terrain type, carrier frequency, angle of incidence, and other parameters. As discussed in greater detail below and in Chapter 10, the ground clutter cross section equals the product of the backscattering coefficient and the ground area resolved by the antenna beam and range resolution distance. Typically, the backscattering coefficient ranges between -10 and -60 dB and a representative value is approximately -24 dB (numerically equal to 4×10^{-3}).

The magnitude of the backscattered return depends on the magnitude of the incident wave. Therefore, most of the ground clutter return is due to the intersection of the antenna's main beam and the ground. The clutter return that competes with the target return is from an area of the ground within the same resolution cell. The cross-range dimension of the area approximately equals the target range multiplied by the antenna beamwidth. If the antenna pattern were the constant theoretical pattern over an angular region and zero elsewhere, this approximation would be exact. Because an actual antenna pattern is considerably different, an accurate computation is more difficult. The intent of this section is only to develop an indication of the severity of the environment, and so the approximation is sufficiently accurate. More exact analysis would change the results by a few dB. The dimension of the resolution cell along the ground in the radial-range direction is determined either by the elevation beamwidth or the radial-range resolution in the direction of the beam center. The smaller of the two effects dominates. For ground-based radars, the latter effect is usually the more pertinent, and the cell length is approximately equal to the range resolution. As an example, a 1 MHz bandwidth radar has a range resolution of 150 m. For an antenna beamwidth of 2° and a target range of 40 km, the cross-range dimension equals 1396 m. For a scattering coefficient of -24 dB, the ground clutter's cross section equals 836 m^2. This is approximately 29 dB greater than the 1 m^2 target. Thus, the target is not detectable when using a single-pulse radar processor. Multiple-pulse processors using delay-line cancellers or doppler filter banks will improve the target detectability.

The ability of the delay-line canceller to cancel the ground clutter depends on the clutter spectrum. If the clutter were a perfectly stationary reflector and the radar transmitter were a truly gated sinusoid, the clutter at the subtracter via the direct and delayed path would be identical and perfect cancellation thereby achieved. However, most clutter has slight internal motion due to the wind. The velocity spread about the average zero velocity limits the achievable cancellation. A typical velocity spread value equals 0.25 m/s. For a radar operating frequency of 3 GHz, this velocity causes a doppler spread of 5 Hz. The output clutter power from any filter can be calculated from

$$p_{co} = \int_{-\infty}^{\infty} C(f)|H(f)|^2 \, df \tag{2.2}$$

where $C(f)$ is the clutter's power density spectrum and $H(f)$ is the filter's transfer function. For simplicity, let us assume that the clutter spectrum is rectangular with the two-sided bandwidth equal to B. For the above example, $B = 5$ Hz. Thus,

$$C(f) = \frac{p_{ci}}{B} \mathrm{Rect}(2f/B) \tag{2.3}$$

where p_{ci} is the input clutter power. The magnitude of the filter transfer function is given by (1.20). Because the clutter spectrum is narrow compared to the PRF, the first term of a series is a reasonable approximation to $|H(f)|^2$, resulting in

$$\begin{aligned} p_{co} &= 2\frac{p_{ci}}{B} \int_{-B/2}^{B/2} (\pi f T)^2 \, df \\ \frac{p_{co}}{p_{ci}} &= \frac{\pi^2}{8} (BT)^2 \end{aligned} \tag{2.4}$$

For $B = 5$ Hz and $T = 1000$ μs, the clutter cancellation ratio equals -45.1 dB. As the input SCR equals -29 dB, the output SCR equals 16.1 dB when the target's doppler frequency is such that the target gain equals 1. The maximum target gain equals 2 so that the maximum output SCR equals 19.1 dB. If the target's doppler frequency is at, or close to, a blind speed, the output SCR is much less than the input SCR.

The maximum SCR improvement factor assumes a true sinusoidal transmitter. Actual transmitters have small random deviations from their nominal frequency, and the above calculation thus is optimistic. Note that the ratio of clutter bandwidth to carrier frequency in the above example is approximately 1×10^{-9}. Thus, the short-term transmitter stability must be significantly less than 1×10^{-9} for the transmitter's instability to have negligible effect on the SCR improvement.

2.8.2 Weather Clutter

The pertinent resolution cell for weather clutter computations is a volume rather than an area. Assuming that the entire main beam is filled with clutter scatterers, the volume of the cell is approximated by

$$V_c \approx \theta_a \theta_e R^2 (c\tau/2) \tag{2.5}$$

For both beamwidths equal to 2°, 40 km range, and 150 m range resolution, the volume equals 2.92×10^8 m^3. The backscattering coefficient for rain depends on the intensity of the rainstorm, r, and the carrier frequency, f_0. A first-order approximation is the reflectivity, η, expressed in dB:

$$\eta = -110 + 38.5 \log f_0 + 16.6 \log r \tag{2.6}$$

where r is in mm/hr and f_0 is in GHz. For a 4 mm/hr storm and a 3 GHz operating frequency, eta is evaluated as −81.6 dB or 6.86×10^{-9}. Thus, the rain's clutter cross section equals 2.0 m^2.

Although this cross section is significantly less than that of the ground, it still can cause a significant degradation in target detectability. Thus, the rain's clutter cross section exceeds that of the example target, and thus the target is not detectable without the use of multiple-pulse coherent processing. Although ground clutter ceases to be a problem at long ranges due to the earth's curvature, the rain's clutter cross section still exists at long range and increases as the square of range. At a range of 100 km, the clutter cross section equals 78 m^2. Also, the frequency of the rain clutter is not centered at zero so that it is often not well cancelled by using a delay-line canceller.

The rain's clutter spectrum depends on the wind. Wind speeds vary significantly with altitude, and velocity reversals sometimes occur within altitude differences of a few kilometers. For initial design purposes, a crude first-order approximation is that wind velocity increases linearly with altitude by the value of 2 m/s per 1 k of altitude increase. As an example, at a radial range of 100 km, the altitude above a flat earth at an elevation angle of 3° is 5.24 km. The corresponding approximate wind velocity equals 10.5 m/s, and the corresponding doppler frequency is 210 Hz. Thus, if a surveillance radar has a 2° elevation beamwidth and the beam is centered 3° above a flat earth, the clutter at the beam center has a doppler frequency of 210 Hz. The clutter at the upper and lower beam edges have doppler frequencies of 280 Hz and 140 Hz, respectively, so that the clutter spectral bandwidth is 140 Hz. Therefore, rain clutter can be much more difficult to cancel because its center frequency is nonzero and its spread is very large.

2.8.3 Noise Jamming

A jammer of power P_J at a range to the radar of R_J induces a received power of

$$P_{rJ} = \frac{P_J G_J A_{rJ}}{4\pi R_j^2} \tag{2.7}$$

where G_J is the gain of the jammer's antenna in the direction of the radar and A_{rJ} is the radar receiving antenna's effective aperture in the direction of the jammer. Similarly to (1.6), the effective antenna aperture is

$$A_{rJ} = G_{rJ}\lambda^2/4\pi \tag{2.8}$$

where G_{rJ} is the radar receiving antenna's gain in the direction of the jammer. For sidelobe jamming, a typical maximum value for G_{rJ} is 0 dB with respect to an ideal isotropic antenna (dBI). A noise jammer develops a thermal-noiselike waveform with its power spread over a bandwidth B_J. When the spectral density of the energy received by the radar due to the noise jammer is greater than that of thermal noise, the presence of the jammer significantly decreases the maximum detection range of the radar. As a specific example, consider an SOJ that induces a noise intensity 16 times greater than that of thermal noise. Because the received target energy varies as the inverse fourth power of range, the maximum target detection range in jamming equals one-half of its maximum detection range in thermal noise. As a numerical example, assume a 10 W jammer at a range of 100 km with a jammer bandwidth of 10 MHz and a jammer antenna gain of 10. The received power density at the radar equals 6.3×10^{-20} W/Hz. This exceeds the spectral density of thermal noise by a factor of approximately 16 so that the detection range is decreased to half its value under thermal-noise conditions.

2.8.4 Target Characteristics and Detection Probability

A major conceptual difference between targets and (distributed) clutter is that the target's physical size is usually much smaller than that of the resolution cell, whereas the clutter is usually much larger than the resolution cell, and therefore, fills it. Often, the target is idealized and may be thought of as an isolated point. However, in addition to distributed clutter, the environment sometimes consists of the combination of distributed clutter and point clutter. The point-clutter concept pertains when the energy reflected by point clutter in the resolution cell is an appreciable fraction of that due to distributed clutter in the resolution cell.

The energy reflected from a point target depends on the incident angle, carrier frequency of the waveform, and polarization. The spatial characteristics of the reflection are similar to the antenna characteristic sketched in Figure 1.2. For an aircraft, the peak reflection occurs broadside to the axis of the fuselage. Like an antenna, the main-lobe width is approximately equal to the wavelength divided by the length of the aircraft. For a carrier frequency of 3 GHz and an aircraft length of 10 m, the main-lobe width is on the order of 1×10^{-3} radians or 0.057°. Outside the very narrow main-lobe region, the total aircraft reflection is due to the phasor sum of individual reflections from all over the aircraft. In general, the combined reflection pattern varies significantly with small angle changes, like the single term due to the fuselage. Thus, small perturbations of the aircraft's flight path due to winds significantly change the radar cross section. As a consequence, a reasonable analytical assumption is that the target cross section is a random variable. The reflected target energy calculated by (1.2) thus also is a random variable.

2.8.4.1 *Detection Theory*

Figure 1.1 showed a block indicated as *target decisions*. The intent of this block is that a "target-present" or "target-absent" decision is made for each resolution cell. This decision is based on each cell's voltage. When the voltage is larger than a preset threshold value, a target-present decision is made. Sometimes this decision is in error because, due to random fluctuations, the voltage can be larger than the threshold, even when the environment is thermal noise only (or the sum of thermal noise and distributed clutter). The probability of this error depends on the ratio of the threshold value and noise power, as given by (1.12).

The probability of target detection depends on the SNR, given by the ratio of (1.2) and (1.3). We assume the target cross section to be a random variable, and thus the SNR is a random variable. For aircraft targets, measurements show that the *probability density function* (pdf) for the single-pulse SNR, γ, is approximately equal to the exponential density. Thus,

$$p_\gamma(\gamma) = \exp(-\gamma/\gamma_a)/\gamma_a \tag{2.9}$$

where γ_a is the average SNR value. When the probability of threshold exceedance is averaged with respect to (2.9), the probability of target detection is given by (1.13).

The multiple-pulse pdf for the SNR depends on the aircraft flight characteristics, time between pulse transmissions, and carrier frequency of each pulse. When the carrier frequency is constant, the usual assumption is that the SNR of each pulse of a dwell of N pulses is identical when the dwell duration is less than about 10 μs. If the carrier frequency of each pulse changes by an amount approximately equal to that which causes a range resolution equal to the target's radial extent, the per-pulse

SNR values are statistically independent and the pdf is given by (2.9). Due to aircraft trajectory changes, the scan-to-scan SNR values are assumed to be statistically independent, even if the carrier frequency is unchanged.

These target models are termed Swerling 1 when the SNR values are identical and Swerling 2 when they are statistically independent. When the pdf for γ is (2.9) convolved with itself (a gamma density), the comparable target models are termed Swerling 3 and Swerling 4. Other mathematical models are used when the measured target statistics are not close to these pdf values.

Chapter 3
Nonadaptive (Conventional) Receiving Antennas

3.1 INTRODUCTION

Antennas are used for energy transmission and reception. The system requirements for the two applications are different, and this chapter considers only receiving antennas.

The antenna characteristics control the magnitudes of the received target signal, interfering clutter, and jamming waveforms. Environmentally independent, deterministic designs that decrease the magntiude of one of the interference signals tend to increase the signal received from other types of interference as well as decrease the signal received from the target. Environmentally dependent adaptive designs can significantly improve the overall performance. To be able to assess the potential improvement quantitatively, and hence justify the increased implementation cost of adaptive antennas, we need to determine the performance attained when using nonadaptive antennas.

There are three types of radar antennas in use. These are linear arrays, planar arrays, and reflectors. All have a gain *versus* angle characteristic as that shown in Figure 1.2. The initial microwave radars used reflectors. For reflector antennas, the high-gain, main-beam angular region is perpendicular to the plane of the reflector. Mechanical scanning of the antenna is used to provide the desired coverage. Array antennas are often used for newly designed radars. The beam is scanned electronically rather than mechanically. The electronic scanning is done by varying the values of phase shifters, giving rise to the alternative name of *phased arrays*.

3.2 LINEAR ARRAYS

A linear array consists of nominally identical, equispaced elements that are located along a line. A plane wave impinging on an antenna excites signals in each antenna element. For an arrival angle, θ, measured from the normal to the line array, and interelement spacing, d, the element-to-element time delay of the received signal equals $d \sin\theta/c$. Choosing the end element (element 1) as a reference and denoting its received signal as $g_1(\theta)x(t)$, the signal received by the nth element equals $g_n(\theta)x[t - (n - 1)d \sin\theta/c]$, where $g_n(\theta)$ denotes the gain pattern of the nth element. For radar applications, neglecting noise, $x(t)$ is essentially a pulsed sinusoid. The antenna pattern is commonly defined in terms of the gain to the sinusoid's carrier frequency portion of the waveform. The change of pattern over the bandwidth of the pulse is usually negligible for nonadaptive antennas. However, it can be significant for adaptive antennas, and its effect will be considered in detail later.

For a sinusoidal waveform of frequency f_0, the effect of time delay is a phase shift with the element-to-element phase shift ϕ equal to $2\pi f_0$ times the element-to-element delay. Thus, $\phi = -2\pi f_0 d \sin\theta/c$. Because the wavelength, λ, is related to the carrier frequency as $f_0 \times \lambda = c$, we have

$$\phi = 2\pi(d/\lambda) \sin\theta \tag{3.1}$$

The phase shift at the nth element equals $-2\pi(n - 1)(d/\lambda) \sin\theta$. The antenna output is formed as the weighted sum of the element voltages. The complex amplitude of the sinusoidal output is given by*

$$e_{\text{out}} = a\Sigma w_n^* g_n(\theta) \exp[-j(n - 1)\phi] \tag{3.2}$$

where a is the amplitude of the sinusoid at each element and w_n is the weight applied to the nth element. The voltage gain equals e_{out}/a. The performance analysis assumes that the elements have identical gain characteristics. This assumption neglects element-to-element mutual coupling, but allows the main details of the pattern to be determined with only small errors, avoiding excessive and complicated analysis.

3.2.1 Array Factor

The common element power gain, $G_e(\theta)$, is the magnitude squared of the elemental voltage gain. Thus, the antenna power gain is the product of two terms:

$$G(\theta) = G_e(\theta)G_a(\theta) \tag{3.3}$$

where $G_a(\theta)$ is defined as the array factor gain:

$$G_a(\theta) = |\Sigma w_n^* \exp[-j(n - 1)\phi]|^2 \tag{3.4}$$

---

*Summations with respect to index n are over the range of n equals 1 to N (the number of array elements).

Let us consider the case of equal-valued real weights. As discussed earlier for the analysis of doppler filters, the evaluation of target detectability is simplified when the output noise power from a circuit equals the input noise power. This is also true for antenna analysis so that the magnitudes of the weight are defined as equal to $1/\sqrt{N}$. Note that this assumed value is for the purposes of analysis and is not indicative of actual implementation. Initially, the phases of the weights are set equal to zero. Section 3.2.2 discusses the use of nonzero phases for beam steering. The summation can be recognized as a geometric series, and the array power gain is

$$G_a(\phi) = N[\sin(N\phi/2)/N \sin(\phi/2)]^2 \tag{3.5}$$

Note that the form of this equation is essentially identical to (1.23), as derived for the doppler filter response. The response curve of Figure 1.6 shows the filter response *versus* doppler frequency. This same curve is the form of the power gain *versus* ϕ. Figure 1.6 shows the response as periodic with the periodicity equal to $1/T$. As discussed in detail below, the comparable concept for antenna theory is that of grating lobes. The next section develops a more detailed analysis of the properties of the response patterns.

As an algebraic derivation of the pattern, first note that G_a is an even function of ϕ. Then note that, for small angles, the sine function can be approximated by the first two terms of a Taylor series:

$$\sin(x) \approx x - \frac{1}{6}x^3 \tag{3.6}$$

Combining (3.5) and (3.6), with a few other approximations, gives

$$G_a \approx N[1 - (N^2 - 1)\phi^2/12] \tag{3.7}$$

The peak occurs when $\phi = 0$ (and therefore $\theta = 0$). The maximum gain equals the number of elements, as can be directly inferred from (3.4) by noting that G_a is the sum of phasors. The sum is maximum when all phases are identical. This occurs for $\phi = 0$, and more generally when ϕ is a multiple of 2π. From (3.7), the gain decreases quadratically as the magnitude of ϕ increases.

From (3.5), the rate of variation of the numerator is N times as rapid as the denominator. Thus, as ϕ increases, the array gain varies because of the combined effect of rapid numerator variation and slow denominator variation. The array factor gain equals zero whenever the numerator equals zero and the denominator does not equal zero. When both numerator and denominator equal zero, from (3.7), the gain equals N. This occurs when ϕ equals multiples of 2π so that the angular spacing between the two major peaks sketched in Figure 3.1 is

$$(d/\lambda)\sin\theta = 1 \tag{3.8}$$

The major peak at zero is denoted as the *main lobe*. Others are termed *grating lobes*. Grating lobes are usually undesirable as they cause significant ambiguity in angle.

The grating lobes' locations depend upon the d/λ ratio. For $d/\lambda = 1$ grating lobes occur when θ equals integer multiples of $\pm\pi/2$. For this case, the array factor has two peaks broadside to the array ($\theta = 0$ and π) and two peaks in the plane of the array. The product of the array and element patterns tends to suppress the grating lobes so that the overall antenna pattern has only a single main lobe. For $d/\lambda > 1$, multiple grating lobes occur. For $d/\lambda < 1$, there are no grating lobes when the weights are real valued. As will be shown later, grating lobes occur when d/λ is greater than one-half and the weight values include phase shifts for beam steering.

The array pattern close to the main-lobe peak is primarily controlled by the numerator of (3.5). The gain equals zero when ϕ equals integral multiples of $2\pi/N$. From (3.1), assuming small angles of arrival, nulls occur at these arrival angles such that

$$\theta = k\lambda/Nd \tag{3.9}$$

for positive integral k. The angle between the main-lobe peak and the first null, θ_n, equals λ/D, where $D = Nd$ is the array width. Thus, this angle equals the beamwidth defined in (1.4).

The sidelobe peaks occur approximately halfway between the nulls. At these points, the numerator of (3.5) equals unity. The approximate gain of the sidelobe peaks relative to the peak main-lobe gain is

$$G_{\mathrm{SL}} = (N\sin\phi)^{-2}; \quad \phi = \pi(|k| + 0.5)/N \tag{3.10}$$

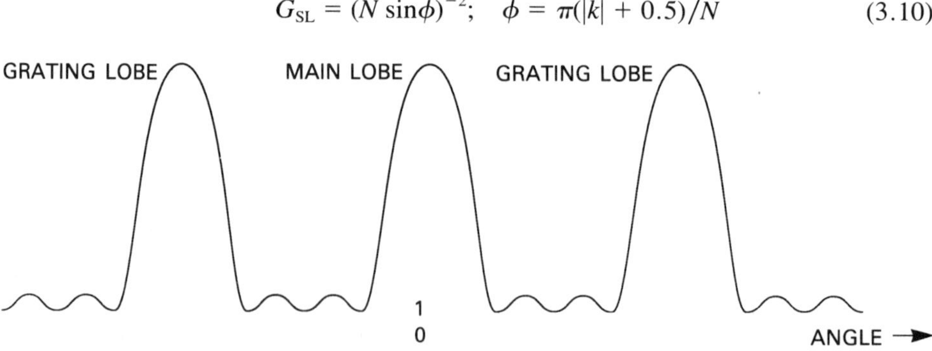

Figure 3.1 Array pattern.

For $k = 1$ and 2, the approximate relative sidelobe gain equals -13.5 and -17.9 dB, respectively.

3.2.2 Beam Steering

The array can be steered so that the beam peak is at any angle of arrival by including phase shifts in the weights indicated by (3.4). Consider the steering weights:

$$\begin{aligned} w_n &= \exp[j(n-1)\phi_s]/\sqrt{N} \\ \phi_s &= 2\pi(d/\lambda)\sin\theta_s \end{aligned} \tag{3.11}$$

Combining (3.4) and (3.11) gives

$$G_a = N\{\sin[N(\phi - \phi_s)/2]/N\sin[(\phi - \phi_s)/2]\}^2 \tag{3.12}$$

so that the peak of the beam is in the direction of the angle θ_s. The angle of the first null, θ_n, measured from the beam peak occurs for

$$\sin(\theta_s + \theta_n) - \sin\theta_s = \lambda/D \tag{3.13}$$

Using the trigonometric identity for the difference of two sine functions gives

$$2\sin(\theta_n/2)\cos(\theta_s + \theta_n/2) = \lambda/D \tag{3.14}$$

For small beamwidths and large scanning angles, this approximates to

$$\theta_n = (\lambda/D)/\cos(\theta_s) \tag{3.15}$$

Thus, the beamwidth increases as the beam is steered from broadside.

The grating lobes occur when the denominator of (3.9) equals zero. Therefore, a grating lobe cannot occur if

$$d/\lambda|\sin\theta - \sin\theta_s| < 1 \tag{3.16}$$

for all angles θ; $\theta \neq \theta_s$. This can always be satisfied if d is very small. However, use of very small element-to-element spacing is undesirable as this increases the antenna beamwidth (and increases the losses associated with mutual coupling). The largest possible spacing that can be used without grating lobes is desirable.

The maximum value of the argument of the absolute value function equals $1 + |\sin\theta_s|$. Thus, the design requirement for the normalized element spacing is

$$d/\lambda < (1 + |\sin\theta_s|)^{-1} \tag{3.17}$$

When d/λ equals one-half, the inequality is always satisfied. For a maximum scan angle of 45°, the normalized spacing can be increased to 0.59.

3.2.3 Sidelobe Reduction Weighting

Equation (3.10) is for the relative peak sidelobe gain. Evaluations have shown that the sidelobes adjacent to the main beam have relative gains of approximately −13.5 dB. These high sidelobe levels are undesirable for several reasons: they create an increased vulnerability to jamming; they also can cause an ambiguity in angle as a large target received through a sidelobe will be mistaken for a target having an angle that is the steered angle. Methods of modifying the antenna response to create decreased sidelobes are of interest.

Equation (3.4) gives the array gain of the antenna in the form of a *discrete Fourier transform* (DFT). Note that this is consistent with the results previously obtained. Specifically, the DFT of a sampled constant-valued function is well known as the bracketed function of (3.5). (The function $[\sin(Nx)/\sin(x)]/N$ is defined as the SNIC function by some authors.) A sketch of this transform relationship is shown in Figure 3.2(a). The SNIC function is similar to the sinc function, defined as $(\sin NX)/NX$. The major difference is that the SNIC function is periodic and the sinc function is not. The sinc function is the Fourier transform of the dotted rectangle function shown in Figure 3.2(a). Thus, known Fourier transform relationships can be used to design antenna patterns with decreased sidelobe levels.

As a specific example, let us consider sampled cosine-squared weighting, indicated by the sketch of Figure 3.2(b). With this weighting, the amplitude of the weights in the array center are large. The weight values taper toward zero at the

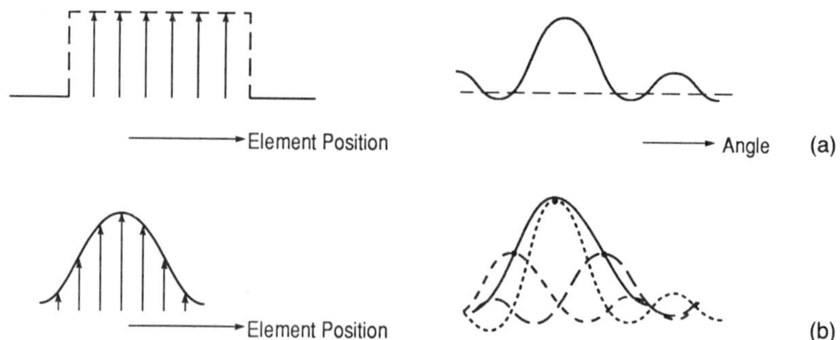

Figure 3.2 Antenna pattern Fourier transform relation.

array edges. Expanding the cosine-squared by standard trigonometric relations gives the continuous weighting function (or illumination function) as

$$I(p) = [1 + \cos 2\pi p/D]/2; \quad -D/2 < p \leq D/2$$

$$= \left[\frac{1}{2} + \frac{1}{4}\exp(j2\pi p/D) + \frac{1}{4}\exp(-j2\pi p/D)\right]\mathrm{Rect}(p/D) \qquad (3.18)$$

for an aperture of width D. The Fourier transform of the final form of (3.18) can be obtained by noting that the transform of a rectangular function is a sinc [(sin$x$)/x] function. Thus,

$$\mathrm{Rect}(p/D) \leftrightarrow D\ \sin(\pi\Phi D)/\pi\Phi D \qquad (3.19)$$

Also, for any function $f(p)$ with a transform of $F(\cdot)$:

$$f(p)\exp(j2\pi p/D) \leftrightarrow F(\Phi - 2\pi/D) \qquad (3.20)$$

In addition, the effect of multiplying a function by a scalar (e.g., $\frac{1}{2}$ or $\frac{1}{4}$) is that the tranform is multiplied by this same scalar.

The right-hand sketch of Figure 3.2(b) shows three dotted sinc functions. These are the transforms of the three portions of the illumination function of (3.18). The rightmost portion of the three curves shows the sidelobes in a phase and amplitude relation that tends to add to a total sidelobe of almost zero. This approximate cancellation continues for the farther removed sidelobe region. Note that the beamwidth of the solid curve is larger than that of a dotted curve. In particular, the width to the first null is doubled. The increased beamwidth is undesirable as it decreases the ability to resolve multiple targets that are closely spaced in angle and increases the received clutter power. In addition, the weighting decreases the peak antenna gain, and thus reduces the target's detectability by decreasing the target SNR. The combination of decreased sidelobes with increased beamwidth and decreased SNR is always true. Some weighting functions give a better performance compromise than others. Table 3.1 summarizes the characteristics of several weighting functions with θ_B defined as λ/D.

Figure 3.3(a) shows the array pattern for a 32-element array using −40 dB Taylor weights. We can see that the maximum relative sidelobe level equals −40 dB and that the other sidelobes are all less than this value. To achieve this pattern, the precise theoretical weight values must be used. Usually, the theoretical values cannot be implemented. Figure 3.3(b) shows the effect of random 5% errors. The next section discusses some theoretical aspects of the effect of errors.

Table 3.1
Comparison of Characteristics of Weighting Functions

Function	*SNR Loss (dB)*	*−3 dB Width (Ratio)*	*Peak Sidelobe (dB)*
Uniform	0.0	0.89 θ_B (1.00)	−13.2
Taylor (N = 8)	1.14	1.25 θ_B (1.41)	−40.0
$\cos^2$	1.76	1.46 θ_B (1.65)	−31.7
$\cos^4$	2.88	1.94 θ_B (2.19)	−47.0

3.2.4 Effect of Weighting Errors

Several architectures are available to form a beam by combining the outputs of the array elements. One is the direct implementation of weighting and phase shifting of the RF or IF signals. Another is to use analog-to-digital (A/D) converters for each array element followed by digital processing. For both architectures, only approximations to the required weights can be implemented. As an example, the available phase shifters can only shift by discrete phases, such as 180°, 90°, 45°, *et cetera*. The infinite precision, exact phase shift values required to steer to particular directions are not implemented. The weighting errors hence cause pattern perturbations. This section develops a quantitative relationship between the weighting errors and pattern perturbations.

A reasonable error model considers both a magnitude error due to receiver or element gain mismatches, and phase errors due to phase quantization and mismatches. The applied elemental weight, $w_{n\varepsilon}$, equals the desired weight multiplied by $(1 + g_{n\varepsilon}) \exp(j\phi_{n\varepsilon})$ where $g_{n\varepsilon}$ is the gain error and $\phi_{n\varepsilon}$ is the phase error. From (3.2), the array factor voltage gain is

$$g_{a\varepsilon} = \sum w_n^*(1 + g_{n\varepsilon}) \exp\{-j[\phi_{n\varepsilon} + (n-1)\phi]\} \tag{3.21}$$

Expanding (3.21) with the approximations that products of errors are negligible and for small arguments, $\exp(-j\phi_{n\varepsilon}) \approx 1 - j\phi_{n\varepsilon}$ gives

$$g_{a\varepsilon} = \sum w_n^* \exp[-j(n-1)\phi] + (g_{n\varepsilon} - j\phi_{n\varepsilon}) w_n^* \exp[-j(n-1)\phi] \tag{3.22}$$

This is the sum of an unerrored and errored term. According to the central limit theorem, the statistical density functions of the error term are approximately Gaussian. To derive the mean and variance of the approximately Gaussian statistics, we have assumed that the errors are zero mean, identically distributed, statistically independent, and uniformly distributed over an interval. As an example, assume that

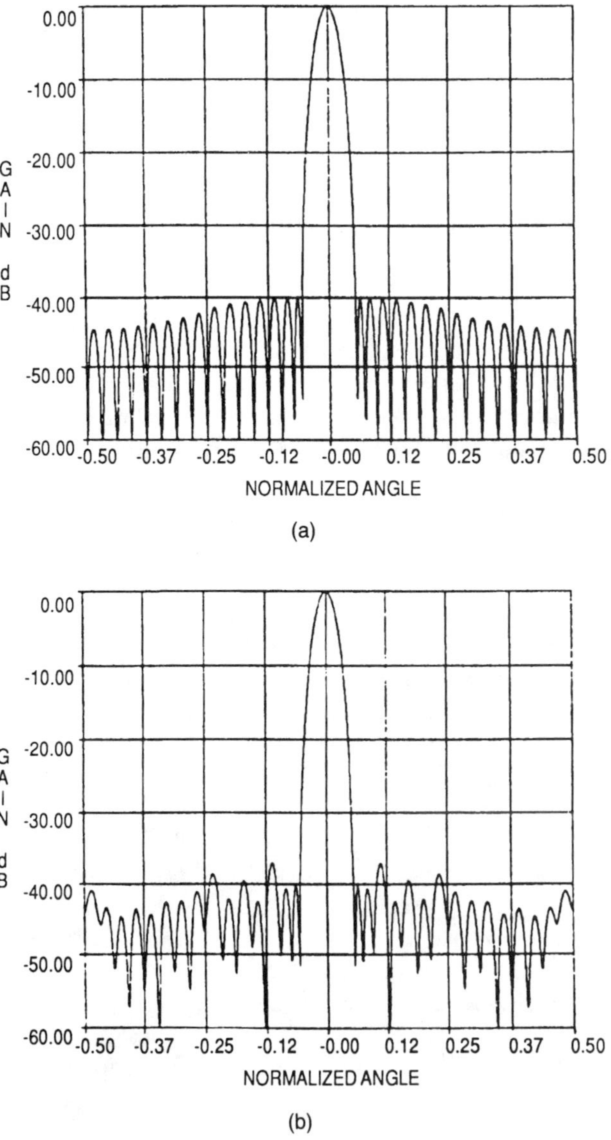

Figure 3.3 32-element line array pattern.

the nominal amplifier gain equals unity and the actual amplitude gain is a random variable uniformly distributed over the interval (1.05, 0.95). This corresponds to a maximum amplifier gain error of approximately ±0.43 dB. The mean of the error term of zero and the variance is

$$\mathrm{var}(g_{a\varepsilon}) = [\sigma_g^2 + \sigma_\phi^2]\Sigma|w_n|^2 \tag{3.23}$$

where σ_g^2 and σ_ϕ^2 denote the variance of the gain and phase errors, respectively. Due to the constraint of unity noise gain, the summation equals unity so that the variance of the array factor voltage gain equals the sum of the variances of the amplifier gain errors.

For amplifier gain errors that are uniformly distributed over the range $(-\varepsilon/2, \varepsilon/2)$, the variance equals $\varepsilon^2/12$. For the previous example of ±0.43 dB amplifier gain errors, the gain error variance equals 8.33×10^{-4}. This corresponds to an array gain level of −30.8 dBI. An array antenna with 64 elements has a peak gain of 18.06 dB. Thus, these errors affect sidelobes at the relative level of −48.8 dB. Similar results can be obtained for the effect of phase errors.

3.3 PLANAR ARRAYS

A line array generates a beam that is directive in only one angular dimension. This beam shape is termed a *fan beam*. To generate a beam that is directive in both angular dimensions, azimuth and elevation, a planar array must be used. One method of building a planar array is to place elements on a rectangular grid. This planar array can be viewed as a line array of line arrays. That is, each "element" of the "second" array is a line array, and the effect of the "element" pattern is important. The generalization of (3.5) is

$$G_a = N_a N_e \left[\frac{\sin(N_a \phi_a/2)}{N_a \sin(\phi_a/2)} \frac{\sin(N_e \phi_e/2)}{N_e \sin(\phi_e/2)} \right] \tag{3.24}$$

where ϕ_a and ϕ_e denote phase shifts in the azimuth and elevation angles, respectively. N_a is the number of elements in each row of the array, and N_e is the number of rows so that the total number of array elements equals $N_a N_e$. The array main lobe can be steered off broadside by phase shifters. The phase shift applied to the nth row, mth column element of the array to steer to the azimuth angle θ_{as} and the elevation angle θ_{es} is

$$\phi_{nm} = -2\pi d[(n - 1)\sin\theta_{es} + (m - 1)\sin\theta_{as}]/\lambda \tag{3.25}$$

Note that steering in elevation is accomplished by phase shifting from row to row, and azimuth steering is accomplished by phase shifting from column to column.

3.4 REFLECTOR ANTENNAS

The most common antenna currently in use is the reflector, as sketched in Figure 3.4. On transmission, the energy flows from the transmitter through the waveguide to the feed. The electromagnetic energy launched by the feed is then reflected and transmitted into space in a direction perpendicular to the plane of the reflector. The reflector shape is usually parabolic to generate a constant phase wavefront, equivalent to that transmitted from a planar phased array steered to broadside. As the beam is always perpendicular to the reflector plane, the beam is steered by physical rotation of the array.

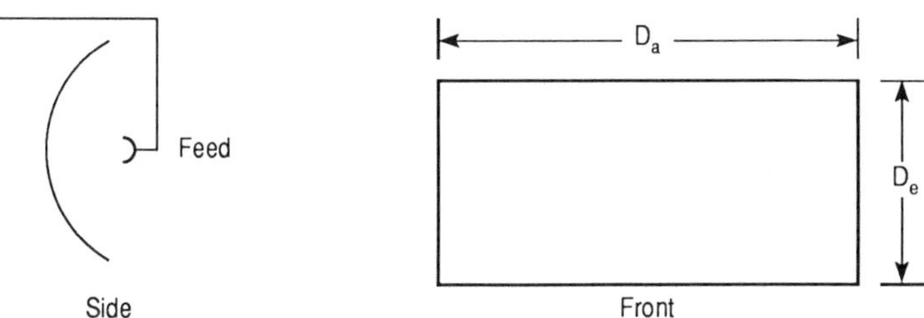

Figure 3.4 Reflector antenna.

Sidelobe level is controlled through the antenna pattern of the feed, and generating very low sidelobe levels is difficult. Some error sources that limit the achievable sidelobe levels are caused by imperfect matching to a parabolic shape and blockage due to the physical existence of the feed.

Chapter 4
Sidelobe Canceller (SLC)

4.1 INTRODUCTION

A fundamental limit to the maximum detection range of a radar or other sensor is thermal noise. For some applications, noiselike artificial (man-made) interference is received at power levels much higher than those of the thermal noise, causing a significant degradation of detection range. Two examples of the noiselike interference are RFI and *jamming*. The difference is the intent of the transmission. RFI is usually transmission with a purpose other than interference, such as from a communication transmitter, and the interference is inadvertent. Jamming, also termed *electronic countermeasures* (ECM), is intentional, and it is intended to decrease the radar's performance so as to degrade the capability of a radar-directed weapon system. Electronic counter-countermeasure (ECCM) techniques are incorporated into the radar to reduce the effect of the jamming.

One common type of jamming is *noise jamming* received through the sidelobes of the radar antenna. An example is given in Section 2.8.3. As indicated in Section 3.2.3, one ECCM technique uses illumination function weighting to reduce the antenna sidelobes. However, if the jammer's power is sufficiently large or its range is sufficiently small, the received jammer noise is much greater than the received thermal noise so that the radar performance is degraded. Another ECCM technique that can be used in conjunction with illumination weighting is *antenna adaptivity*. This concept develops an antenna pattern with a gain that is adapted to be very small in the direction of the jammer or jammers. Thus, the jammer is "cancelled" and, ideally, the radar performance is only limited by thermal noise, even in a jamming environment.

The first adaptive antenna technique developed was the *sidelobe canceller* used with reflector antennas. It is an effective ECCM technique in a single-jammer environment. The *multiple sidelobe canceller* technique was developed to cope with a multiple-jammer environment. As discussed, these techniques are applicable for both reflector and array antennas.

4.2 SIDELOBE CANCELLATION CONCEPT

The sidelobe canceller uses an auxiliary antenna in addition to the main antenna. Combining the outputs from the two antennas creates an overall antenna that "cancels" the jammer. The quotation marks are necessary because exact cancellation does not occur. However, the cancellation concept is approximately valid and is useful to obtain a physical interpretation of the system performance.

A sketch of a conceptual sidelobe canceller is given in Figure 4.1. The auxiliary antenna is physically close to the reflector, but outside of the reflector area to prevent blockage of the received energy. In an array, the auxiliary antenna is often implemented as one or more of the elements of the array of the main antenna.

Let us denote the voltage gains of the main and auxiliary antennas in the direction of the jammer as $g_s(\theta_J)$ and $g_1(\theta_J)$ respectively.* The subscript s is intended to emphasize that the jammer is received in the main antenna via a sidelobe. We

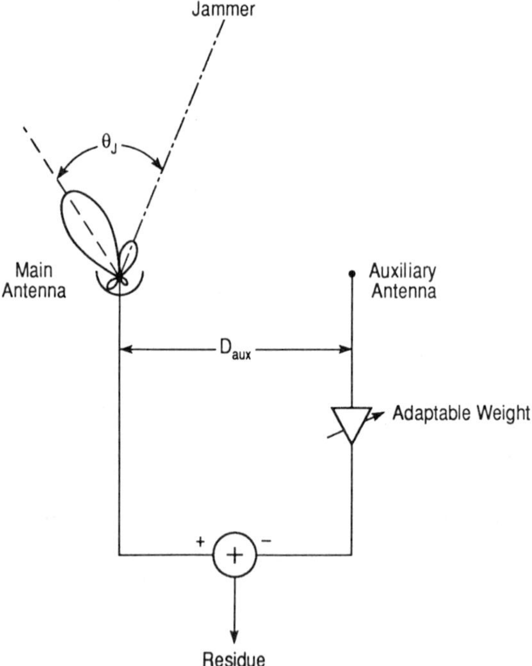

Figure 4.1 Conceptual sidelobe canceller antenna system.

*The results of this and the following sections can be generalized to the two-dimensional angle space. The discussion is limited to a single dimension only for the convenience of our presentation.

denote the complex envelope of the voltage received by the main antenna due to the jammer, $y_{0J}(t)$, as

$$y_{0J}(t) = g_s(\theta_J)j(t) \tag{4.1}$$

where $j(t)$ is the complex envelope due to the jammer. That received by the auxiliary antenna is

$$y_{1J}(t) = g_1(\theta_J)j(t - t_1)\exp(-j\omega_0 t_1) \tag{4.2}$$

where t_1 is the propagation delay between the two antennas:

$$t_1 = d_{0,1}\sin\theta_J/c \tag{4.3}$$

where $d_{0,1}$ is the spacing between the centers of the two antennas and c is the velocity of propagation. Note in (4.2) that the propagation delay appears as both a delay in the jammer complex envelope and a phase shift due to the delay of the carrier. Initially, we assume that the product of bandwidth and time delay is very small so that $j(t - t_1) \approx j(t)$. Then, the antenna system output, $r_J(t)$, for an adaptive weight value, w_1, due solely to the jammer, is

$$r_J(t) = j(t)[g_s(\theta_J) - w_1^* g_1(\theta_J)\exp(-j\omega_0 t_1)] \tag{4.4}$$

The asterisk denotes the complex conjugate. (The terminology of "adaptive weight" is equivalent to "adaptive gain.") If

$$w_1 = [g_s(\theta_J)/g_1(\theta_J)]^* \exp(-j\omega_0 t_1) \tag{4.5}$$

the jammer-induced residue output equals zero so that the jammer is cancelled. Thus, the combined antenna pattern, due to the adaptive combination of the main and auxiliary antennas, has a null at the jammer angle.

Although the jammer-induced residue would equal zero, the total residue would not be zero because of the effects of thermal noise. Specifically, the total voltage of each antenna is the sum of the jammer-induced component and thermal noise. The thermal noise is due to noise received from external sources plus that generated in the receiver circuitry. Usually, the total noise power, denoted as p_{noise}, is the same for all antennas. The total output residue power is the sum of that due to the jammer (zero for this condition) and that due to the thermal noise. The thermal noise power of the two antennas adds (it is assumed to be statistically independent), and that of the auxiliary antenna is modified due to the adaptive weight. Thus, the output residue power, p_{res}, is

$$p_{\text{res}} = p_{\text{noise}}[1 + G_s(\theta_J)/G_1(\theta_J)] \tag{4.6}$$

where G_s denotes the sidelobe power gain of the main antenna and $G_1(\theta_J)$ denotes the power gain of the auxiliary antenna.

Note that the residue power is greater than the thermal noise power. Thus, although adaptive antennas form an important ECCM technique, they do not return the radar performance to the level that it achieves in a nonjamming environment. As a numerical example, assume that, in a nonjamming environment, the ratio of target reflected energy to thermal noise power, SNR = 20 dB, so that the target is detectable with very high probability. Let us assume a jamming environment such that the power received from the jammer through the antenna sidelobe, the jamming-to-noise power ratio (JNR), is 50 dB above the thermal noise power. The ratio of signal energy to jamming plus noise (SJNR) is approximately −30 dB, and the probability of target detection is approximately zero. The adaptive antenna system using the weight value given in (4.5) cancels the jammer, but, as given by (4.6), the residue power is greater than the noise power. If $G_s(\theta_J) = G_1(\theta_J)$, the residue power is 3 dB higher than the thermal noise power, the SNR is reduced from its nonjamming environment value of 20 dB to 17 dB and the target detection probability is reduced. Note that a smaller performance loss occurs if the auxiliary antenna gain is much higher than the main antenna sidelobe gain. Thus, a good low-sidelobe design for the main antenna is desired even though the SLC ECCM technique is implemented.

4.3 RESIDUE POWER MINIMIZATION

As shown in the preceding section, the residue power is higher than the thermal noise power. Because residue power is the limiting factor in target detectability, determining which weight value causes the minimum is important. As shown, the minimum does not occur at the same weight that causes exact jammer cancellation.

Let us denote the complex envelopes of the thermal noises for the main and auxiliary antennas as $n_0(t)$ and $n_1(t)$, respectively. The total voltages of the main and auxiliary antennas, $y_0(t)$ and $y_1(t)$, respectively, equal the sum of thermal noise plus the jammer-induced voltages given by (4.1) and (4.2). Thus, for the narrowband condition,

$$\begin{aligned} y_0(t) &= n_0(t) + g_s j(t) \\ y_1(t) &= n_1(t) + g_1 j(t) \exp(-j\omega_0 t_1) \end{aligned} \tag{4.7}$$

The dependence of g_s and g_1 on θ_J has been suppressed. The residue is

$$r(t) = n_0(t) - w_1^* n_1(t) + j(t)[g_s - w_1^* g_1 \exp(-j\omega_0 t_1)] \tag{4.8}$$

The residue power is

$$p_{res} = (1 + |w_1|^2)p_{noise} + p_J[G_s + |w_1|^2 G_1 - w_1 g_s g_1^* \exp(j\omega_0 t_1) - w_1^* g_s^* g_1 \exp(-j\omega_0 t_1)] \quad (4.9)$$

where p_J denotes the jammer power as received by an antenna of unit gain.

The complex-valued weight that minimizes the residue power can be found by partial differentiation with respect to the real and imaginary parts of w_1. An equivalent, but much simpler, procedure is to compute the partial derivative with respect to w_1^* (or w_1) while ignoring the relation between w_1 and w_1^*. The weight value for a zero derivative is easily found as

$$w_{opt} = p_J g_s^* g_1 \exp(-j\omega_0 t_1)/(p_{noise} + p_J G_1) \quad (4.10)$$

Because we can see in (4.9) that the residue power tends toward infinity as a function of the magnitude of w_1, the zero derivative condition corresponds to the minimum, and not the maximum, residue power. Comparing (4.10) and (4.5), we can see that both have the same phase angle but differing magnitude. The magnitudes become equal as the jammer power tends toward infinity. Thus, for very large jammer power, minimizing the residue power is approximately equivalent to cancelling the jammer.

Substituting the optimum weight value from (4.10) into (4.9) gives the minimum residue power as

$$p_{min} = p_{noise}\left[1 + \frac{p_J^2 G_s G_1 + p_J p_{noise} G_s}{(p_{noise} + p_J G_1)^2}\right] \quad (4.11)$$

We can see that as p_J tends toward infinity, (4.11) approaches (4.6). In general, the residue power obtained by exact jammer cancellation is only slightly greater than that obtained by using the residue power minimizaton weight. Also, note that, for narrowband jamming, the minimum residue power is approximately independent of the jammer power when the latter is much greater than the thermal noise power.

As previously indicated, the presence of a noise jammer causes a performance loss, even if a sidelobe canceller is implemented, because the residue power always is greater than the thermal noise power. Additionally, there are other reasons for loss. An important one is that the residue power varies as a function of the unknown jammer characteristics, and a fixed detection threshold cannot be used because it would tend to cause many false alarms. A *constant false alarm rate* (CFAR) circuit is necessary and it always causes an additional performance loss.

Another source of loss is the fact that computing the optimum weight via (4.10) is not possible in an actual implementation because the various parameters, particularly the sidelobe gain, g_s, are unknown. An estimation algorithm must be used, and this is another source of loss.

An additional effect that is usually negligible for a single canceller, but can be significant for an MSLC, is that the gain in the direction of the target for the adapted system can be less than that of the main antenna alone. For such systems, maximizing the output SJNR is often desirable rather than simply minimizing the output residue power and ignoring the effect on antenna gain in the direction of the target.

4.4 CANCELLATION DEGRADATION EFFECTS

In Section 4.2, we computed the weight value that achieved perfect cancellation of very narrowband jamming. Precise cancellation is only possible for zero-bandwidth (sinusoidal waveform) jamming. For a zero-bandwidth waveform, the propagation time delay between the main and auxiliary antennas causes the waveforms at the subtracter point to differ by only a single phase shift. The adaptive weight can compensate for the phase shift, and cancellation thus occurs. For nonzero-bandwidth waveforms, the propagation delay causes the waveforms at the subtracter point to differ by other than a complex-valued scalar. As the adaptive weight only corrects for gain factors, precise cancellation cannot occur. As the bandwidth increases, the residue power due to this effect becomes larger than the minimum residue power for the narrowband optimum weight.

Another cancellation limitation that occurs for nonzero bandwidth is due to differing amplifier transfer functions in the main and auxiliary legs. When the two transfer functions have different variations as a function of frequency, the same input waveform to the two channels causes different output waveforms. Because they differ, perfect cancellation cannot be achieved. As indicated in the following sections, good cancellation can be achieved only when the transfer functions differ by less than tenths or hundredths of a dB.

4.4.1 Propagation Delay Effect on Cancellation

The nonzero-bandwidth effect will be analyzed by neglecting thermal noise and solely focusing on the jammer. We shall show that for this assumption the output residue power increases as a function of bandwidth and jammer power, as indicated by the sketch of Figure 4.2. Also given in the figure is the curve for zero-bandwidth residue power, shown as a horizontal line because the output residue power is independent of jammer power. The intersection of the lines for the cases of zero and nonzero bandwidth indicate that, for jammer powers greater than the intersection value, the residue power increases with jammer power and depends on the waveform bandwidth. For jammer powers less than the intersection value, the output residue power is approximately independent of jammer power, as given by (4.6). The dotted line indicates the condition of power out being equal to power in. The curves for nonzero bandwidth are parallel to this line. Thus, for each nonzero bandwidth, the output power is a constant number of dB less than the input power, this difference being the cancellation ratio. Note that the cancellation ratio is a valid performance measure

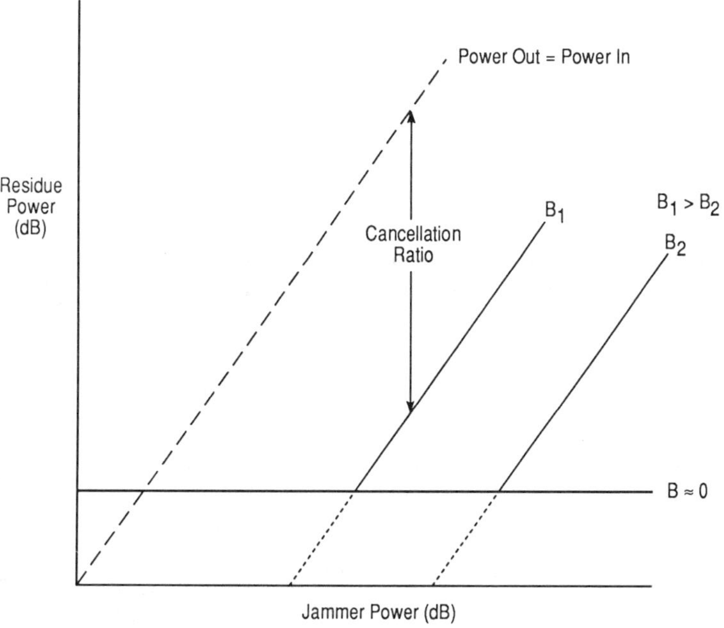

Figure 4.2 Bandwidth effect on sidelobe canceller.

only when the bandwidth is sufficiently large for the residue power due to this effect to exceed that due to the thermal noise.

From (4.1) and (4.2), the residue power is

$$r(t) = g_s j(t) - w_1^* g_1 j(t - t_1) \exp(-j\phi); \quad \phi = \omega_0 t_1 \tag{4.12}$$

The residue power is the statistical expectation of $|r(t)|^2$. Thus,

$$\begin{aligned} p_{\text{res}} &= p_J[G_s - w_1 g_s g_1^* p \exp(j\phi) \\ &\quad - w_1^* g_s^* g_1 \rho^* \exp(-j\phi) + w_1 w_1^* G_1] \\ E[j(t) j^*(t - t_1)] &= p_J \rho \end{aligned} \tag{4.13}$$

where $E(\cdot)$ denotes the statistical expectation operator and ρ represents the normalized correlation coefficient between the delayed and undelayed waveforms. As discussed in Subsection 4.4.1.1, the value of ρ depends on the product of the radar bandwidth and propagation delay. When the bandwidth equals zero, the product equals zero and ρ equals unity. For nonzero products, the magnitude of ρ is less than unity.

The optimum weight for minimum residue power is found by evaluating the partial derivative of (4.13) with respect to w^* and setting the derivative equal to zero. Thus,

$$\begin{aligned}\frac{\partial p_{\text{res}}}{\partial w_1^*} &= p_J[-g_s^* g_1 \rho \exp(-j\phi) + w_1 G_1] \\ w_{\text{opt}} &= \rho[g_s/g_1 \exp(j\phi)]^*\end{aligned} \tag{4.14}$$

Note that (4.14) reduces to (4.6) when ρ equals unity. Combining (4.13) and (4.14) gives the minimum residue power for the nonzero bandwidth (neglecting thermal noise) as

$$p_{\min} = p_J G_s(1 - |\rho|^2) \tag{4.15}$$

Note that for zero bandwidth, ρ equals unity and $p_{\min}$ equals zero. For nonzero bandwidth, $p_{\min}$ is larger than zero.

4.4.1.1 Propagation Delay Cancellation Limitation and Radar System Parameters

For the usual ECM or ECCM condition, the transmitted noise jammer bandwidth is much greater than the radar receiver bandwidth. Thus, the received jammer waveform has a bandwidth equal to the radar bandwidth. A consequence is that radar system design is a constant trade-off between competing desires. As an example, the desire for increased range resolution indicates a need for large radar signal bandwidth. However, the need to maintain good performance in a noise jammer environment indicates a need to obtain good jammer cancellation, and therefore small radar signal bandwidth. This section modifies the cancellation equation to express the result in terms of radar system parameters to highlight the pertinent system trade-off parameters.

There is a relation between the normalized correlation coefficient and transmitted radar waveform. For a matched filter processor, the radar receiver transfer function is the complex conjugate of the spectrum of the radar waveform. Thus, for a transmitted waveform spectrum of $E(f)$, the receiver transfer function equals $E^*(f)$. When the noise jammer bandwidth is much greater than the radar bandwidth, the output jammer power spectrum is proportional to the squared magnitude of the receiver transfer function, or $|E(f)|^2$. The jammer autocorrelation coefficient is the Fourier transform of the power spectrum. However, the Fourier transform of the squared magnitude of the receiver transfer function also equals the complex envelope of the radar signal waveform at the matched filter output. Thus, as an example, assume that the transmitted radar waveform is a pulse of duration τ. The matched filter output is a triangular function of total duration 2τ. The value of the normalized correlation coefficient equals the value of the triangular function at the lag time equal to the propagation delay between the two antennas. Thus,

$$\rho = 1 - |t_1/\tau| \tag{4.16a}$$

For a *linear frequency modulation* (LFM) waveform of two-sided bandwidth, B, the magnitude of the transmitted spectrum, or matched filter transfer function, is the rectangular function Rect(f/B). The autocorrelation function is the Fourier transform, and evaluates to

$$\rho = \sin(\pi B t_1)/\pi B t_1 \approx 1 - (\pi B t_1)^2/6 \tag{4.16b}$$

The approximation is via a power series expansion of the sine function and is valid for values of ρ close to unity. Also note that

$$1 - \rho^2 = (\pi B t_1)^2/3 \tag{4.17}$$

Let us assume a main antenna width equal to D and that the auxiliary antenna is placed on the edge of the main antenna. From (4.3),

$$t_1 = D \sin\theta_J/(2c) \tag{4.18}$$

If we combine this result with the expression relating the antenna width and beamwidth (1.4), we obtain

$$t_1 = \sin\theta_J/(2f_0\theta_a) \tag{4.19}$$

Combining (4.15), (4.17), and (4.19) gives

$$p_{\min}/p_{\text{main}} = [\pi(B/f_0)\sin\theta_J/\theta_a]^2/12 \tag{4.20}$$

where p_{main} is the jammer power received by the main antenna via a sidelobe gain of G_s. We can observe that the cancellation ratio, defined as the ratio of the power received by the main antenna prior to adaptation to the residue power, decreases as the radar bandwidth increases and as antenna beamwidth decreases. Thus, for both range and angle, improved resolution degrades the ECCM performance of the sidelobe canceller. As a numerical example of the achievable cancellation ratio for typical system parameters, we shall assume an antenna beamwidth of 1/30 radian (1.91°), a bandwidth of 3 MHz, a carrier frequency of 3 GHz, and a jammer angle of 30°. Evaluating (4.20) shows that the cancellation ratio equals 37.3 dB. Thus, for received jammer power less than about 37 dB, the residue power is close to the thermal noise level and approximately independent of jammer power. However, for received jammer power greater than about 37 dB, the residue power increases linearly (in dB) with jammer power. If the system bandwidth increases to 10 MHz, the transition power level decreases to 26.8 dB.

4.4.2 Channel Mismatch Effect on Jammer Cancellation

The propagation delay limitation on jammer cancellation depends on the jammer's angle of arrival. When the jammer angle is zero, there is no propagation delay. However, perfect jammer cancellation occurs only if the transfer functions in the main antenna and auxiliary antenna paths are identical. When the transfer functions differ by other than a scalar multiple, the two waveforms at the subtracter inputs are not identical and perfect cancellation will not occur. The cancellation ratio limitation due to this effect will be analyzed by neglecting thermal noise and propagation delay. The result will be an equation for cancellation ratio as a function of channel matching. As in the previous section, this limitaton is negligible when the propagation delay limitation is more severe and it is of primary importance when the limitation is more severe than that due to propagation delay.

Equation (4.15) is general in the sense that it is valid for any two inputs that are adaptively combined to compute the residue. This can be derived by defining the main antenna voltage as $y_M(t)$ and auxiliary antenna voltage as $y_A(t)$. The residue thus is

$$r(t) = y_M(t) - w^* y_A(t) \tag{4.21}$$

The residue power is

$$p_{\text{res}} = p_{\text{main}} + |w|^2 p_{\text{aux}} - w^* \rho p_{ma} - w \rho^* p_{ma} \tag{4.22}$$

where p_{main} and p_{aux} are the total powers received by the main and auxiliary antennas respectively,

$$p_{\text{ma}} = \sqrt{p_{\text{main}} p_{\text{aux}}}$$

and ρ is the normalized correlation coefficient:

$$\rho = E[\, y_A(t) y_M^*(t)] / p_{ma} \tag{4.23}$$

Differentiation of (4.22) with respect to w^* shows the weight value that minimizes the residue power to be

$$w_{\text{opt}} = \rho \sqrt{p_{\text{main}} / p_{\text{aux}}} \tag{4.24}$$

and the combination of this relation with (4.22) shows that the minimum residue power is

$$p_{\min} = p_{\text{main}}(1 - |\rho|^2) \tag{4.25}$$

Note that this is equivalent to (4.15).

Let us denote the impulse response of the main channel as $h_M(t)$ and that of the auxiliary channel as $h_A(t)$. The input to each channel is $j(t)$. Thus,

$$\begin{aligned} y_M(t) &= \int j(t - x)h_M(x)\,\mathrm{d}x \\ y_A(t) &= \int j(t - x)h_A(x)\,\mathrm{d}x \end{aligned} \tag{4.26}$$

When the bandwidth of the noise jammer is much greater than the channel bandwidth, we have

$$\begin{aligned} E[\,y_A(t)y_M^*(t)] &= p_J \int h_A(x)h_M^*(x)\,\mathrm{d}x \\ p_{\text{main}} &= p_J \int |h_M(x)|^2\,\mathrm{d}x \\ p_{\text{aux}} &= p_J \int |h_A(x)|^2\,\mathrm{d}x \end{aligned} \tag{4.27a}$$

Although valid, these equations are of only marginal utility, as it is easier to visualize (and specify) channel mismatches in the frequency domain than the impulse-function domain.

According to Parsavel's theorem, these integrals can be transformed into the frequency domain. Thus,

$$\begin{aligned} E[\,y_A(t)y_M^*(t)] &= p_J \int H_A(f)H_M^*(f)\,\mathrm{d}f \\ p_{\text{main}} &= p_J \int |H_M(f)|^2\,\mathrm{d}f \\ p_{\text{aux}} &= p_J \int |H_A(f)|^2\,\mathrm{d}f \end{aligned} \tag{4.27b}$$

where $H_M(f)$ and $H_A(f)$ are the Fourier transforms of the respective impulse functions. As an example, assume that both filters are nominally rectangular bandpass functions of unity gain over the bandwidth, B. Due to errors, their gains vary linearly with frequency, but have opposite slopes. Thus,

$$\left.\begin{aligned} H_M(f) &= 1 - \alpha f \\ H_A(f) &= 1 + \alpha f \end{aligned}\right\} -B/2 < f < B/2 \tag{4.28}$$

Then,

$$\begin{aligned} p_{\text{main}} &= p_J B(1 + \alpha^2 B^2/12) = p_{\text{aux}} \\ E[y_A(t) y_M^*(t)] &= p_J B(1 - \alpha^2 B^2/12) \\ \rho &= \frac{1 - \alpha^2 B^2/12}{1 + \alpha^2 B^2/12} \approx 1 - \alpha^2 B^2/6 \end{aligned} \tag{4.29}$$

and combining these relations with (4.25) gives

$$p_{\min}/p_{\text{main}} = \alpha^2 B^2/3 \tag{4.30}$$

The peak transfer function error, ε_p, equals $\alpha B/2$, and in terms of the peak error:

$$p_{\min}/p_{\text{main}} = \frac{4}{3}\,\varepsilon_P^2 \tag{4.31}$$

Transfer function errors are commonly expressed in decibels. For an error of ε_{dB}:

$$p_{\min}/p_{\text{main}} = \frac{4}{3}\,(10^{0.05\varepsilon_{\text{dB}}} - 1)^2 \tag{4.32}$$

As an example, if ε_{dB} equals 0.1 dB, $p_{\min}/p_{\text{main}}$ equals -37.5 dB. For a peak error of 0.5 dB, it equals -23.3 dB. By comparing these values with those obtained for the propagation-delay analysis, we can see that amplifier mismatches of more than a few tenths of a decibel would be the limiting factor for achievable jammer cancellation.

Similar results are obtained for other forms of the transfer function. As an example, assume that the transfer function error is a sinusoidal ripple. The equivalent of (4.28) is

$$H_M(f) = 1 + \varepsilon_P \sin(\theta_M + 2\pi f/f_M)$$
$$H_A(f) = 1 + \varepsilon_P \sin(\theta_A + 2\pi f/f_A) \quad (4.33)$$

where θ_M and θ_A are independent random variables, uniformly distributed over 2π. The respective ripple frequencies are f_M and f_A. Note that

$$E(\cos\theta_M) = E(\sin\theta_M) = E(\cos\theta_A) = E(\sin\theta_A) = 0 \quad (4.34)$$

Using (4.34) in (4.27b) shows that

$$p_{\text{main}} = p_{\text{aux}} = p_J B(1 + \varepsilon_P^2/2)$$
$$\rho p_{ma} = p_J B \quad (4.35)$$

so that

$$\rho = (1 + \varepsilon_P^2/2)^{-1} \quad (4.36)$$

Assuming that ε_P is small, we have

$$p_{\text{min}}/p_{\text{main}} = \varepsilon_P^2 \quad (4.37)$$

which is very close to the result of (4.31).

Chapter 5
Multiple-Sidelobe Canceller (MSLC)

5.1 INTRODUCTION

The previous chapter discussed an adaptive antenna system that uses a single auxiliary antenna to cancel a single jammer. Adaptivity can be generalized to cancel multiple jammers by use of multiple auxiliaries. As shown, the use of at least J auxiliaries are necessary to cancel J independent jammers. As for the single-jammer case, the performance attained depends on the radar bandwidth and channel-to-channel match. In general, the performance improves as the number of auxiliaries increases beyond J.

5.2 MSLC CONCEPT

The initial discussion of the sidelobe canceller in Chapter 4 showed that an adaptive weight could be used to cancel a narrowband jammer received by a sidelobe of the main antenna. Similarly, we shall show that J narrowband jammers can be cancelled by adaptively weighting the outputs of J auxiliaries. Initially, for simplicity, the case of two narrowband jammers will be considered.

Figure 5.1 shows a two-auxiliary cancellation system. Let us denote the voltage received by the main antenna solely due to the jammers as $y_{0J}(t)$. Then, the generalization of (4.1) is

$$y_{0J}(t) = g_{01}j_1(t) + g_{02}j_2(t) \tag{5.1}$$

where g_{01} and g_{02} denote the main antenna's sidelobe gains in the directions of the two jammers, and $j_1(t)$ and $j_2(t)$ denote the complex envelopes of the jammers received by a unit gain antenna.

The voltages received by auxiliary antennas 1 and 2 are

$$\begin{aligned} y_{1J}(t) &= g_{11}j_1(t)\exp(-j\omega_0 t_{11}) + g_{12}j_2(t)\exp(-j\omega_0 t_{12}) \\ y_{2J}(t) &= g_{21}j_1(t)\exp(-j\omega_0 t_{21}) + g_{22}j_2(t)\exp(-j\omega_0 t_{22}) \end{aligned} \tag{5.2}$$

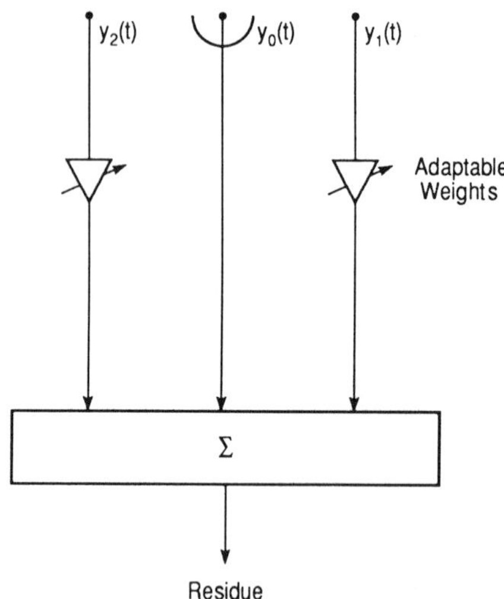

Figure 5.1 Conceptual block diagram of two-auxiliary MSLC.

The values g_{nm} are gains of auxiliary antenna, n, in the direction of jammer, m. Similarly, t_{nm} is the time delay of reception of jammer m at auxiliary n, relative to the time of reception by the main antenna.

The residue for jammers only is obtained as

$$r_J(t) = y_{0J} - w_1^* y_{1J} - w_2^* y_{2J} \tag{5.3}$$

This residue is zero when

$$\begin{aligned} w_1^* g_{11} \exp(-j\phi_{11}) + w_2^* g_{21} \exp(-j\phi_{21}) &= g_{01} \\ w_1^* g_{12} \exp(-j\phi_{12}) + w_2^* g_{22} \exp(-j\phi_{22}) &= g_{02} \end{aligned} \tag{5.4}$$

where

$$\phi_{nm} = \omega_0 t_{nm} \tag{5.5}$$

In matrix notation, (5.4) is

$$\mathbf{GW}^* = \mathbf{G}_0 \tag{5.6}$$

where the n-mth entry of the 2×2 matrix $\mathbf{G}$ equals g_{nm} [$\exp(-j\phi_{nm})$] and similar entries for the 2×1 column vectors $\mathbf{G}_0$ and $\mathbf{W}^*$. In general, the matrix $\mathbf{G}$ is nonsingular so that cancellation of two jammers with arbitrary angles of arrival is possible by using two auxiliaries. More generally, (5.6) can be written for J jammers and J auxiliaries, showing that cancellation of an arbitrary number of narrowband jammers with arbitrary angles of arrival can be accomplished by using an adaptive configuration, where the number of auxiliary antennas equals the number of narrowband jammers.

Note that exact jammer cancellation is usually not the preferred technique because target detection probability is maximized by minimizing residue power. As shown previously, the residue power for a single jammer is minimized by a slightly different weight value than that which achieves jammer cancellation. As shown later, this is also true for multiple jammers. The exact jammer cancellation technique is currently emphasized because the results are easily derived and approximately identical to the residue power minimization technique for large jammer powers.

5.3 MINIMUM REQUIRED NUMBER OF AUXILIARY ANTENNAS

For special conditions, multiple jammer cancellation can occur when the number of auxiliary antennas is less than the number of jammers. As an example, consider a two-jammer environment and a single auxiliary antenna. A single auxiliary antenna is equivalent to the case where one of the adaptable weights is constrained to always equal zero. Thus, from (5.4), two jammers can be cancelled when

$$w_1^* = g_{01} \exp(j\phi_{11})/g_{11} = g_{02} \exp(j\phi_{12})/g_{12} \tag{5.7}$$

Thus, for particular sets of jammers' angles of arrival, the main antenna sidelobe gains, relative time delays, and auxiliary antenna gains may be that (5.7) is satisfied. In general, however, this does not occur.

Another possibility that should be considered is whether the single auxiliary antenna system indicated in Figure 5.2 could be used to cancel two jammers. For this system, (5.1) and the first equation of (5.2) are still applicable. The second equation of (5.2) becomes

$$y_{2J}(t) = g_{11}j_1(t) \exp[-j\omega_0(t_{11} + t_d)] + g_{12}j_2(t) \exp[-j\omega_0(t_{12} + t_d)] \tag{5.8}$$

To be able to cancel two jammers with arbitrary angles of arrival, the matrix:

$$\mathbf{G} = \begin{pmatrix} g_{11} \exp(-j\omega_0 t_{11}) & g_{12} \exp(-j\omega_0 t_{12}) \\ g_{11} \exp[-j\omega_0(t_{11} + t_d)] & g_{12} \exp[-j\omega_0(t_{12} + t_d)] \end{pmatrix} \tag{5.9}$$

must be nonsingular. However, we can see that the elements of the second row equal the elements of the first row multiplied by the common factor $\exp(-j\omega_0 t_d)$. Thus, the matrix is singular and the circuit of Figure 5.2 is not capable of cancelling two jammers with arbitrary angles of arrival.

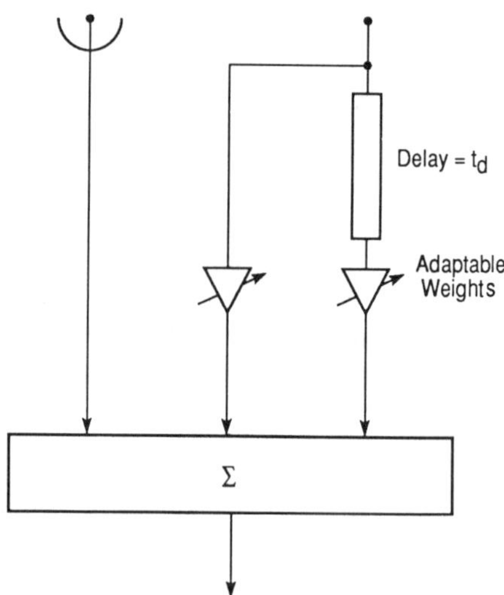

Figure 5.2 Conceptual block diagram of conjectured canceller.

In general, the minimum number of auxiliary antennas equals the number of jammers to be cancelled.

5.4 TWO-JAMMER CANCELLATION PERFORMANCE AS A FUNCTION OF NUMBER AND SPACING OF AUXILIARY ANTENNAS

The performance will be calculated under the condition of very large power (approximately infinite) and very narrowband jammers. The adapted gain of the main antenna and auxiliaries always will have exact nulls at the jammers' angles of arrival. As the jammers are very narrowband, they are cancelled and the residue power is due only to the combination of the thermal noises of the antennas. We shall assume that the gain pattern of each auxiliary antenna is omnidirectional.

5.4.1 Two Auxiliary Antennas

The first case to be considered is for two auxiliary antennas equispaced and bracketing the main antenna as shown in Figure 5.3(a). The matrix formulation of (5.4) for this case is

$$\begin{pmatrix} \exp(j\phi_1) & \exp(-j\phi_1) \\ \exp(j\phi_2) & \exp(-j\phi_2) \end{pmatrix} \begin{pmatrix} w_1^* \\ w_2^* \end{pmatrix} = \begin{pmatrix} g_{01} \\ g_{02} \end{pmatrix} \tag{5.10}$$

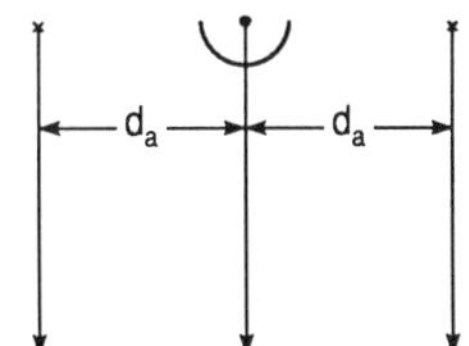

(a) Two Bracketing Auxiliaries

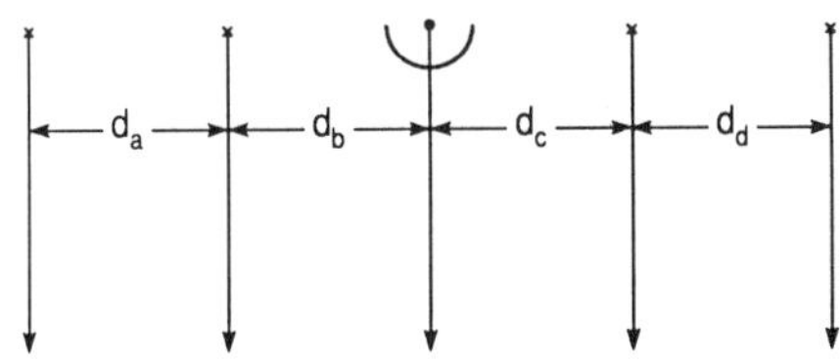

(b) Four Bracketing Auxiliaries

Figure 5.3 Auxiliary antenna configurations.

where ϕ_1 and ϕ_2 denote the phase shifts induced by the two jammers. The phase shifts are $\pm\phi_1$ and $\pm\phi_2$ because the antennas are symmetrically spaced around the main antenna. The solution of (5.10) is

$$\begin{pmatrix} w_1^* \\ w_2^* \end{pmatrix} = \begin{pmatrix} \exp(-j\phi_2) & -\exp(-j\phi_1) \\ -\exp(j\phi_2) & \exp(j\phi_1) \end{pmatrix} \begin{pmatrix} g_{01} \\ g_{02} \end{pmatrix} \Big/ \det \qquad (5.11)$$
$$\det = 2j \sin(\phi_1 - \phi_2)$$

The residue noise power due to the thermal noise addition from the three antennas is

$$p_{\text{res}} = p_{\text{noise}}(1 + |w_1|^2 + |w_2|^2) \qquad (5.12)$$

Note that $|w_1|^2 + |w_2|^2$ is the squared length of the adaptive weight vector and equals $W^H W$, where the superscript H denotes the transposed conjugate. For the residue power to be small, the squared length of the weight vector must be small. Manipulating (5.11) shows that the residue power is

$$p_{\text{res}} = p_{\text{noise}}\{1 + 2[|g_{01}|^2 + |g_{02}|^2 - (g_{01}^* g_{02} + g_{01} g_{02}^*) \cos(\phi_1 - \phi_2)]/|\det|^2\} \qquad (5.13)$$

The residue power becomes infinite when the determinant equals zero and the bracketed term is nonzero. For two very closely spaced jammers, ϕ_1 tends toward ϕ_2, but g_{01} tends toward g_{02} so that the bracketed term also tends toward zero. When ϕ_1 and

ϕ_2 differ by a nonzero integer multiple of π, the bracketed term is nonzero and the determinant equals zero so that the residue power tends toward infinity. For angles of arrival of θ_1 and θ_2, and any nonzero integer k, this occurs when

$$k\pi = \phi_1 - \phi_2 = 2\pi d_a[\sin\theta_1 - \sin\theta_2]/\lambda \tag{5.14}$$

For a dish antenna, with d_a equal to half the dish diameter, after trigonometric manipulation and approximation, this reduces to

$$k = [(\theta_1 - \theta_2)/\theta_b] \cos\theta_{av} \tag{5.15}$$

where the beamwidth equals θ_b, and θ_{av} is the average of θ_1 and θ_2. This result indicates that cancellation of two jammers separated "exactly" by integer multiples of a beamwidth causes large residue powers.

A more quantitative measure of the residue power as a function of the jammers' angles of arrival can be obtained by modeling the jammers' sidelobe voltage gains as zero-mean random variables. This is a reasonable assumption because, as can be seen by examining the sidelobe gain pattern of Figure 2.1, the effect of implementation errors is to cause a noiselike sidelobe gain pattern.

A convenient way to evaluate (5.12) is as a function of three regions of the difference between ϕ_1 and ϕ_2. For one region, the difference between ϕ_1 and ϕ_2 is large, corresponding to angle-of-arrival differences of more than approximately one beamwidth. Thus, the angles of arrival are in different sidelobes so that the gains g_{01} and g_{02} are reasonably modeled as independent random variables. Averaging (5.13) shows that the average residue power is

$$p_{res} = p_{noise}[1 + G_s/\sin^2(\phi_1 - \phi_2)] \tag{5.16}$$

where G_s denotes the average absolute sidelobe power gain. As a numerical example, the average absolute sidelobe power gain tends to be on the order of one-half. Hence, the residue power is at least 10 dB above the thermal noise power if the magnitude of the phase angle difference is less than 13.6°. If the phase angle difference is a uniformly distributed random variable, the probability of the normalized residue power exceeding 10 dB is 0.15.

A second region is where the two jammers are at almost the same angle so that they are nearly equivalent to a single jammer. For this case, the bracketed term of (5.13) tends toward zero so that the residue power does not become infinite, although the difference between ϕ_1 and ϕ_2 tends toward zero. If g_{01} equals g_{02}, we have

$$p_{\text{res}} = p_{\text{noise}}[1 + G_s(1 - \cos\varepsilon)/\sin^2\varepsilon] \tag{5.17}$$

where ε equals $\phi_1 - \phi_2$. Small-angle approximations for cosine and sine then give

$$p_{\text{res}} = p_{\text{noise}}(1 + G_s/2) \tag{5.18}$$

for two very closely spaced jammers.

This result is exactly equal to the residue power obtained for a single jammer. To show this, let us consider cancellation of a single jammer with a single omnidirectional auxiliary antenna. The jammer voltage received at the main antenna differs from that received at the auxiliary antenna by the factor g_0 and a phase shift due to time-of-arrival delay. Thus, the magnitude of the weight value used by the single auxiliary equals $|g_0|$ and the residue power is

$$p_{\text{res}} = p_{\text{noise}}(1 + |g_0|^2) \tag{5.19}$$

For two auxiliaries and a single jammer, symmetry implies that the weights' magnitude of both auxiliary outputs equals $|g_0|/2$. Thus,

$$|w_1|^2 + |w_2|^2 = 2G_s/4 \tag{5.20}$$

which result is equivalent to (5.18).

The third region connects the previous two regions. The gains g_{01} and g_{02} are neither statistically independent nor identical. A reasonable model is that they are correlated random variables and the normalized correlation coefficient, ρ, depends on ε. The exact dependence of ρ on ε probably differs for different antennas, different deterministic weights, and the errors of the deterministic weights. A first-order approximation is that ρ is unity for $\varepsilon = 0$, and varies with the magnitude of ε. We will assume that $\rho = 0$ for $\varepsilon \geqslant \pi/2$. A convenient assumption is

$$\begin{aligned} E(g_{01}g_{02}^*) &= G_s\rho \\ \rho &= \cos\varepsilon; \quad |\varepsilon| < \pi/2 \\ \rho &= 0; \quad |\varepsilon| > \pi/2 \end{aligned} \tag{5.21}$$

Combining (5.13) and (5.21) gives

$$p_{\text{res}} = p_{\text{noise}}[1 + G_s(1 - \rho\cos\varepsilon)/\sin^2\varepsilon] \tag{5.22}$$

Note that $\rho = 0$ gives (5.16), and ρ equals unity produces (5.17). Hence, (5.22) reduces to

$$\begin{aligned} p_{res} &= p_{noise}[1 + G_s]; \quad |\varepsilon| < \pi/2 \\ &= p_{noise}[1 + G_s/\sin^2\varepsilon]; \quad |\varepsilon| > \pi/2 \end{aligned} \tag{5.23}$$

Thus, given the assumption of (5.21), two narrowband jammers spaced within a fraction of a sidelobe can be cancelled with a resulting residue power slightly larger than that of a single jammer. Although this statement is exactly true for only the particular assumption of (5.21), it is approximately true for similar assumptions.

In summary, cancellation of two narrowband jammers by two auxiliary antennas can be done with relatively low residue power, except when the two jammers have particular geometric spacings. The probability that the normalized residue power exceeds 10 dB is approximately 0.15.

5.4.2 Four Auxiliary Antennas

This section derives an equation for the residue power for the four-auxiliary-antenna system of Figure 5.3(b), where the adaptive weights are set to cancel exactly two very narrowband jammers. As is shown, the residue power is less than that of the two-auxiliary system. In addition, we can see that substantial performance improvement is attained by nonuniformly spacing the four auxiliary antennas.

For four uniformly spaced auxiliary antennas, the equivalent of (5.10) is

$$\begin{pmatrix} \exp(j\phi_1) & \exp(j\phi_1/3) & \exp(-j\phi_1/3) & \exp(-j\phi_1) \\ \exp(j\phi_2) & \exp(j\phi_2/3) & \exp(-j\phi_2/3) & \exp(-j\phi_2) \end{pmatrix} \begin{pmatrix} w_1^* \\ w_2^* \\ w_3^* \\ w_4^* \end{pmatrix} = \begin{pmatrix} g_{01} \\ g_{02} \\ g_{03} \\ g_{04} \end{pmatrix} \tag{5.24}$$

The four-auxiliary-antenna system is obtained from the two-auxiliary system by placing two additional antennas between the original auxiliary locations and the main antennas. Thus, the phase shifts for the first and last auxiliary are the same value for the two- and four-auxiliary systems.

This system of equations is not square. It is termed an *underdetermined equation set* and has an infinite number of solutions. Note that, as for the two-auxiliary cancellation configuration, the residue power for the four-auxiliary case is due to thermal-noise combination from the main antenna and four auxiliaries. The residue power is minimized when the length of the weight vector is minimized. Thus, although (5.24) has an infinite number of solutions, the unique solution with minimum length minimizes the residue power. Finding the minimum length solution of an underdetermined equation set is a classic problem. The solution is given by the Penrose pseudoinverse. Specifically, (5.24) can be written in matrix notation as

$$\boldsymbol{GW}^* = \mathbf{G}_0 \tag{5.25}$$

This matrix form is identical to (5.6), but now $\mathbf{G}$ is a 2×4 matrix, $\mathbf{W}$ is a 4×1 column vector, and $\mathbf{G}_0$ is a 2×1 column vector. The minimum-weight-length solution is given by

$$\mathbf{W}^* = \mathbf{G}^{\mathbf{H}}(\mathbf{G}\mathbf{G}^{\mathbf{H}})^{-1}\mathbf{G}_0 \tag{5.26}$$

Note that $\mathbf{G}\mathbf{G}^{\mathbf{H}}$ is a square matrix and generally will have an inverse. Also note that if $\mathbf{G}$ is a square matrix, the pseudoinverse solution reduces to the conventional solution.

By using (5.26), we can then show that

$$\begin{aligned} \mathbf{W}^{\mathbf{H}}\mathbf{W} &= \mathbf{G}_0^{\mathbf{H}}(\mathbf{G}\mathbf{G}^{\mathbf{H}})^{-1}\mathbf{G}_0 \\ p_{\text{res}} &= p_{\text{noise}}[1 + \mathbf{G}_0^{\mathbf{H}}(\mathbf{G}\mathbf{G}^{\mathbf{H}})^{-1}\mathbf{G}] \end{aligned} \tag{5.27}$$

From (5.24), we can show that

$$\begin{aligned} \mathbf{G}\mathbf{G}^{\mathbf{H}} &= 2\begin{pmatrix} 2 & \cos\varepsilon + \cos(\varepsilon/3) \\ \cos\varepsilon + \cos(\varepsilon/3) & 2 \end{pmatrix} \\ \varepsilon &= \phi_2 - \phi_1 \end{aligned} \tag{5.28}$$

Note that when $\varepsilon = 0$, each matrix entry is equal to 2 so that the matrix is singular. However, $\varepsilon = 0$ corresponds closely to the single-jammer case and, as in Section 5.3, this does not cause large residue. However, the matrix is also singular when ε equals any integral multiple of 3π. These cases correspond to large residue power. Again, modeling g_{01} and g_{02} as independent random variables gives

$$p_{\text{res}} = p_{\text{noise}}\{1 + 2G_s/[4 - (\cos\varepsilon + \cos(\varepsilon/3))^2]\} \tag{5.29}$$

Figure 5.4 compares the residue power for the two and four equispaced auxiliary canceller configurations as a function of the differential phase shift ε, for ε in the region where g_{01} and g_{02} are modeled as independent random variables. The solid curve is for the two-auxiliary configuration, and the dashed curve is for the four-auxiliary configuration. We can see that the four-auxiliary system has substantially smaller residue power. This is especially true in the regions of 180° and 360°, where the residue power of the two-auxiliary configuration becomes infinite, whereas that of the four-auxiliary configuration does not. As discussed previously, the probability of the normalized residue power exceeding 10 dB is approximately 0.15 for the two-auxiliary-antenna configuration. The finite residue power at 180° and 360° implies

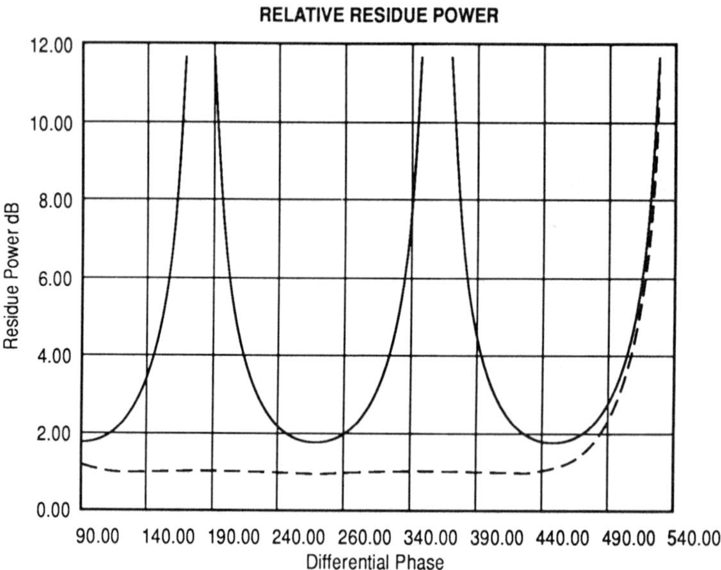

Figure 5.4 Performance comparison for two and four uniformly spaced auxiliary MSLC.

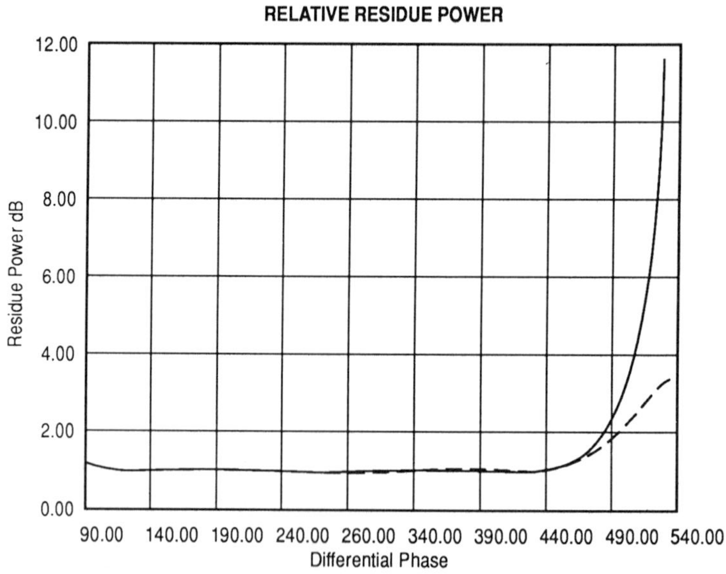

Figure 5.5 Performance comparison for uniformly and nonuniformly spaced four-auxiliary MSLC.

that the comparable probability is reduced to 0.05 for the uniformly spaced four-auxiliary-antenna configuration.

Even better performance can be obtained by nonuniform spacing of the auxiliaries. The solid curve of Figure 5.5 is a repeat of the four-auxiliary configuration of Figure 5.4. The dashed curve is for a nonuniformly spaced four-auxiliary system. Specifically, the two auxiliaries closest to the main antenna are displaced ±30% from their uniformly spaced positions. We can see that a major effect of nonuniform spacing is that the infinite power condition at 540° has been alleviated. Finite residue power requires that the matrix determinant be nonzero. In theory, this can be attained over all angles by using spacings between auxiliaries that are relatively prime.

5.5 RESIDUE MINIMIZATION

The preceding material analyzed the residue power when the weights used to combine the auxiliary and main antenna voltages were chosen to cancel narrowband jammers exactly. As previously indicated in the single-jammer analysis (Chapter 4), target detectability is maximized by minimizing residue power, and the values of the weights for residue power minimizaton are not quite identical to those used for exact jammer cancellation. Also, the MSLC technique can only perform exact jammer cancellation for the very narrowband case, and the nonzero radar bandwidth usually has significant effect on the residue power. This section derives equations and analyzes the attained performance for the residue power minimizaton criterion for the multiple-jammer and multiple-auxiliary-antenna scenario. In general, as will be shown, the equations are such that developing simple physical interpretations is difficult. Thus, predicting the performance that will be attained for a particular configuration and jammer geometry is likewise difficult. A useful result of the prior sections on exact cancellation of two narrowband jammers is that a relatively simple interpretation of the effect of the number and spacing of auxiliary antennas can be developed. As shown, these interpretations can be used to explain the results obtained for the nonzero bandwidth residue minimization criterion.

5.5.1 Optimum Weights Derivation

The voltage at the main antenna for J jammers plus thermal noise is

$$y_0(t) = n_0(t) + \sum_{k=1}^{J} g_{0k} j_k(t) \tag{5.30}$$

where the thermal noise is $n_0(t)$, $j_k(t)$ denotes the complex envelope of the kth jammer, and g_{0k} is the voltage gain of the main antenna at the jammer's angle of arrival.

The J jammer waveforms are not necessarily due to J physically distinct jammers. Some of these may result from reflections due to multipath effects.

The voltage at the nth auxiliary antenna is

$$y_n(t) = n_n(t) + \sum_{k=1}^{J} g_{nk} \exp(-j\phi_{nk}) j_k(t - t_{nk}) \tag{5.31}$$

where $n_n(t)$ denotes the thermal noise, g_{nk} is the voltage gain of the nth auxiliary antenna in the direction of the kth jammer, t_{nk} is the propagation delay with respect to the main antenna's time of arrival, and ϕ_{nk} is the phase shift of the carrier induced by the time delay. The residue is obtained as the weighted sum of the main and auxiliary antenna voltages:

$$r(t) = y_0(t) - \sum_{n=1}^{N} w_n^* y_n(t) \tag{5.32}$$

5.5.1.1 Statistical Review

Our goal is to determine the weights that minimize the residue power. As shown, the optimum weights are a function of the interantenna statistical correlations. This section gives a short review of statistical correlation.

The residue power is the statistical expectation of $|r(t)|^2$. Thus,*

$$\begin{aligned} p_{\text{res}} = E\{&|y_0(t)|^2 - \Sigma[w_n y_0(t) y_n^*(t) + w_n^* y_0^*(t) y_n(t)] \\ &+ \Sigma\Sigma w_n^* w_m y_n(t) y_m^*(t)\} \end{aligned} \tag{5.33}$$

where $E(\cdot)$ denotes the statistical expectation operator. In this equation, the weight values are fixed, but unknown. The random processes occur as pairwise products. The covariances are defined as $E[y_n(t)y_m^*(t)]$. In general, the covariances depend on time. For the MSLC application, the statistics are time-independent and

$$r_{nm} = E[y_n(t) y_m^*(t)] \tag{5.34}$$

From (5.31),

$$\begin{aligned} r_{nm} = E\Bigg\{&\left[n_n(t) + \sum_{k=1}^{J} g_{nk} \exp(-j\phi_{nk}) j_k(t - t_{nk})\right] \\ &\left[n_m^*(t) + \sum_{k=1}^{J} g_{mk}^* \exp(j\phi_{mk}) j_k^*(t - t_{mk})\right]\Bigg\} \end{aligned}$$

*The summation limits for indices n or m are all 1 through N.

This can be simplified by term-by-term multiplication and then applying the theorem that the expectation of a sum is the sum of the expectations. As an example,

$$E|j_1(t) + j_2(t)|^2 = E[j_1(t)j_1^*(t)] + E[j_1(t)j_2^*(t)] + E[j_2(t)j_1^*(t)] + E[j_2(t)j_2^*(t)] \quad (5.35)$$

A further simplification is obtained by recognizing that the thermal noise of any channel is statistically independent of any jamming waveform. Because both are zero-mean random processes:

$$E[n_n(t)j_k^*(t - t_{mk})] = 0 \quad (5.36)$$

for all subscripts n, k, and m. Similarly, the thermal noise of different channels is statistically independent, so that

$$\begin{aligned} E[n_n(t)n_m^*(t)] &= 0; \quad n \neq m \\ &= p_{\text{noise}}; \quad n = m \end{aligned} \quad (5.37)$$

For certain applications, some of the J jammer waveforms are dependent. For simplicity, this section will only consider the statistically independent case. Thus,

$$\begin{aligned} E[j_k(t - t_{nk})j_p^*(t - t_{mp})] &= 0; \quad k \neq p \\ &= p_k\rho(t_{mk} - t_{nk}); \quad k = p \end{aligned} \quad (5.38)$$

where $\rho(\cdot)$ is the normalized correlation coefficient. As discussed in Subsection 4.3.1.1, the value of the normalized correlation coefficient depends on the value of the time delay, $t_{mk} - t_{nk}$, and the radar waveform. For a delay of t_d and a simple pulse transmission of duration τ, we have

$$\begin{aligned} \rho(t_d) &= 1 - |t_d/\tau|; \quad |t_d| < \tau \\ &= 0; \quad |t_d| > \tau \end{aligned} \quad (5.39)$$

For an LFM waveform of two-sided bandwidth, B:

$$\rho(t_d) = \sin(\pi B t_d)/(\pi B t_d) \quad (5.40)$$

For the very narrowband waveform assumption, the normalized correlation coefficient approaches unity. As for the single-jammer canceller and multiple jammers, the normalized correlation coefficient must be close to 1 to obtain high cancellation ratios. This requirement, in turn, implies that high cancellation ratios cannot be obtained unless the product of the time delay and radar waveform bandwidth is small.

Combining the preceding results gives

$$r_{nm} = p_{\text{noise}}\delta_{nm} + \sum_{k=1}^{J} p_k g_{nk} g_{mk}^* \exp[-j(\phi_{nk} - \phi_{mk})]\rho(t_{mk} - t_{nk}) \qquad (5.41)$$

where $\delta_{nm} = 0$ for $n \neq m$, and $\delta_{nm} = 1$ for $n = m$. Note that for $n = m$, r_{nn} is real. Also, note that r_{mn} is the complex conjugate of r_{nm}.

5.5.1.2 Matrix Representation

Manipulation of the residue power equations is simplified by the use of matrix equations. Define the N-dimensional column vectors $\mathbf{W}$ and $\mathbf{Y}$ as

$$\begin{aligned} \mathbf{W}^{\mathrm{T}} &= (w_1, w_2, \ldots, w_N) \\ \mathbf{Y}^{\mathrm{T}} &= (y_1(t), y_2(t), \ldots, y_N(t)) \end{aligned} \qquad (5.42)$$

where superscript $\mathbf{T}$ denotes the transposed conjugate. The inner product between the vectors $\mathbf{W}$ and $\mathbf{Y}$ is defined as

$$\boldsymbol{\alpha} = \sum_{n=1}^{N} w_n^* y_n = \mathbf{W}^{\mathbf{H}}\mathbf{Y} \qquad (5.43)$$

where superscript $\mathbf{H}$ denotes the complex conjugate. Because $\boldsymbol{\alpha}$ is a scalar, it equals its transpose. Thus,

$$\begin{aligned} \boldsymbol{\alpha} &= \boldsymbol{\alpha}^{\mathrm{T}} = (\mathbf{W}^{\mathbf{H}}\mathbf{Y})^{\mathrm{T}} = \mathbf{Y}^{\mathrm{T}}\mathbf{W}^* \\ \boldsymbol{\alpha}^* &= \mathbf{Y}^{\mathbf{H}}\mathbf{W} \end{aligned} \qquad (5.44)$$

Combining (5.32) and (5.43) gives the residue voltage as

$$r(t) = y_0(t) - \mathbf{W}^{\mathbf{H}}\mathbf{Y} \qquad (5.45)$$

Combining (5.44) and (5.45), the residue power is

$$p_{\text{res}} = E[r(t)r^*(t)] = E[y_0(t) - \mathbf{W}^{\mathbf{H}}\mathbf{Y}][y_0^*(t) - \mathbf{Y}^{\mathbf{H}}\mathbf{W}] \qquad (5.46)$$

The covariance matrix, $\mathbf{R}$, is a square $N \times N$ matrix, the nth row, mth column entry of which is r_{nm}. Note that the nth row, mth column of the matrix product $\mathbf{YY^H}$ is $y_n(t)y_m^*(t)$. Thus,

$$E(\mathbf{YY^H}) = \mathbf{R} \tag{5.47}$$

Similarly, $y_0^*(t)\mathbf{Y}$ is an N-dimensional vector, the nth component of which equals $y_0^*(t)y_n(t)$. Thus, the vector $\mathbf{R}_0$ is

$$\mathbf{R}_0 = E[\,y_0^*(t)\mathbf{Y}] \tag{5.48}$$

and the nth entry of $\mathbf{R}_0$ equals r_{n0}, the expectation of the product $y_n(t)y_0^*(t)$.

The terms of the covariance matrix, $\mathbf{R}$, are the pairwise correlations of the auxiliary antennas. The terms of the correlation vector are the pairwise correlations of the auxiliary and main antennas. Each of these matrices can be decomposed into sums of terms for thermal noise and each jammer. From (5.41), $\mathbf{R}$ can be decomposed into $J + 1$ matrices. One is for thermal noise; the others are for the J jammers. Because the thermal noise is statistically independent and of the same power level at each antenna, its contribution to $\mathbf{R} = p_{\text{noise}}\mathbf{I}$, where $\mathbf{I}$ denotes the $N \times N$ identity matrix. Thus,

$$\mathbf{R} = p_{\text{noise}}\mathbf{I} + \sum_{k=1}^{J} p_k J_k \tag{5.49}$$

where J_k is an $N \times N$ matrix. From (5.41), the nth row, mth column entry of J_k is

$$(\mathbf{J}_k)_{nm} = g_{nk}g_{mk}^* \exp[-j(\phi_{nk} - \phi_{mk})]\rho(t_{mk} - t_{nk}) \tag{5.50}$$

There is a somewhat similar decomposition for $\mathbf{R}_0$. One difference is that there is no thermal-noise term, as the thermal noise is independent at all antennas. Therefore,

$$\mathbf{R}_0 = \sum_{k=1}^{J} p_k \mathbf{J}_{k0} \tag{5.51}$$

where $\mathbf{J}_{k0}$ is an N-dimensional vector and its nth entry is given by

$$(\mathbf{J}_{k0})_n = g_{nk}g_{0k}^* \exp(-j\phi_{nk})\rho(-t_{nk}) \tag{5.52}$$

For the very narrowband waveform case, all of the normalized correlation coefficients equal unity and $\mathbf{J}_k$ is equal to a matrix outer product. Thus, we define the vector $\mathbf{S}_k$, the nth component of which is $g_{nk} \exp(-j\phi_{nk})$. Then,

$$\mathbf{J}_k = \mathbf{S}_k\mathbf{S}_k^{\mathbf{H}} \tag{5.53}$$

5.5.1.3 Optimum Weight Vector

In matrix notation, the residue power for an arbitrary weight vector $\mathbf{W}$ is

$$p_{\text{res}} = E|y_0(t)|^2 - \mathbf{R}_0^{\mathbf{H}}\mathbf{W} - \mathbf{W}^{\mathbf{H}}\mathbf{R}_0 + \mathbf{W}^{\mathbf{H}}\mathbf{R}\mathbf{W} \tag{5.54}$$

The first term is the total power received by the main antenna and will be denoted as p_{main}.

The weights for minimum residue power can be obtained by differentiating with respect to w_1^*, w_2^*, through w_N^* and setting the N derivatives simultaneously equal to zero. Note, as an example, that

$$\frac{\partial}{\partial w_1^*}(\mathbf{W}^{\mathbf{H}}\mathbf{A}) = \frac{\partial}{\partial w_1^*}(\Sigma w_n^* a_n) = a_n \tag{5.55}$$

Thus,

$$\frac{\partial(p_{\text{res}})}{\partial w_k^*} = -(\mathbf{R}_0)_k + (\mathbf{R}\mathbf{W})_k \tag{5.56}$$

Setting the N partial equation to zero gives the optimum weight vector matrix equation:

$$\mathbf{R}\mathbf{W}_{\text{opt}} = \mathbf{R}_0 \tag{5.57}$$

Substituting this into (5.54) shows that the residue power, when using the optimum weight vector, is

$$p_{\min} = p_{\text{main}} - \mathbf{R}_0^{\mathbf{H}}\mathbf{R}^{-1}\mathbf{R}_0 = p_{\text{main}} - \mathbf{R}_0^{\mathbf{H}}\mathbf{W}_{\text{opt}} \tag{5.58}$$

For any other weight vector, $\mathbf{W} = \mathbf{W}_{\text{opt}} + \Delta$, substitution into (5.54) shows that

$$p_{\text{res}} = p_{\min} + \Delta^{\mathbf{H}}\mathbf{R}\Delta \tag{5.59}$$

5.6 SIMPLIFIED EXAMPLES OF OPTIMUM WEIGHT VECTOR PERFORMANCE

This section derives equations for the output residue power when using optimum weights for some simple jammer scenarios. We do so to gain additional insight into behavior as a function of radar system parameters such as the number of auxiliary elements and the radar bandwidth. The following section gives additional results for more complex environments, evaluated by computing the residue power from (5.58), using the covariance matrix and correlation vector for the specified environment.

5.6.1 Thermal-Noise Environment

The interference consists solely of zero-mean noise, which is statistically independent from antenna to antenna, and common power denoted as p_{noise}. Then, the correlation vector equals the zero vector and the covariance matrix equals the identity matrix multiplied by p_{noise}. From (5.57), the optimum weight vector equals the zero vector. Thus, in the absence of jamming, the output consists only of the main antenna voltage. From (5.58), the minimum residue power equals the unadapted main antenna power, which equals the thermal-noise level.

5.6.2 Single Narrowband Jammer

From (5.49) and (5.53), the covariance matrix is

$$\mathbf{R} = p_{\text{noise}}\mathbf{I} + p_1\mathbf{S}_1\mathbf{S}_1^{\mathbf{H}} \tag{5.60}$$

We shall assume that each auxiliary antenna is identical and omnidirectional so that the nth component of $\mathbf{S}_1 = \exp(-j\phi_{n1})$. Similarly, from (5.51), the correlation vector is

$$\mathbf{R}_0 = g_{01}^* p_1 \mathbf{S}_1 \tag{5.61}$$

For N auxiliary antennas, $\mathbf{R}$ is a square $N \times N$ matrix. Both $\mathbf{R}_0$ and $\mathbf{S}_1$ are N-dimensional vectors. A convenient approach is to define p_{noise} as equal to unity and to interpret p_1 as the *jammer-to-noise ratio* (JNR). Then, by direct multiplication, we can verify that

$$\mathbf{R}^{-1} = \mathbf{I} - [(p_1/(1 + p_1N)]\mathbf{S}_1\mathbf{S}_1^{\mathbf{H}} \tag{5.62}$$

and, from (5.57), the optimum weight vector is

$$\mathbf{W}_{\text{opt}} = g_{01}^*\mathbf{S}_1 p_1/(1 + p_1N) \tag{5.63}$$

This weight vector steers the auxiliary array to the jammer angle and develops a beam that peaks at the jammer angle. This beam is termed a *perturbation* or *retrodirective beam*. The main antenna power, prior to adaptation, is the sum of the thermal-noise power plus the received jammer power. Thus,

$$p_{\text{main}} = 1 + |g_{01}|^2 p_1 = 1 + G_s p_1 \tag{5.64}$$

where G_s denotes the main antenna's power gain in the jammer direction. From (5.58),

$$p_{\min} = 1 + G_s p_1/(1 + p_1 N) \tag{5.65}$$

If $p_1 = 0$, being the thermal-noise-only environment of Section 5.6.1, the minimum residue power equals the thermal-noise power of unity. The residue power monotonically increases with jammer power, but, for large jammer power, it asymptotically approaches

$$(p_{\min})_{\text{aym}} \sim 1 + G_s/N \tag{5.66}$$

Thus, the residue power decreases as the main antenna sidelobe gain decreases and decreases as the number of elements in the auxiliary array increases.

5.6.3 Multiple Orthogonal Direction Vector Jammers

From (5.49) and (5.53), the covariance matrix is

$$\mathbf{R} = \mathbf{I} + \Sigma p_k \mathbf{S}_k \mathbf{S}_k^{\mathbf{H}} \tag{5.67}$$

and, for omnidirectional auxiliary antennas, the nth component of $\mathbf{S}_k$ equals $\exp(-j\phi_{nk})$. From (5.51) and (5.52), the correlation vector is

$$\mathbf{R}_0 = \Sigma p_k g_{0k}^* \mathbf{S}_k \tag{5.68}$$

In general, to obtain a closed-form expression for the optimum weight vector is difficult because of the difficulty of obtaining an expression for the inverse of $\mathbf{R}$. When the directional vectors are mutually orthogonal, there is a simple expression for the inverse, and therefore for the optimum weight vector.

There are many conditions for orthogonal directional vectors. As an example, consider the case where the array of auxiliary antennas is an equispaced line array with interelement spacing equal to d_x. For a jammer at spatial angle θ_1, the nth component of the direction vector equals $\exp[-j2\pi d_x(n-1)\sin\theta_1/\lambda]$. For a second jammer at θ_2 the inner product for the two directional vectors is

$$\mathbf{S}_2^{\mathbf{H}}\mathbf{S}_1 = \Sigma \exp[-j2\pi d_x(n-1)(\sin\theta_1 - \sin\theta_2)/\lambda] \tag{5.69}$$

which equals zero when $Nd_x(\sin\theta_1 - \sin\theta_2)/\lambda$ equals an integer. Note that the beamwidth of an array of total width Nd_x is approximately equal to λ/Nd_x. Thus, two directional vectors are orthogonal when the differential spatial angle is approximately equal to the auxiliary array beamwidth.

For K orthogonal directional vector jammers, with $K < N$, the largest K eigenvalues equal $1 + p_kN$ with associated unit length eigenvectors of $\mathbf{S}_k/N^{1/2}$. The other $N - K$ eigenvalues equal unity with eigenvectors orthogonal to the set of $\mathbf{S}_k$. Let us denote the nth eigenvalue of R as d_n. Then,

$$\mathbf{R}^{-1} = \Sigma\, d_n^{-1}\mathbf{S}_n\mathbf{S}_n^{\mathbf{H}}/N^{1/2} \tag{5.70}$$

Combining (5.57), (5.67), (5.68), and (5.70), and recognizing that $\mathbf{S}_n^{\mathbf{H}}\mathbf{S}_m$ equals either zero or N gives

$$\mathbf{W}_{\text{opt}} = \Sigma\, g_k^*\mathbf{S}_k[p_k/(1 + p_kN)] \tag{5.71}$$

and combining this with (5.58) gives

$$p_{\text{res}} = p_{\text{main}} - N\,\Sigma\, |g_k|^2 p_k^2/(1 + p_kN) \tag{5.72}$$

For equal-power interference sources and an average antenna sidelobe gain of G_a, this reduces to

$$p_{\text{res}} = p_{\text{main}} - NKG_a p^2/(1 + pN) \tag{5.73}$$

and for pN much greater than unity this is approximately

$$p_{\text{res}} = p_{\text{main}} - KG_a p + KG_a/N \tag{5.74}$$

Note that p_{main} is the sum of thermal noise (currently defined to be unit power) plus the sidelobe received interference. Thus,

$$p_{\text{res}} = (1 + KG_a/N)p_{\text{noise}} \tag{5.75}$$

which generalizes to non-unity thermal-noise power. Note that the residue power does not depend on the power of the interferers, but rather the number of the interferers.

5.6.4 Nonzero Bandwidth Jammers

One way of viewing the degradation of cancellation for nonzero bandwidth is via the ambiguity of frequency and spatial angle. Thus, the relative phase shift of a received waveform for a carrier frequency of f_0 and angle of arrival θ depends on the product of $f_0 \sin\theta$. Thus, a change of f_0 is ambiguous with a change of angle of arrival, and a point source of nonzero bandwidth is equivalent to a spatially distributed zero-bandwidth source.

A qualitative indication of the effect of system bandwidth on cancellation as a function of the number of jamming sources and auxiliary antennas can be obtained by first considering the case where the number of jammers equals the number of auxiliary antennas and the adaptive weight vector is set so that the response of the adapted antenna pattern is exactly equal to zero at the jammer angles. In the neighborhood of the kth jammer, the adapted gain pattern is approximately

$$g_a(\theta) \approx (\theta - \theta_k)\dot{g}_a(\theta_k); \quad \theta \approx \theta_k \tag{5.76}$$

Assume that, due to the radar system bandwidth, the source of power p is distributed over an angle β. Partial differentiation of the product $f_0 \sin\theta$ indicates that for a bandwidth, B, the equivalent angle spread equals $B \tan(\theta_k)/f_0$. The resulting residue power is approximately

$$\begin{aligned} p_{\text{res}}(k) &\approx [|\dot{g}_a(\theta_k)|^2 p/\beta] \int_{-\beta/2+\theta_k}^{\beta/2+\theta_k} (\theta - \theta_k)^2 \, d\theta \\ &\approx |\dot{g}_a(\theta_k)|^2 p\beta^2/12 \end{aligned} \tag{5.77}$$

For N auxiliary antennas and N jammers, assuming $\dot{g}_a(\theta_k)$ is the common value of $\dot{g}_a$ and neglecting the dependence of β on θ_k gives a total residue power of

$$p_{\text{res}} = pN|\dot{g}_a|^2(B/f_0)^2/12 \tag{5.78}$$

The last equation states that the residue increases as the square of the relative bandwidth.

Now, consider the case of N auxiliary antennas and $N/2$ jammers. Assume that the optimum adaptive weight is approximately that which creates a double zero at each jammer angle. This clearly is a way of decreasing the sensitivity to bandwidth, although not necessarily in an optimal way. For this case, the equivalent of (5.74) is

$$g_a(\theta) = (\theta - \theta_k)^2\ddot{g}_a(\theta_k); \quad \theta \approx \theta_k \tag{5.79}$$

and the resulting residue power is

$$p_{\text{res}} = p(N/2)|\ddot{g}_a|^2(B/f_0)^4/80 \tag{5.80}$$

Comparing (5.78) and (5.80) and recognizing that B/f_0 usually is very small compared to unity indicates that substantially superior cancellation is achieved when the number of auxiliary antennas exceeds the number of jammers.

5.7 OPTIMUM WEIGHT VECTOR PERFORMANCE

The previous section derived expressions for the optimum weight vector and resulting residue power in terms of the covariance matrix of the auxiliary antennas and the vector of covariances between the main and auxiliary antennas. The values of the covariances depend on many parameters such as the following:

1. Number of interferers;
2. Interferers' spatial angles and powers;
3. Main antenna's characteristics;
4. Number of auxiliary antennas;
5. Spacing of auxiliary antennas;
6. Bandwidth.

The results of Section 5.4 on exact jammer cancellation are useful for predicting performance trends as a function of some parameters. As examples, the section shows that the residue power decreased when the number of auxiliary antennas increased from two to four. Residue power also decreased when using four nonuniformly spaced auxiliary antennas, rather than uniformly spaced auxiliary antennas. However, the results of Section 5.4 did not indicate performance trends as a function of number of jammers, jammer power, radar bandwidth, or channel-to-channel mismatch. The performance trends as a function of these parameters require evaluation of (5.58).

This section gives the results of an optimum weight performance analysis to obtain quantitative MSLC performance data for better specifying radar system requirements. The performance measure is the ratio of the adapted residue power when interferers are present to that when interferers are absent (thermal noise). This increase in residue power is a system degradation, and the allowable system degradation determines the required number of auxiliary elements, allowable radar signal bandwidth, *et cetera*.

The evaluation was performed for a main antenna configured as a 64-element, equispaced, linear array with element-to-element spacing equal to one half-wavelength ($\lambda/2$). The auxiliary antennas are not array elements. The initial portion of the evaluation is for a four-auxiliary-antenna array with both uniformly and nonuniformly spaced auxiliaries. For both cases, the overall length of the auxiliary array equals that of the main antenna array. Many different interferer angles were considered. Specifically, for each selection of parameters 1 through 6, some 501 sets of interferer angles were chosen by a random mechanism. For each angle set, the adapted residue power was computed. The results for the 501 sets were combined to obtain a cumulative probability function of the adapted residue power.

5.7.1 Performance Results

The evaluation was for a main antenna implemented as a 64-element array. The MSLC performance depends on the number of interferers, their powers, and angular locations. For the four-auxiliary MSLC, the number of interferers considered is either one, two, or four. For each number of interferers, each interferer has the same power, but the total interference power is constant. For the first scenario, the power is such that the ratio of interferer power to noise power for the four-interferer scenario equals 30 dB at an array element. The ratio equals 33 dB for the two-interferer scenario and 36 dB for the single-interferer case.

The performance was evaluated by placing the interferers at independent random spatial angles constrained to be within a 50° wide sidelobe region bracketing the main beam. For the *known* spatial angles and *known* powers, the optimum MSLC weights and resulting residue power is computed. The residue power is normalized

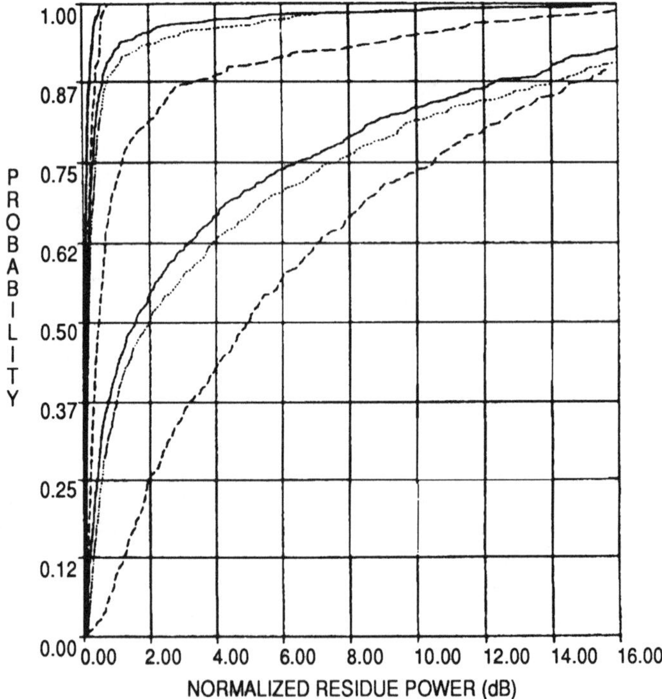

Figure 5.6 Effect of number of jammers on MSLC, 30 dB jammers, four uniformly spaced auxiliaries.

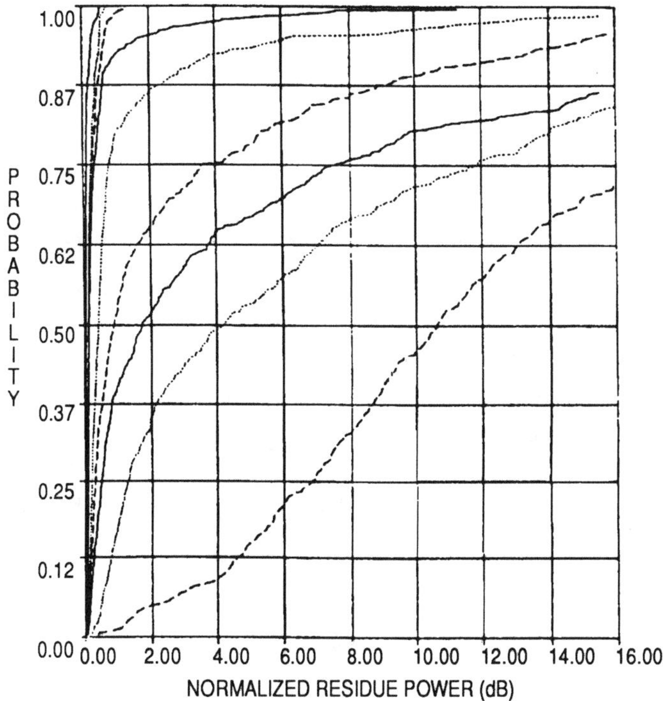

Figure 5.7 Effect of number of jammers on MSLC, 40 dB jammers, four uniformly spaced auxiliaries.

to the level achieved in the absence of interferers (the thermal-noise-only case). The 501 sets of random interferer locations were chosen for each condition. The 501 data points were combined to compute the cumulative probability function of the normalized residue power. This result is shown in Figure 5.6.

The nine curves of Figure 5.6 are for three different bandwidth conditions, and the number of interferers equal to one, two, and four. The fractional bandwidths (ratio of the smaller of the radar signal bandwidth or interferer bandwidth to the radar carrier frequency) are very narrow (essentially zero), 0.001 and 0.004. As an example, 1 MHz bandwidth and a radar carrier frequency of 1 GHz gives a fractional bandwidth of 0.001. The three solid curves are for the very narrow bandwidth case and one, two, and four interferers. The normalized residue power increases as the number of interferers increases. We can see that, for four very narrowband jammers, the probability that the residue power is at least 12 dB greater than thermal noise equals 0.13. For two jammers, the 0.13 probability is for a residue increase of about 0.8 dB. The dotted curves are for a normalized bandwidth of 0.001. The dashed

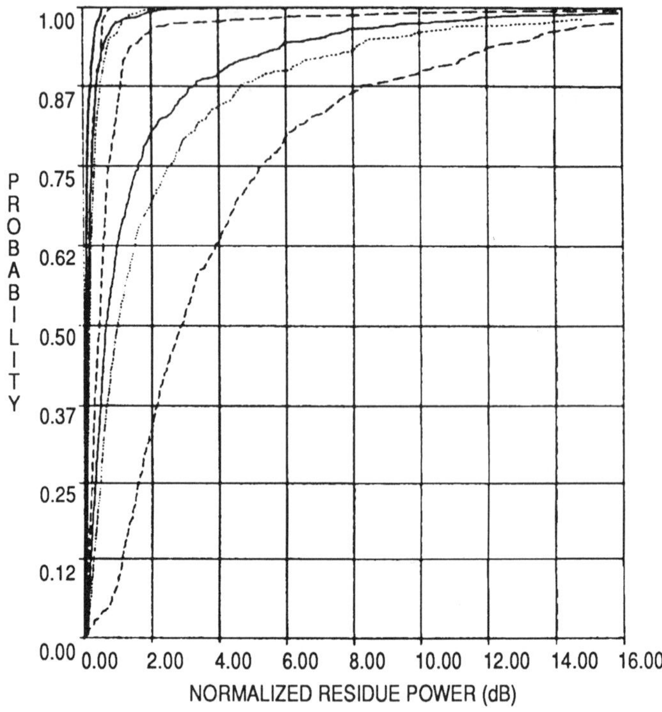

Figure 5.8 Effect of number of jammers on MSLC, 30 dB jammers, four nonuniformly spaced auxiliaries.

curves are for a normalized bandwidth of 0.004. We can see that the normalized residue power increases as the bandwidth increases.

Figure 5.7 differs from Figure 5.6 in that each jammer power has increased by 10 dB and the sets of jammer angles are independently selected. We can see that, for each curve, the normalized residue power increases, but not by an amount equal to 10 dB. The increase of normalized residue power is greatest for the four-jammer case of bandwidth equal to 0.004. This result is similar to the effect shown by Figure 4.2. Specifically, the bandwidth value of 0.004 is less than the transition value. For this example, we can show that the normalized residue power increases by 10 dB for the case of the normalized bandwidth equal to 0.01.

The effect of nonuniform spacing of the auxiliary elements is shown by Figures 5.8 and 5.9. The only change as compared to Figures 5.6 and 5.7 is that the auxiliary elements are nonuniformly spaced. Each set of jammer angles is repeated. We can see that the normalized residue power is substantially decreased. The effects of

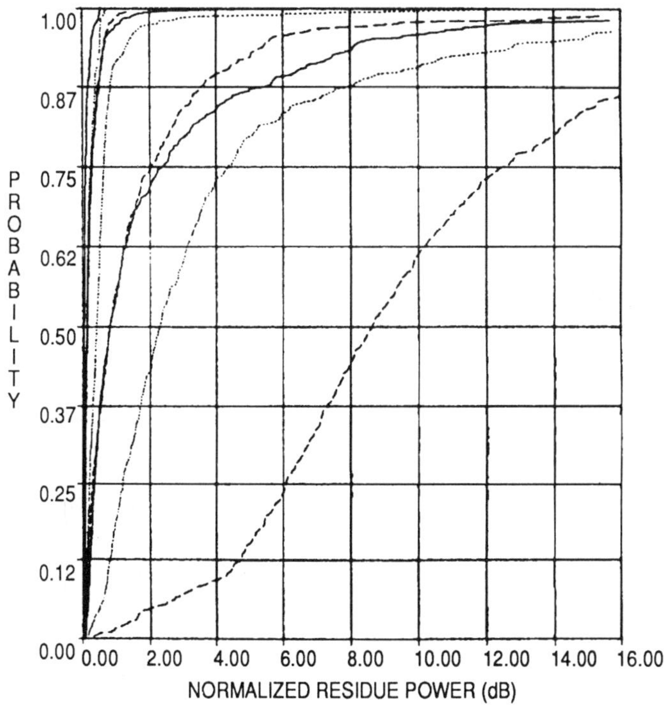

Figure 5.9 Effect of number of jammers on MSLC, 40 dB jammers, four nonuniformly spaced auxiliaries.

channel-to-channel mismatch are indicated by Figures 5.10 and 5.11 for the 30 dB and 40 dB jammer cases. The channel-to-channel mismatch equals 0.1 dB. Comparing Figures 5.8 and 5.10, we can see that the 0.1 dB mismatch has an almost negligible effect on performance for the 30 dB jammer case. However, from a comparison of Figures 5.9 and 5.11, we observe that the 0.1 dB mismatch has a significant effect when the jammer powers are 40 dB. Also note that the degradation is relatively large when good cancellation is obtained for a perfect match, and small when the cancellation was poor. Specifically, note that the normalized residue power noticeably increases for the single-jammer case at all bandwidths. However, the performance for the largest bandwidth, four-jammer scenario is essentially unaffected by the channel mismatch. The achieved cancellation for this case is obviously bandwidth-limited rather than mismatch-limited.

Figures 5.12 and 5.13 are for eight nonuniformly spaced auxiliary antennas and number of jammers equal to two, four, and eight. The jammer powers are 40

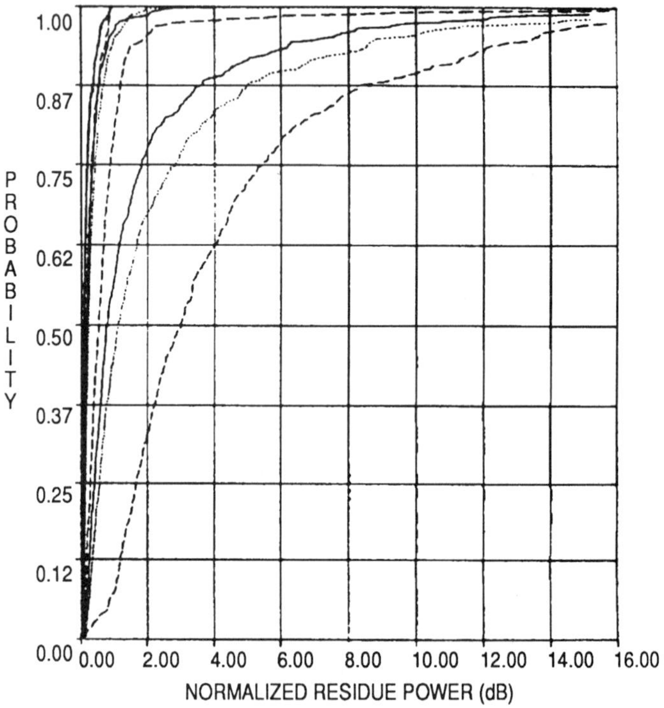

Figure 5.10 Effect of number of jammers on MSLC, 30 dB jammers, four nonuniformly spaced auxiliaries, 0.1 dB channel mismatch.

dB. The normalized bandwidths are the same values as previously used. Figure 5.12 is for perfect channel matching. Figure 5.13 is for a channel mismatch of 0.1 dB. The four nonuniformly spaced auxiliary with perfect channel match is shown in Figure 5.9. Comparing Figures 5.9 and 5.12 shows that, for two and four jammers, the normalized residue power is substantially decreased by increasing the number of auxiliary elements. Note that the normalized residue power is essentially identical when the number of jammers equals one-quarter or one-half the number of auxiliary elements. To a first-order approximation, the normalized residue power depends on the ratio of the number of jammers to the number of auxiliary elements. When the number of jammers equals the number of auxiliary elements, the effect of bandwidth is much greater on the eight-auxiliary system than on the four-auxiliary system. Similarly, comparing Figures 5.11 and 5.13, we can see that, for the same number of jammers and channel mismatch, the normalized residue power is much less for the

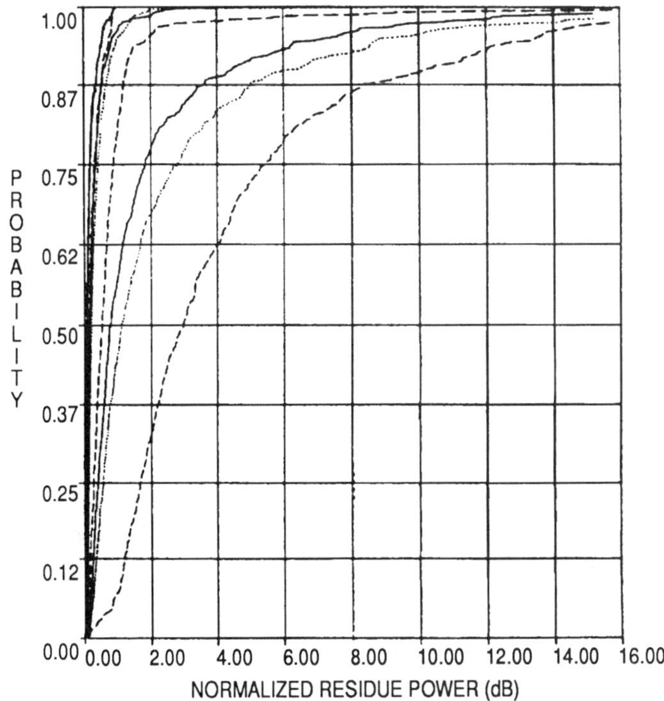

Figure 5.11 Effect of number of jammers on MSLC, 40 dB jammers, four nonuniformly spaced auxiliaries, 0.1 dB channel mismatch.

eight-auxiliary system than for the four-auxiliary system. However, the effect of channel mismatch is much greater for the eight-auxiliary system when the number of jammers is eight as compared to the four-auxiliary system when the number of jammers equals four.

Figures 5.14 and 5.15 are identical to the preceding two except that the array has been increased to 128 elements. The normalized location of the auxiliary elements is unchanged. The effect of the increased array size is that the normalized residue power decreases for the very narrowband waveforms and increases for the large bandwidth waveforms.

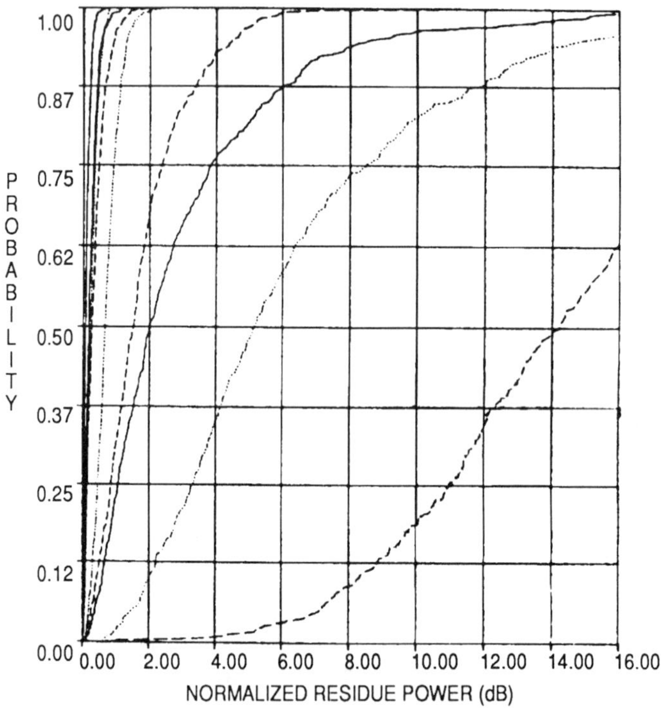

Figure 5.12 Effect of number of jammers on MSLC, 40 dB jammers, eight nonuniformly spaced auxiliaries, 0.0 dB channel mismatch.

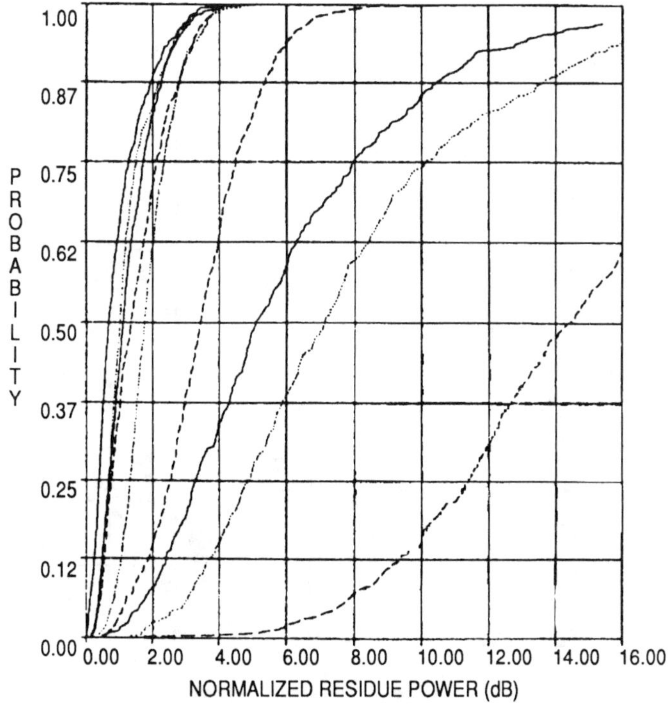

Figure 5.13 Effect of number of jammers on MSLC, 40 dB jammers, eight nonuniformly spaced auxiliaries, 0.1 dB channel mismatch.

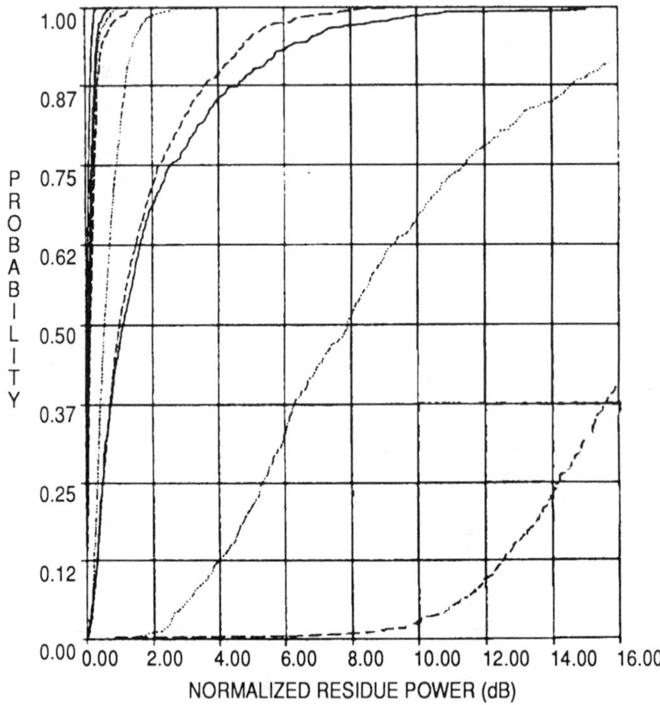

Figure 5.14 Effect of number of jammers on MSLC, 128-element array, 0.0 dB channel mismatch.

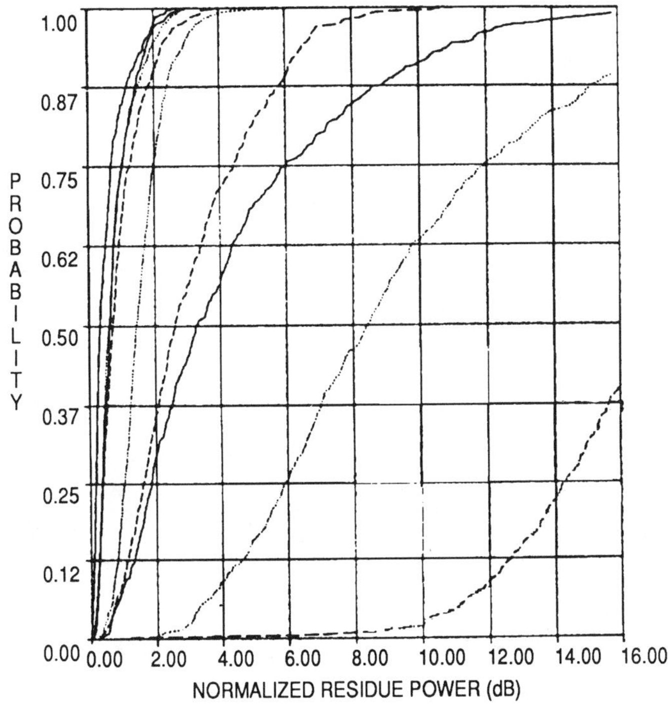

Figure 5.15 Effect of number of jammers on MSLC, 128-element array, 0.1 dB channel mismatch.

Chapter 6
Optimum MSLC Weight Vector Estimation

6.1 INTRODUCTION

Optimum weights are derived in terms of the values of the correlations among the antennas. Knowledge of these correlations is equivalent to knowing the exact angular locations and powers of the jammers. In addition, these correlations require exact knowledge of the gains of the antennas in the directions of the jammers. For any practical application, none of these values is known so that we need to develop algorithms that can estimate the optimum weights. The main use of the theory of optimum weights is that its performance represents an unachievable upper bound that can be used to judge the efficacy of candidate algorithms. The ratio of the two performances is a measure of the efficiency of the estimation algorithm.

Many algorithms have been suggested that directly or indirectly estimate the covariance matrix and compute the estimated weights. Early work, in classified form, resulted in the *Howells-Applebaum correlation feedback loops* [1,2]. This processor was used in radar for cancelling multiple sidelobe jammers. An essentially identical algorithm, usually denoted as the LMS (least mean squares) algorithm, was developed by Widrow and applied to communication problems [3].

The algorithm has been termed a *stochastic gradient descent algorithm*. A significant similarity to the gradient descent algorithm is that the convergence time, as measured by the number of iterations required to approximate the optimum weight solution, can be very long. For some environmental conditions, the convergence time is so long compared to the time scale of a practical problem that the convergence time is essentially infinite.

There are other algorithms that converge much more rapidly than the LMS algorithm. Algorithms that directly estimate the covariance matrix and then compute the estimated weights have been considered. Reed, Mallett, and Brennan [4] analyzed the performance of the adaptive processor when the matrix estimator is the maximum likelihood estimate of a Hermitian matrix. They showed that, for an N-weight adaptive processor, performance within 3 dB of the theoretical optimum can be achieved when the number of independent vector measurements used to estimate the covariance matrix equals $2N - 3$. The somewhat remarkable nature of this result is that, whereas the convergence time of the LMS algorithm strongly depends on the particular covariance matrix, the convergence time of the processor based on the Hermitian covariance matrix maximum likelihood estimator is independent of the matrix and always equals $2N - 3$.

This chapter discusses and analyzes the performance of these and other algorithms. The performance and implementation costs are compared. One significant aspect of the implementation cost is the number of analog-to-digital (A/D) converters that are used. Algorithms that directly estimate covariance matrices require A/D conversion at each antenna. The matrix estimation and weight computation is then performed by the use of special-purpose digital computers. These techniques, denoted as *digital techniques,* are relatively expensive to implement. The LMS technique can be implemented either digitally, by A/D conversion at each antenna, or by all-analog circuitry techniques. The latter is considerably less expensive to implement and is often preferred, even when its performance is inferior to that of other techniques. We also discuss variations of the analog LMS algorithm that have improved convergence properties.

6.2 ANALOG IMPLEMENTATION OF THE LMS ALGORITHM

Figure 6.1 is a conceptual block diagram of the LMS algorithm. It is also denoted as a *correlation feedback loop* because the implementation is a feedback circuit that drives the output until the weight values are such that the residue voltage is uncorrelated with each of the auxiliary inputs. As will be shown in Section 6.2.2, the weights that minimize the residue power are those that cause the residue voltage to be uncorrelated with the auxiliary antenna voltages.

As indicated in Figure 6.1, in complex-envelope notation, the antenna voltages are at the radian carrier frequency ω_0. The implementation for an N-auxiliary-antenna system requires N correlation feedback loops. Only one is shown. The auxiliary voltages are down-converted (mixed) with an offset oscillator with a radian frequency equal to $\omega_0 - \omega_{IF}$. The filter passes the difference frequency.

6.2.1 Multiplication by Mixing and Filtering

For each mixer shown in the block diagram (Figure 6.1), the combination of mixing and filtering is intended to implement the mathematical operation of multiplication of the two inputs. The design requires careful choice of mixers, intermediate frequencies, and filter characteristics to obtain a "multiplier" with small errors.

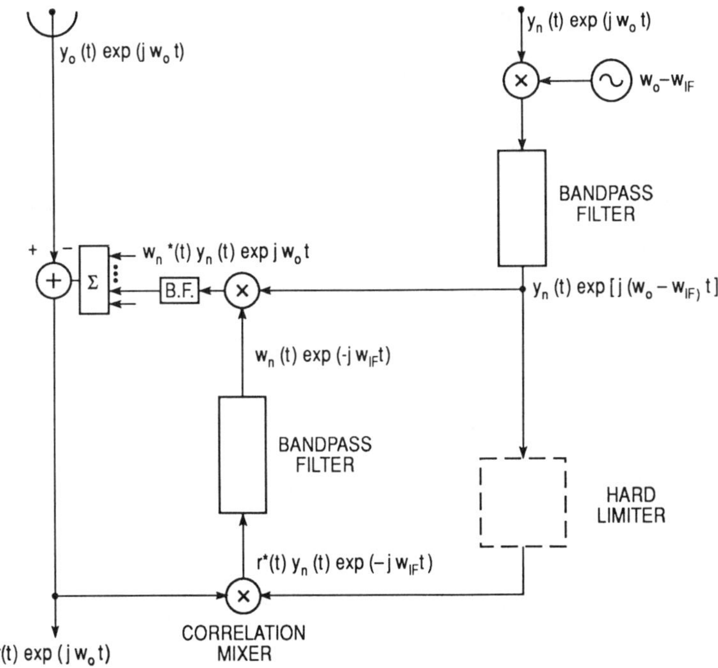

Figure 6.1 Conceptual block diagram of the LMS (correlation feedback loop) algorithm.

The nonlinear input-output characteristic of the mixer can be represented by a power series as

$$e_{\text{out}} = \Sigma_k \alpha_k e_{\text{in}}^k \tag{6.1}$$

The input, e_{in}, is the sum of the two individual inputs:

$$e_{\text{in}} = r_1(t)\cos[\omega_1 t + \phi_1(t)] + r_2(t)\cos[\omega_2 t + \phi_2(t)] \tag{6.2}$$

where $r(t)$ denotes envelope modulation and $\phi(t)$ denotes phase modulation. For $k > 1$ in (6.1), the nonlinearity creates output terms with radian frequencies that differ from those of the input, ω_1 and ω_2. For $k = 2$, the cross term of the squaring creates sum-and-difference frequency terms. The difference term is the desired output:

$$e_2 = r_1(t)r_2(t)\cos[(\omega_2 - \omega_1)t + \phi_2(t) - \phi_1(t)] \tag{6.3}$$

We define the complex envelopes:

$$\begin{aligned} a_1(t) &= r_1(t)\exp[j\phi_1(t)] \\ a_2(t) &= r_2(t)\exp[j\phi_2(t)] \end{aligned} \tag{6.4}$$

The equivalent of (6.3) is that the complex envelope of the output equals $a_2(t)$ times the complex conjugate of $a_1(t)$. Thus, the action of each mixer-filter is such that if the complex representation of the two inputs are denoted as $a_1(t)\exp(j\omega_1 t)$ and $a_2(t)\exp(j\omega_2 t)$, the complex representation of the output is $a_1^*(t)a_2(t)\exp[j(\omega_2 - \omega_1)t]$.

Note that the desired output of (6.3) cannot be attained exactly because the squared terms create frequencies at $2\omega_2$ and $2\omega_1$. The filter, after the mixer, attenuates these terms, but cannot totally eliminate them. Similarly, frequency terms due to the other exponents of (6.1) create error terms, and the designer must choose mixer and filter characteristics so that the multiplication approximation is done with minimal error.

6.2.2 Residue-Auxiliary Correlation

The correlaton of the residue and the nth auxiliary voltage is defined as

$$c_{rn} = E[r^*(t)y_n(t)] \tag{6.5}$$

The vector of correlations, $\mathbf{C}$, the nth component of which is c_{rn}, is

$$\mathbf{C} = E[r^*(t)\mathbf{Y}] \tag{6.6}$$

For the optimum weights, the residue is given by (5.45). Thus,

$$\mathbf{C} = E[y_0^*(t)\mathbf{Y}] - E[\mathbf{Y}\mathbf{Y}^{\mathbf{H}}\mathbf{W}] \tag{6.7}$$

and combining this with (5.47) and (5.48) gives

$$\mathbf{C} = \mathbf{R}_0 - \mathbf{R}\mathbf{W} \tag{6.8}$$

From (5.57), the vector $\mathbf{C} = 0$. Thus, if the correlation feedback circuit exactly converges to the condition that the residue is uncorrelated with each auxiliary voltage, the obtained weights equal the optimum weights.

6.2.3 MSLC Weight Computation Using LMS Algorithm

To continue our discussion of the block diagram, the residue is mixed with each of the down-converted auxiliary voltages. The dashed box denotes a hard limiter. This modification will be discussed in Section 6.2.4. The filter following the correlation mixer is narrowband and centered at the IF. The output of the narrowband filter of each correlation feedback loop is the weight value for the auxiliary antenna. Each weight is mixed with its associated auxiliary voltage and then bandpass filtered. Note that each weight is time-dependent, and thus not equal to the optimum weight. In addition, as each is a function of random processes, the weights are random processes. However, we shall show that the average value of the weights converge to a close approximation of the optimum weights.

In terms of the complex envelopes, we can see from Figure 6.1 that

$$r(t) = y_0(t) - \Sigma_m w_m^*(t) y_m(t)$$
$$w_n(t) = \int h(t - x) r^*(x) y_n(x)\, \mathrm{d}x \tag{6.9}$$

where $h(\cdot)$ denotes the impulse response of the narrowband filter after the correlation mixer. Often, this is a single-pole filter, the operation of which is described by the differential equation:

$$T[\mathrm{d}w_n(t)/\mathrm{d}t] + w_n(t) = g r^*(t) y_n(t) \tag{6.10}$$

where T is the filter time constant and g is the dc gain. Combining (6.10) with the first part of (6.9) gives

$$T[\mathrm{d}w_n(t)/\mathrm{d}t] + w_n(t) = g[y_0^*(t) y_n(t) - \Sigma w_m(t) y_m^*(t) y_n(t)] \tag{6.11}$$

As $w_n(t)$ is a random process, (6.11) is a stochastic difference equation. The solution is the joint probability density function for the set of $w_n(t)$. This cannot be solved by any known techniques. Instead, the moments of $w_n(t)$ can be obtained. This section will derive the first moment, or mean, of $w_n(t)$. We shall show that the mean is time-dependent. Thus, if the interference environment initially solely consists of thermal noise, the weights are values appropriate for a thermal-noise environment. If the environment suddenly changes to thermal noise plus one or more jammers, the feedback action causes the weights to change to values appropriate for jammer cancellation. The change of weight values is not instantaneous. The weight values are time-dependent and the time period required to change from the values appropriate to a thermal-noise environment to the jammer cancellation values is termed the *convergence time*.

6.2.3.1 Single Auxiliary Convergence Time Analysis

The convergence time analysis of the multiple-auxiliary LMS algorithm is complicated by the coupling of the weight values due to the summation in (6.11). To develop a better understanding of the convergence action, the single auxiliary is initially analyzed. For this case, (6.11) becomes

$$T[dw_1(t)/dt] + w_1(t)[1 + g|y_1(t)|^2] = gy_0^*(t)y_1(t) \tag{6.12}$$

Taking the mean value of both sides of (6.12) gives

$$\begin{aligned} &T[dw_a(t)/dt] + w_a(t) + gE[w_1(t)|y_1(t)|^2] = g\rho\sqrt{p_M p_A} \\ &w_a(t) = E[w_1(t)] \\ &p_M = E|y_0(t)|^2 \\ &p_A = E|y_1(t)|^2 = r_{11} \\ &\rho\sqrt{p_M p_A} = E[y_0^*(t)y_1(t)] = r_{10} \end{aligned} \tag{6.13}$$

where p_M and p_A are the total interference powers of the main antenna and auxiliary antennas, respectively, and ρ denotes the correlation coefficient of the main and auxiliary antennas. We shall assume that the filter bandwidth is very narrow compared to the radar bandwidth. Then, the present value of the weight is approximately uncorrelated with the present value of the auxiliary voltage. Thus,

$$E[w_1(t)|y_1(t)|^2] = w_a(t)p_A = w_a(t)r_{11} \tag{6.14}$$

Combining (6.13) and (6.14) produces the first-order constant coefficient differential equation:

$$T[dw_a(t)/dt] + w_a(t)\,[1 + gp_A] = g\rho\sqrt{p_M p_A} \tag{6.15}$$

Assuming the average weight value equals zero at t equals zero, the solution is

$$w_a(t) = g\rho\sqrt{p_M p_A}/(1 + gp_A)\{1 - \exp[-(1 + gp_A)t/T]\} \tag{6.16}$$

The steady-state solution is

$$w_s = g\rho\sqrt{p_M p_A}/(1 + gp_A) \tag{6.17}$$

By comparing this relation to (4.24), we can see that, when the filter gain, g, is large, the steady-state, average-weight value approximately equals the optimum weight.

The transient time constant equals $T/(1 + gp_A)$. During the time of the transient decay, the residue power is greater than that achieved at steady state. By neglecting the difference between the steady-state average-weight and the optimum weight, and assuming the normalized correlation coefficient equals one, the transient time weight error is

$$\delta(t) = \sqrt{p_M/p_A}\, \exp(-gp_A t/T) \tag{6.18}$$

From (5.59), this weight error causes the residue power to equal the minimum residue power plus the following:

$$\begin{aligned} p_{\text{add}} &= p_M \exp(-2gG_{\text{SL}}p_M t/T) \\ p_A &= G_{\text{SL}}\, p_M \end{aligned} \tag{6.19}$$

The sketch of Figure 6.2 depicts the convergence as a parametric function of jammer power.

As indicated by the sketch, the residue power is initially approximately equal to the thermal-noise power. Then, the environment changes, and, from (6.19) at $t = 0$, the additional residue power equals the main antenna power. As the weight approaches its steady-state value, the residue power (in dB) decreases linearly with

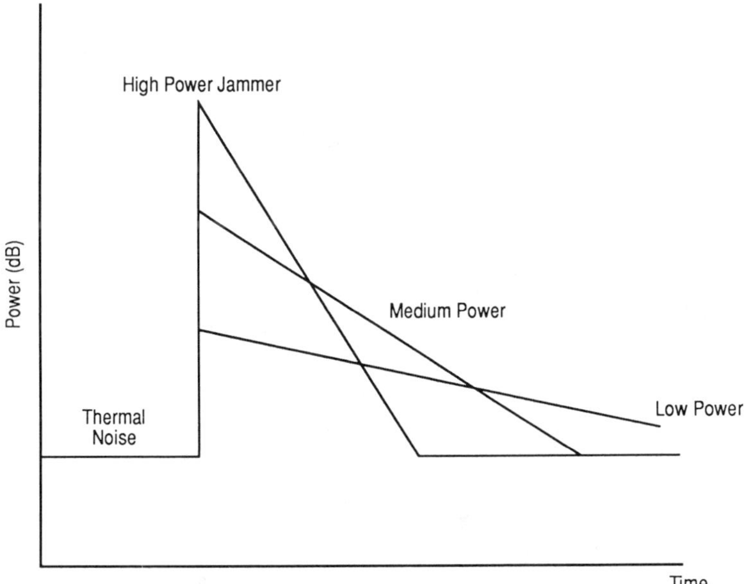

Figure 6.2 Single-jammer convergence characteristics.

a slope that depends on the main antenna power. As we can see, the convergence is fast for large jammer power and slow for small jammer power.

There are two operational constraints on the acceptable values of the time constant. First, the time constant should be longer than the time duration comparable to a range cell so that there would be little tendency for self-cancellation of a target return. Second, the time constant should be much shorter than the radar beam's dwell period. For many radars, this implies that the convergence time should be greater than a few microseconds, but much less than a few milliseconds. However, variations in the jamming threat can easily cause the received jammer power, and therefore the time constant, to vary by four or more orders of magnitude. Thus, long convergence time might limit the applicability of this algorithm.

Several modifications to the LMS algorithm have been proposed to decrease the variability of the convergence time. The sections to follow analyze two modified LMS algorithms with convergence times that are less dependent on the power level of the auxiliary antenna. Both modify the auxiliary antenna input to the correlation mixer.

6.2.4 MSLC Weight Computation Using LMS Algorithm with Hard Limiter

This technique modifies the auxiliary antenna input to the correlation mixer by use of an IF hard limiter. An IF hard limiter is a device with an input-output characteristic such that when the input is $r(t)\cos[\omega_0 t + \phi(t)]$, the output is $L\cos[\omega_0 t + \phi(t)]$, where L is a constant amplitude. Thus, its action is such that all amplitude modulation is removed by the limiting action. As previously discussed for the LMS algorithm, the resulting equation is a stochastic difference equation and the only available analysis is by moments. In addition, the convergence analysis is complicated by the coupling of the weight values. To develop an understanding of the action of the limiter, the convergence analysis is performed for the single-auxiliary-antenna system.

6.2.4.1 Single-Auxiliary Convergence Time Analysis

Equation (6.12) is modified to

$$\begin{aligned} &T[dw_1(t)/dt] + w_1(t)[1 + g_L y_1^*(t) y_L(t)] = g_L y_0^*(t) y_L(t) \\ &y_L(t) = L y_1(t)/|y_1(t)| \end{aligned} \tag{6.20}$$

where $y_L(t)$ denotes the hard-limited version of $y_1(t)$. Taking the average value of both sides of (6.20) again produces a first-order constant coefficient differential equation. The time constant, T_e, and steady-state weight value, w_e, are

$$\begin{aligned} &T_e = T/\{1 + g_L E[y_1^*(t) y_L(t)]\} \\ &w_s = g_L E[y_0^*(t) y_L(t)]/\{1 + g_L E[y_1^*(t) y_L(t)]\} \end{aligned} \tag{6.21}$$

To evaluate the indicated expectations, we need to specify the statistics of $y_0(t)$ and $y_1(t)$. For jamming cancellation, a reasonable assumption is that all voltages have a zero-mean joint Gaussian distribution. Then, the required expectations can be evaluated by transformation to a set of independent zero-mean Gaussian distributed unit variance complex random variables, $x_1(t)$ and $x_0(t)$. Define

$$\begin{aligned} y_1(t) &= \sqrt{p_A}x_1(t) \\ y_0(t) &= \sqrt{p_M}[\rho^* x_1(t) + \sqrt{(1-|\rho|^2)}x_0(t)] \end{aligned} \tag{6.22}$$

To prove the validity of this decomposition, note that, because $x_1(t)$ and $x_0(t)$ are zero-mean Gaussian processes, $y_1(t)$ and $y_0(t)$ are zero-mean Gaussian processes. Also, the respective variances of $y_1(t)$ and $y_0(t)$ are p_A and p_M, and the normalized correlation coefficient is ρ. Thus, all moments of $y_0(t)$ and $y_1(t)$ are satisfied and the decomposition is valid. Then,

$$E[y_1^*(t)y_L(t)] = L\sqrt{p_A}E|x_1(t)|$$

To evaluate the required expectation, note that the probability density function for $x_1(t)$ is a zero-mean complex Gaussian with variance r_{11} and, for the current case, the variance equals 1. Specifically, the probability density $p(x_1)$ is

$$\begin{aligned} p(x_1) &= \exp(-|x_1|^2/r_{11})/\pi r_{11} \\ p(r,\theta) &= r\exp(-r^2/r_{11})\pi r_{11}; \quad r \geqslant 0, -\pi \leqslant \theta \leqslant \pi \\ x_1 &= r\exp(j\theta) \end{aligned} \tag{6.23}$$

and the two densities are equivalent with the second expressing the complex-valued variable in polar coordinates. Thus, for arbitrary μ,

$$E(|x|^\mu) = \int_0^{2\pi}\int_0^\infty [r^{1+\mu}\exp(-r^2/r_{11})/\pi r_{11}]\,dr\,d\theta$$

This evaluates to

$$E(|x|^\mu) = r_{11}^{\mu/2}\Gamma(1+\mu/2) \tag{6.24}$$

For the required expectation, $\mu = 1$, so that $E|x_1(t)| = \sqrt{\pi}/2$. Similarly,

$$E[y_0^*(t)y_L(t)] = L\sqrt{p_M}E|x_1(t)|$$

Substituting these moments into (6.21) gives

$$\begin{aligned} T_e &= T/(1 + g_1\sqrt{p_A}) \\ w_s &= g_1\rho\sqrt{p_M}/(1 + g_1\sqrt{p_A}) \\ g_1 &= g_L\sqrt{\pi}/2 \end{aligned} \tag{6.25}$$

Thus, the convergence time varies as the square root of the auxiliary antenna's power rather than directly as for the nonlimiter LMS algorithm. When g_1 is large, the steady-state mean value of the weight is still approximately equal to the optimum weight.

6.2.5 Weight Computation for Single Auxiliary System Using Normalized LMS Algorithm (NLMS)

The use of the hard limiter changes the auxiliary input to the correlation mixer from $y_1(t)$ to $y_1(t)/|y_1(t)|$. The NLMS algorithm changes the auxiliary input to $y_1(t)/|y_1(t)|^2$. As shown below, this algorithm causes the convergence time for the single-jammer and single-auxiliary-antenna implementation to be totally independent of the auxiliary power.

The circuitry required to implement the algorithm is *instantaneous automatic gain control* (IAGC), where the IAGC action is to control the gain such that it is inversely proportional to $|y_1(t)|^2$. The details of the implementation depend upon the IAGC circuitry. The desired IAGC action can be obtained by gain control of only the auxiliary-antenna input to the correlation mixer. Alternatively, the IAGC action can be split so that the residue and auxiliary antenna voltages are each divided by $|y_1(t)|$.

In either implementation, the equivalent of (6.12) is

$$T[\mathrm{d}w_1(t)/\mathrm{d}t] + w_1(t)[1 + g_N] = g_N y_0^*(t) y_1(t)/|y_1(t)|^2 \tag{6.26}$$

Taking the average value of both sides yields

$$T[\mathrm{d}w_a(t)/\mathrm{d}t] + w_a(t)[1 + g_N] = g_N E[y_0^*(t) y_1(t)/|y_1(t)|^2] \tag{6.27}$$

The effective time constant and steady-state, average-weight values are

$$\begin{aligned} T_e &= T/(1 + g_N) \\ w_s &= [g_N/(1 + g_N)]E[y_0^*(t) y_1(t)/|y_1(t)|^2] \end{aligned} \tag{6.28}$$

Thus, the time constant is independent of jammer power. The expectation can be evaluated by use of (6.22). Thus,

$$E[y_0^* y_1/|y_1|^2] = E\{[\rho\sqrt{p_M p_A} x_1^* + \sqrt{(1 - |\rho|^2)} x_0^*] x_1/p_A|x_1|^2\} \tag{6.29}$$

Because x_0 and x_1 are independent and zero mean, this evaluates to $\rho\sqrt{p_M/p_A}$, which equals the optimum weight value. Thus, when g_N is large, the steady-state, average-weight value is approximately equal to the optimum weight.

6.3 DIGITAL IMPLEMENTATION OF THE LMS ALGORITHMS

To obtain improved performance, some radar system designs use an A/D converter for the main antenna and each auxiliary. The desired performance improvements are usually related to aspects other than adaptivity. However, as the radar data are discrete rather than continuous, modifications are required to implement antenna adaptivity. These systems implement the correlation feedback loop via digital filtering and multiplying.

There are several implementation techniques for the A/D converter. All can be viewed as converting a bandpass waveform into a pair of real-valued discrete sequences, and the pair of sequences can be viewed as a single complex-valued sequence. Thus, consider a bandpass waveform:

$$e(t) = x_r(t) \cos w_0 t - x_i(t) \sin w_0 t \tag{6.30}$$

The ideal A/D converter samples the input at equispaced time instants. The sampling period is denoted as T_s. At time kT_s, where k is an integer, the voltage outputs are $x_r(kT_s)$ and $x_i(kT_s)$. These real voltages can be regarded as the complex envelope:

$$y(k) = x_r(kT_s) + jx_i(kT_s) \tag{6.31}$$

For either thermal noise or jamming, the statistics of $x_r(t)$ and $x_i(t)$ are such that they are zero-mean Gaussian processes with common variance. Then, $y(k)$ is a zero-mean complex Gaussian sequence with variance equal to twice the power of $e(t)$. The factor of 2 in power occurs because of a mathematical peculiarity in the definition of complex random processes and causes no analytical difficulty.

The sampling period is usually approximately equal to the reciprocal of the radar bandwidth. Then, samples of $e(t)$ at different time instants are approximately statistically independent. The samples are exactly independent when $e(t)$ is strictly band limited with bandwidth less than or equal to $1/T_s$. The following analyses assume independence at different time instants. Thus, for N auxiliary antennas, the A/D output sequences are $y_0(k)$ from the main antenna and $y_1(k)$ through $y_N(k)$ for the N auxiliary antennas. For the same value of k, these sequences are correlated, but $y_n(k)$ and $y_m(1)$ are always independent for $k \neq 1$ for any values of n and m.

6.3.1 MSLC Digital LMS Algorithm

A conceptual block diagram of the multiple-loop LMS algorithm is given in Figure 6.3, which shows the details of three correlation feedback loops. For an N-auxiliary-antenna implementation, N parallel loops are required. The actual implementation

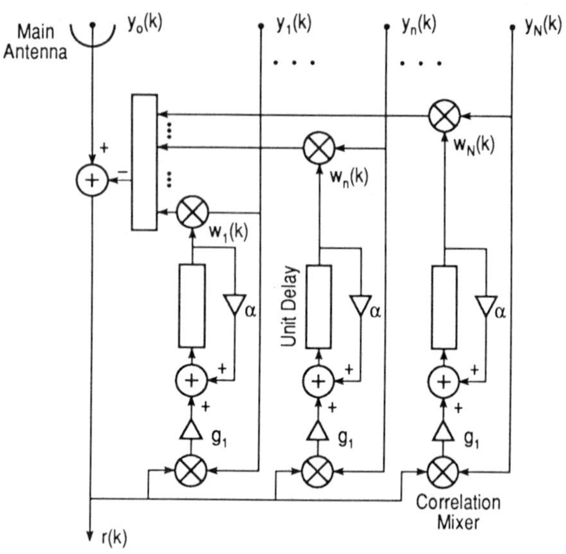

Figure 6.3 Conceptual block diagram of digital LMS algorithm.

can be via microprocessor or a dedicated special-purpose computer. The implementation shows a single-pole digital filter after the correlation mixer. In general, other filters can be used, but the analysis of this section will only consider the single-pole filter.

As can be seen directly from the figure, the adaptive weight generated by the nth loop satisfies the difference equation:

$$\begin{aligned} w_n(k+1) &= \alpha w_n(k) + g r^*(k) y_n(k); \quad n = 1, N \\ r(k) &= y_0(k) - \Sigma_m w_m^*(k) y_m(k) \end{aligned} \tag{6.32}$$

If we compare this to (6.9), we can see that the form of the weight update and residue equations are essentially identical. These equations combine into N first-order, coupled, stochastic difference equations. Initially, the solution is obtained for the simpler case of a single auxiliary antenna. Chapter 7 to follow derives the solution for the more general case.

6.3.2 Digital LMS Single-Auxiliary Analysis

For a single-auxiliary antenna, (6.32) combines into

$$w_1(k+1) = [\alpha - g|y_1(k)|^2] w_1(k) + g y_0^*(k) y_1(k) \tag{6.33}$$

As for the analog LMS algorithm analysis, we derive the time-varying dependence of the average value of the adaptive weight.

Note that the current value of $w_1(k)$ does not depend on the current values of $y_0(k)$ and $y_1(k)$, but only on previous values. Because the sampling rate is such that the current value of $y_1(k)$ is statistically independent of the previous values, $w_1(k)$ is statistically independent of $y_1(k)$. Thus, taking the average value of both sides of (6.33) gives

$$w_a(k+1) = [\alpha - gp_A]w_a(k) = g\rho\sqrt{p_M p_A} \tag{6.34}$$

Assuming that the initial value of the average weight equals zero, the solution of this first-order constant coefficient difference equation is

$$w_a(k) = [1 - (\alpha - gp_A)^k]g\rho\sqrt{p_M p_A}/(1 - \alpha + gp_A) \tag{6.35}$$

This equation indicates that there are two significant differences in the convergence properties of the digital LMS algorithm as compared to the analog LMS algorithm. For the analog algorithm, the feedback circuit is stable for all positive values of the filter gain, g. For the digital algorithm, convergence occurs only if

$$-1 < \alpha - gp_A < 1 \tag{6.36}$$

or, equivalently,

$$(\alpha - 1)/p_A < g < (1 + \alpha)/p_A \tag{6.37}$$

The second difference is that, at convergence, the average steady-state weight value is

$$w_s = g\rho\sqrt{p_M p_A}/(1 - \alpha + gp_A) \tag{6.38}$$

Thus, for α equal to unity, the average steady-state weight is exactly equal to the optimum weight. This result contrasts with the analog implementation where the steady-state weight is only approximately equal to the optimum weight. This difference in steady state behavior occurs because the dc gain of the narrowband filter after the correlation multiplier is always finite for the analog implementation, but is infinite when $\alpha = 1$ in the digital implementation.

When $\alpha = 1$, the stability constraint on g is

$$0 < g < 2/p_A \tag{6.39}$$

Thus, g must be positive and less than a value that is a function of the auxiliary power. The value of p_A is constrained by the power level that saturates the auxiliary antenna A/D converter, and this, in turn, specifies the maximum allowable value of g. For a specified value of g, the functional form of (6.35) is very much like exponential decay. Specifically, for α equal to unity and gp_a less than unity, (6.35) may be rewritten as

$$w_a(k) = \{1 - \exp[k \ln(1 - gp_A)]\} g\rho\sqrt{p_M/p_A} \tag{6.40}$$

and the effective time constant is

$$T_e = -1/\ln(1 - gp_A) \tag{6.41}$$

For gp_A close to unity, the effective time constant is short. As p_A decreases, the time constant increases and tends toward infinity as p_A tends toward zero. Thus, the weight value for the digital implementation, like that of the analog implementation, converges exponentially with a time constant that varies inversely with the auxiliary power level.

6.3.3 Single-Auxiliary Analysis of Digital LMS Algorithm with Hard Limiting

When the auxiliary input to the correlation multiplier is modified by limiting, the modification of (6.33) is very much like that caused by limiting in the analog implementation. Specifically, the modified stochastic difference equation is

$$w_1(k + 1) = [1 - g_L y_1^*(k) y_L(k)] w_1(k) = g_L y_0^*(k) y_L(k) \tag{6.42}$$

There are two possible methods of implementing a limiting action. The first is like the analog limiter implementation in that

$$y_L(k) = y_1(k)/|y_1(k)| \tag{6.43}$$

The other is a digital limiter; thus,

$$\begin{aligned} y_L(k) &= \mathrm{Re}[y_1(k)]/|\mathrm{Re}[y_1(k)]| + j\,\mathrm{Im}[y_1(k)]/|\,\mathrm{Im}[y_1(k)]| \\ &= \mathrm{sgn}\{\mathrm{Re}[y_1(k)]\} + j\,\mathrm{sgn}\{\mathrm{Im}[y_1(k)]\} \end{aligned} \tag{6.44}$$

where sgn(·) denotes the sign of the argument. The latter form is quite simple to implement, whereas the first is relatively complicated.

The first type of limiting is mathematically identical to the limit action analyzed in Section 6.2.5. The previous result is

$$E[y_L(k) y_n^*(k)] = (\gamma/\sqrt{p_A}) E[y_1(k) y_n^*(k)]; \quad n = 0, 1 \tag{6.45}$$

where γ is a constant. We can show that this result is also correct for the second type of limiting, but with a different value for γ. Thus, the action of the limiter is essentially identical to that obtained by the analog implementation of Secton 6.2.5. The digital limiter causes the convergence time constant to vary inversely as the square root of the auxiliary antenna power, as for the analog limiter implementation of the LMS algorithm.

6.3.4 Single-Auxiliary Analysis of Digital NLMS Algorithm

The digital version of the NLMS algorithm for a single auxiliary antenna uses a weight update equation:

$$w_1(k + 1) = w_1(k) + [g_N/|y_1(k)^2]r^*(k)y_1(k) \tag{6.46}$$

As for the analog implementation, the modification as compared to the LMS weight update equation (6.32) results in a convergence time that is independent of the auxiliary power. Thus, combining (6.32) and (6.46), and solving for the average weight value, results in

$$w_a(k + 1) = (1 - g_N)w_a(k) + g_N E[y_0^*(k)y_1(k)/|y_1(k)|^2] \tag{6.47}$$

and, from (6.29), the expectation equals $\rho\sqrt{p_M/p_A}$. The solution of the constant coefficient difference equation is

$$\begin{aligned} w_a(k) &= [1 - (1 - g_N)^k]\rho\sqrt{p_M/p_A} \\ 2 &> g_N > 0 \end{aligned} \tag{6.48}$$

and the inequality constraint on the gain constant is required to ensure stability. We can see that the convergence time constant is independent of the auxiliary power level and the steady-state, average-weight value equals the optimum weight.

6.4 ADDITIONAL RESIDUE POWER DUE TO WEIGHT NOISE FOR SINGLE AUXILIARY SYSTEMS

As previously indicated, the adaptive weight computed by any of the preceding algorithms is a random process. The analyses demonstrated that for the discrete implementations the average value of the adaptive weight converged to the optimum weight. However, because the weight is a random process, even after convergence, the weight value will randomly deviate from the optimum value. As shown by (5.59), a weight vector deviation Δ from the optimum weight vector causes the residue power to be larger than the minimum residue power by

$$p_{\text{add}} = \boldsymbol{\Delta}^{\mathbf{H}}\mathbf{R}_{\Delta} = \text{tr}\mathbf{R}_{\Delta}\boldsymbol{\Delta}^{\mathbf{H}} \tag{6.49}$$

As weight vector deviation is random, the average added power is obtained by taking the expectation of (6.49) with respect to Δ. Thus,

$$\begin{aligned} p_{\text{add}} &= \text{tr}\ \mathbf{RR}_{\Delta} \\ \mathbf{R}_{\Delta} &= E(\Delta\Delta^{\mathbf{H}}) \end{aligned} \tag{6.50}$$

and $\mathbf{R}_{\Delta}$ is the covariance matrix of the weight noise. For a single-weight canceller system and auxiliary power of p_A, this reduces to

$$p_{\text{add}} = p_A E(|\delta_1|^2) \tag{6.51}$$

This section derives the additional residue power due to weight noise for the digital implementations of the LMS, LMS with limiter, and NLMS algorithms. We shall show that, although the LMS with limiter and NLMS algorithms have superior convergence properties, this is balanced by additional residue power induced by weight noise. Thus, the choice of the preferred algorithm is not clear-cut, but requires a performance trade-off.

6.4.1 LMS Algorithm

For a single-auxiliary antenna, the residue voltage is

$$\begin{aligned} r(k) &= y_0(k) - w_1^*(k)y_1(k) \\ w_1(k) &= w_{\text{opt}} + \delta_1(k) \\ r(k) &= r_0(k) - \delta_1^*(k)y_1(k) \\ r_0(k) &= y_0(k) - w_{\text{opt}}^* y_1(k) \end{aligned} \tag{6.52}$$

where $r_0(k)$ is defined as the residue voltage obtained by using the optimum weight value. The weight-update equation is

$$w_1(k+1) = w_1(k) + gr^*(k)y_1(k) \tag{6.53}$$

and can be written in terms of $r_0(k)$ as

$$w_1(k+1) = w_1(k) + gr_0^*(k)y_1(k) - g\delta_1(k)|y_1(k)|^2 \tag{6.54}$$

Subtracting the optimum weight from both sides results in

$$\delta_1(k + 1) = \delta_1(k) - g|y_1(k)|^2\delta_1(k) + gr_0^*(k)y_1(k) \tag{6.55}$$

Let us define the autocorrelation of the weight error as

$$E[\delta_1(k)\delta_1^*(k)] = r_\delta(k) \tag{6.56}$$

From (6.55), an update equation for the autocorrelation of the weight vector is

$$\begin{aligned} r_\delta(k + 1) = E\{&[\delta_1(k) - g|y_1(k)|^2\delta_1(k) + gr_0^*(k)y_1(k)] \\ &[\delta_1^*(k) - g|y_1(k)|^2\delta_1^*(k) + gr_0(k)y_1^*(k)]\} \end{aligned} \tag{6.57}$$

Note that $w_1(k)$ depends on only previous data, and the A/D sampling frequency is chosen such that current data are statistically independent of previous data. Thus, $y_1(k)$ and $\delta_1(k)$ are statistically independent, so that

$$E[|y_1(k)|^2|\delta(k)|^2] = p_A r_\delta(k) \tag{6.58}$$

Also,

$$E[\delta_1(k)r_0(k)y_1^*(k)] = E[\delta_1(k)]E[r_0(k)y_1^*(k)] = 0 \tag{6.59}$$

because, according to Section 6.2.2, the residue, using the optimum weight, is uncorrelated with the auxiliary input. Therefore, (6.57) reduces to

$$r_\delta(k + 1) = (1 - 2gP_A + g^2E|y_1(k)|^4)r_\delta(k) + g^2p_{\min}p_A \tag{6.60}$$

For complex-valued, zero-mean Gaussian variates, a useful result of (6.24) is

$$E|z|^{2n} = n!\, E(|z|^2)^n \tag{6.61}$$

so that

$$E|y_1(k)|^4 = 2p_a^2 \tag{6.62}$$

Thus, (6.60) is a constant coefficient difference equation, which converges if

$$0 < gp_A(1 - gp_A) < 1 \tag{6.63}$$

At steady state,

$$r_{\delta s} = (gp_{\min}/2)/(1 - gp_A) \tag{6.64}$$

From (6.51), the additional steady-state residue power due to weight noise is

$$p_{\text{add}} = p_{\min}(gp_A/2)/(1 - gp_A) \tag{6.65}$$

Note that for small values of gp_A the additional steady-state residue power due to weight noise is small. However, as gp_A increases toward unity, the additional power tends toward infinity. Thus, although the first moment analysis leading to (6.39) indicates that gp_A equal to unity is satisfactory, the second moment analysis indicates gp_A should be much less than unity.

6.4.2 LMS Algorithm with Limiter

The weight-error update equation is modified to

$$w_1(k + 1) = w_1(k) + g_L r^*(k) y_1(k)/|y_1(k)| \tag{6.66}$$

The equivalent of (6.55) is modified to

$$\delta_1(k + 1) = [1 - g_L|y_1(k)|]\delta_1(k) + g_L r_0^*(k) y_1(k)/|y_1(k)| \tag{6.67}$$

From (6.67), an update equation for the autocorrelation of the weight error is

$$r_\delta(k + 1) = (1 - 2g_L\sqrt{p_A \pi/2} + g_L^2 p_A) r_\delta(k) + g_L^2 p_{\min} \tag{6.68}$$

At steady state, the solution of the constant coefficient difference equation is

$$r_{\delta s} = g_L p_{\min}/(2\sqrt{p_A \pi/2} - g_L p_A) \tag{6.69}$$

and the additional residue power due to weight noise at convergence is

$$p_{\text{add}} = p_{\min}(g_L\sqrt{p_A})/(2\sqrt{\pi/2} - g_L\sqrt{p_A}) \tag{6.70}$$

The interpretation of (6.70) is identical to the interpretation of (6.65). Thus, for small values of the effective weight update gain, $g_L\sqrt{p_A}$ for this implementation, the additional steady-state residue power is small. As the effective update gain increases, the additional residue power tends toward infinity.

6.4.3 NLMS Algorithm

The good convergence properties of this algorithm are offset by the additional residue power due to increased steady-state weight noise. As shown below, this effect is exceptionally large for the single auxiliary antenna. For this case, additional residue power due to the steady-state weight noise tends toward infinity for all gain values. There are two caveats to this result. One is that the infinite power only occurs for the single-auxiliary-antenna case. The second is that the infinite power is really an infinite variance, and infinite variance does not necessarily imply poor performance. As an example, the tracking error for a monopulse radar has an infinite variance, although the tracking performance is satisfactory. Infinite variance occurs whenever the tails of the probability density function decrease at a slow rate.

6.4.3.1 Infinite Variance Interpretation

As a specific example of a case where infinite variance occurs and the possible misinterpretations, *assume* the probability density function for the steady-state weight error. (The density function is actually unknown, which is why the analysis evaluates first and second moments.) Thus,

$$p(\delta) = (\delta_s/2)/(\delta_s + |\delta|)^2; \quad |\delta| \geqslant 0 \tag{6.71}$$

We can see that the value of the density is maximum at δ equals zero and the function is even around zero. The probability that the magnitude of the error is less than δ_M is

$$p_M = \delta_s \int_0^{\delta_M} d\delta/(\delta_s + \delta)^2 = 1 - \delta_s/(\delta_s + \delta_M) \tag{6.72}$$

At $\delta_M = 0$, the value of p_M is zero. As δ_M tends toward infinity, the value of p_M tends toward unity. The value of $p_M = 1/2$ for $\delta_M = \delta_s$. Thus, the probability properties of this *assumed* probability density function are reasonable. However, the second moment is infinite, and therefore the variance is infinite.

Infinite variance is indicative of an increased probability that the error exceeds δ_M. To illustrate this, consider the alternative assumption for the error probability density function:

$$p(\delta) = (3\delta_s^3/2)/(\delta_s + |\delta|)^4; \quad |\delta| \geqslant 0 \tag{6.73}$$

The variance of this density function is finite. The probability that the magnitude of the error is less than δ_M is

$$p_M = 1 - [\delta_s/(\delta_s + \delta_M)]^3 \tag{6.74}$$

The value of $p_M = 1/2$ for $\delta_M = 0.26\delta_s$. Thus, this density is much more concentrated than the prior one. Smaller variance is indicative of smaller spread, but infinite variance does not imply infinite spread.

6.4.3.2 NLMS Weight-Error Autocorrelation Evaluation

By using the procedures of the preceding sections, the weight-error update equation is

$$\delta(k + 1) = (1 - g_N)\delta(k) + g_N r_0^*(k) y_1(k)/|y_1(k)|^2 \tag{6.75}$$

The autocorrelation update equation is

$$r_\delta(k + 1) = (1 - g_N)^2 r_\delta(k) + g_N^2 p_{\min} E(1/|y_1(k)|^2) \tag{6.76}$$

The expectation is evaluated by substituting $\mu = -2$ into (6.24). Because the γ function is infinite for zero argument, the expectation is infinite, and the weight-noise variance for the single auxiliary antenna NLMS algorithm is likewise infinite.

REFERENCES

[1] P.W. Howells (1965), Intermediate Frequency Sidelobe Canceller, U.S. Patent 3,202 990, August 24, 1965 (Filed May 4, 1959).

[2] S.P. Applebaum (1966), "Steady State and Transient Performance of the Sidelobe Canceller," Special Projects Laboratory, Syracuse Research Corporation, Syracuse, NY, April 1966.

[3] B. Widrow, *et al.*, "Adaptive Antenna Systems," *Proc. IEEE,* Vol. 55, No. 12, December 1967.

[4] I.S. Reed, J.D. Mallett, and L.E. Brennan, "Rapid Convergence Rate in Adaptive Arrays," *IEEE Trans. Aerospace and Electronics Systems,* Vol. AES-10, No. 6, November 1974.

Chapter 7
Multiple Auxiliary Antenna Algorithm Analysis

7.1 INTRODUCTION

This chapter analyzes the convergence and weight-noise properties of digital implementation of the LMS and NLMS algorithms for multiple-auxiliary-antenna systems. Only digital algorithms are considered because, like the single-auxiliary-antenna case, the form of analysis and results are similar for the analog and digital algorithms. The LMS algorithm with limiter is not considered because its properties can be extrapolated from those of the LMS alogrithm as was shown for the single-auxiliary analysis.

7.2 LMS ALGORITHM

From (6.32), the update equation for the nth weight of an N-auxiliary-antenna system is

$$w_n(k + 1) = w_n(k) + gy_0^*(k)y_n(k) - g\Sigma_m w_m(k)y_m^*(k)y_n(k) \tag{7.1}$$

and the filter feedback gain, α, is specified as equal to unity. Let us define the vectors $W(k)$ and $Y(k)$ with components $w_n(k)$ and $y_n(k)$, respectively. The N-weight update equations of (7.1) can be written in vector format as

$$\mathbf{W}(k + 1) = \mathbf{W}(k) + gy_0^*(k)\mathbf{Y}(k) - g\mathbf{Y}(k)\mathbf{Y}^{\mathbf{H}}(k)\mathbf{W}(k) \tag{7.2}$$

This equation is an N-dimensional, stochastic difference equation. As for the single-auxiliary-antenna analysis, the preferred analytical result is to obtain the joint-time-dependent probability density function for $\mathbf{W}(k)$. However, this analytic result cannot be obtained. Instead, a first moment analysis is performed to determine the algorithm convergence properties and a second moment analysis is performed to determine the additional residue power due to weight noise at steady state.

7.2.1 MSLC LMS Convergence Analysis

Denote the average value of $w_n(k)$ as $w_{an}(k)$ and the vector of average values as $\mathbf{W}_a(k)$. As defined in (5.34), we have

$$
\begin{aligned}
r_{n0} &= E[y_n(k)y_0^*(k)] \\
r_{nm} &= E[y_n(k)y_m^*(k)]
\end{aligned}
\tag{7.3}
$$

Then, the average value of both sides of (7.1) is

$$w_{an}(k+1) = w_{an}(k) + gr_{n0} - g\Sigma_m w_{am}(k)r_{nm} \tag{7.4}$$

The N-dimension version of (7.4) is

$$\mathbf{W}_a(k+1) = \mathbf{W}_a(k) + g\mathbf{R}_0 - g\mathbf{R}\mathbf{W}_a(k) \tag{7.5}$$

where $\mathbf{R}_0$ and $\mathbf{R}$ are as defined in (5.47) and (5.48). Specifically, $\mathbf{R}$ is an $N \times N$ matrix, with its nth row, mth column entry equal to r_{nm}. $\mathbf{R}_0$ is an N-dimensional vector, with its nth entry equal to r_{n0}. Note that if (7.5) converges, $\mathbf{W}_a(k) = \mathbf{W}_a(k + 1) = \mathbf{W}_s$, where $\mathbf{W}_s$ is a vector. The steady-state solution is

$$\mathbf{R}\mathbf{W}_s = \mathbf{R}_0 \tag{7.6}$$

Comparing this result to (5.57) shows that the LMS algorithm converges such that the average weight vector equals the optimum weight vector. Thus, the result of convergence to the optimum weight, proved in Chapter 6 for a single-auxiliary-antenna system, is generally true for an N-auxiliary-antenna system.

Deriving the time-dependent properties of the average weight vector of (7.5) is more difficult than that for the single-auxiliary antenna because the equations are coupled. Thus, from (7.4), the difference equation for the nth weight is a function of all the other weights. The equations would be uncoupled if r_{nm} were nonzero for only $n = m$. This is equivalent to $\mathbf{R}$ being a diagonal matrix in (7.5). The solution of (7.5) can be more easily obtained by use of a change of coordinates such that the matrix $\mathbf{R}$ in the new coordinate system is diagonal.

7.2.2 Eigenvector Review

The standard method of determining the required coordinate change is via the theory of eigenvectors. Consider the matrix equation:

$$\mathbf{R}\mathbf{X} = \mathbf{Y} \tag{7.7}$$

A geometrical interpretation of this equation is that the vector **X** is transformed into the vector **Y** by the transformation matrix **R**. In general, there is no straightforward relation between **X** and **Y**. However, for each square matrix **R**, there are some vectors that transform into a vector that is simply a scalar multiple of the original vector. Thus, there is a vector $\mathbf{F}_1$, termed an *eigenvector,* such that

$$\mathbf{RF}_1 = d_1\mathbf{F}_1 \tag{7.8}$$

where $\mathbf{d}_1$ is a scalar, which is called the *eigenvalue*. To determine an eigenvector, note that this equation is equivalent to

$$(\mathbf{R} - d_1\mathbf{I})\mathbf{F}_1 = \mathbf{0} \tag{7.9}$$

where **I** denotes the identity matrix and **0** denotes the zero vector. Because a vector can be transformed into the zero vector only if the transformation matrix is singular and the determinant of a singular matrix equals zero, eigenvalues values are values of d that solve the equation:

$$\det(\mathbf{R} - dI) = \mathbf{0} \tag{7.10}$$

where det(·) denotes the determinant. For an Nth-order matrix, (7.10) is an Nth-degree polynomial in d. The roots of the polynomial are the eigenvalues. After an eigenvalue is found, the eigenvector can, in concept, be evaluated by using (7.9). An example of the procedure is given later.

For the MSLC analysis, **R** is a covariance matrix. It has special properties, and, in turn, its eigenvalues have special properties. First, the matrix is its own transpose conjugate (Hermitian). Thus, each matrix component satisfies the relation:

$$r_{nm} = r^*_{mn} \tag{7.11}$$

and the equivalent matrix equation is

$$\mathbf{R} = \mathbf{R}^{\mathbf{H}} \tag{7.12}$$

Equation (7.11) directly follows from the definition in (5.34). When **R** is Hermitian, it has N orthogonal eigenvectors and the eigenvalues are real-valued. In addition, when **R** is a covariance matrix the eigenvalues are non-negative. When the covariance matrix is due to the sum of thermal noise and jammers, as in (5.49), each eigenvalue is positive and equal to or greater than p_{noise}.

Although eigenvectors can be of arbitrary length, considering only unit-length eigenvectors is convenient. Then,

$$\mathbf{F}_n^{\mathbf{H}}\mathbf{F}_m = \begin{cases} 1, & n = m \\ 0, & n = m \end{cases} \tag{7.13}$$

Define the square matrix $\mathbf{F}$ with its nth column equal to the nth unit length eigenvector $\mathbf{F}_n$:

$$\mathbf{F} = (\mathbf{F}_1, \mathbf{F}_2, \ldots, \mathbf{F}_N) \tag{7.14}$$

Note that, from (7.13),

$$\mathbf{F}^{\mathbf{H}}\mathbf{F} = \mathbf{I} \tag{7.15}$$

Premultiplying by $\mathbf{F}^{-\mathbf{H}}$ and postmultiplying by $\mathbf{F}^{\mathbf{H}}$ also proves that

$$\mathbf{F}\mathbf{F}^{\mathbf{H}} = \mathbf{I} \tag{7.16}$$

Generalizing (7.8) by using (7.14) gives

$$\begin{aligned} &\mathbf{R}(\mathbf{F}_1, \mathbf{F}_2, \ldots, \mathbf{F}_N) = (d_1\mathbf{F}_1, d_2\mathbf{F}_2, \ldots, d_N F_N) \\ &\mathbf{R}\mathbf{F} = \mathbf{F}\mathbf{D} \\ &\mathbf{R} = \mathbf{F}\mathbf{D}\mathbf{F}^{\mathbf{H}} \end{aligned} \tag{7.17}$$

where $\mathbf{D}$ is a diagonal matrix with entries d_n.

Two useful theorems concerning eigenvalues are

$$\begin{aligned} \det(\mathbf{R}) &= \prod_n d_n \\ \operatorname{tr}(\mathbf{R}) &= \sum_n d_n \end{aligned} \tag{7.18}$$

The first theorem can be proved by noting that

$$
\begin{aligned}
\det(R) &= \det(\mathbf{FDF^H}) = \det(\mathbf{F})\det(\mathbf{D})\det(\mathbf{F^H}) \\
&= \det(\mathbf{D})\det(\mathbf{FF^H}) = \det(\mathbf{D})\det(\mathbf{I}) \\
&= \prod_n d_n
\end{aligned}
\tag{7.19}
$$

To prove the second theorem, note that

$$
\begin{aligned}
\mathrm{tr}(\mathbf{R}) &= \mathrm{tr}(\mathbf{FDF^H}) = \mathrm{tr}(\mathbf{DF^HF}) \\
&= \mathrm{tr}(\mathbf{DI}) = \mathrm{tr}\mathbf{D} = \sum_n d_n
\end{aligned}
\tag{7.20}
$$

7.2.3 Decoupled MSLC LMS Convergence Equations

Substituting (7.16) and (7.17) into (7.5) gives

$$
\mathbf{W}_a(k+1) = \mathbf{F}(\mathbf{I} - g\mathbf{D})\mathbf{F^H}\mathbf{W}_a(k) + g\mathbf{R}_0 \tag{7.21}
$$

Premultiplying by $\mathbf{F^H}$ and using (7.15) results in

$$
\begin{aligned}
\mathbf{V}_a(k+1) &= (\mathbf{I} - g\mathbf{D})\mathbf{V}_a(k) + g\mathbf{F^H}\mathbf{R}_0 \\
\mathbf{V}_a(k) &= \mathbf{F^H}\mathbf{W}_a(k)
\end{aligned}
\tag{7.22}
$$

Because the matrix $\mathbf{I} - g\mathbf{D}$ is diagonal, (7.22) represents N decoupled, first-order, constant coefficient difference equations. Assuming that the initial value of $\mathbf{W}$ is zero, the solution for the nth component of $\mathbf{V}_a(k)$ is

$$
v_{an}(k) = [1 - (1 - gd_n)^k](\mathbf{F}_n^{\mathbf{H}}\mathbf{R}_0/d_n) \tag{7.23}
$$

and the convergence is stable if

$$
0 < gd_n < 2 \tag{7.24}
$$

Because each eigenvalue is positive, similarly to (6.39), g must be positive and less than $2/d_n$. From (7.18), a conservative constraint is

$$
0 < g < 2/\mathrm{tr}\mathbf{R} \tag{7.25}
$$

From (7.6) the steady-state solution, for $\mathbf{W}_s$, is the optimum weight vector. Therefore, the steady-state value of $\mathbf{V}_a$ is the transformed optimum weight. Its nth-component transient deviation from optimum is $-\mathbf{F}_n^{\mathbf{H}}\mathbf{R}_0\,(1 - gd_n)^k/d_n$. From (5.59), the additional power due to weight errors is

$$p_{\text{add}} = \Delta^{\mathbf{H}}\mathbf{R}\Delta \tag{7.26}$$

Combining this result with (7.17) gives

$$p_{\text{add}} = \Delta^{\mathbf{H}}\mathbf{F}\mathbf{D}\mathbf{F}^{\mathbf{H}}\Delta \tag{7.27}$$

Note that, because $\mathbf{F}^{\mathbf{H}}\mathbf{W}_a(k) = \mathbf{V}_a(k)$, $\mathbf{F}^{\mathbf{H}}\Delta$ equals the difference between $\mathbf{V}_a(k)$ and the optimum or steady-state $\mathbf{V}_a$. Let us define

$$\delta v_n = \mathbf{F}_n^{\mathbf{H}}\Delta \tag{7.28}$$

Then,

$$p_{\text{add}} = \Sigma|\delta v_n|^2 d_n \tag{7.29}$$

so that the transient portion of the average residue power as

$$p_{\text{add}} = \Sigma_n(|\mathbf{F}_n^{\mathbf{H}}\mathbf{R}_0|^2/d_n)(1 - gd_n)^{2k} \tag{7.30}$$

The additional residue power thus can be expressed as the sum of N components, which decay with time (index k). The power associated with each component is q_n, where

$$q_n = |\mathbf{F}_n^{\mathbf{H}}\mathbf{R}_0|^2/d_n \tag{7.31}$$

and an alternative form of (7.30) is

$$p_{\text{add}} = \Sigma q_n(1 - g_1 d_n)^{2k} \tag{7.32}$$

The decay rate of a component depends on the eigenvalue. The decay time constant can be put in more familiar terms by noting that

$$\begin{aligned}(1 - g\mathbf{d}_n)^{2k} &= \mathrm{e}^{2k\ln(1-g\mathbf{d}_n)} = \mathrm{e}^{-k/K_n}\\ K_n^{-1} &= -2\ln(1 - gd_n)\end{aligned} \tag{7.33}$$

where K_n is defined as the time constant. When the eigenvalue is small, this may be approximated as

$$K_n \approx (2gd_n)^{-1} \tag{7.34}$$

Small eigenvalues therefore cause large time constants.

The optimum residue power may also be expressed in terms of eigenvalues and eigenvectors by noting that

$$\mathbf{R}^{-1} = \mathbf{FD}^{-1}\mathbf{F}^{\mathbf{H}} \tag{7.35}$$

so that

$$\begin{aligned}\mathbf{R}_0^{\mathbf{H}}\mathbf{R}^{-1}\mathbf{R}_0 &= \mathbf{R}_0^{\mathbf{H}}\mathbf{FD}^{-1}\mathbf{F}^{\mathbf{H}}\mathbf{R}_0 \\ &= \Sigma|\mathbf{F}_n^{\mathbf{H}}\mathbf{R}_0|^2/d_{\mathrm{n}} \\ &= \Sigma \mathrm{q}_{\mathrm{n}};\ q_n = |\mathbf{F}_n^{\mathbf{H}}\mathbf{R}_0|^2/d_n \end{aligned} \tag{7.36}$$

Thus, (5.58) becomes

$$p_{\mathrm{min}} = p_{\mathrm{main}} - \Sigma q_n \tag{7.37}$$

Combining (7.30) and (7.37) results in

$$p_{\mathrm{res}} = p_{\mathrm{main}} - \Sigma q_n[1 - (1 - gd_n)^{2k}] \tag{7.38}$$

Note that the transient behavior of the residue power depends on the time constants of each component, K_n and the power associated with that component q_n. A small eigenvalue does not necessarily imply slow convergence of the residue power. If the time constant is very long, but the value of q_n is very small, the convergence of the residue is approximately independent of the convergence of that component.

7.2.4 LMS Algorithm Convergence Example

To illustrate the analytical techniques and show the sensitivity of performance to the scenario, this section will perform a detailed analysis of a two-auxiliary-antenna system for several jammer scenarios. In general, computing the algorithm convergence properties for arbitrary jammer scenarios and number of auxiliary antennas is tedious because of the computational difficulty of computing eigenvalues and eigenvectors for matrices of order greater than 2. Computer subroutines are available to compute the eigenstructure of larger ordered matrices. We shall also discuss the results for a four-auxiliary system, as computed by using eigenstructure subroutines.

The analysis of the two-auxiliary-antenna system will consider three cases. The first two cases are for a single jammer of nonzero bandwidth. For the first case, the two auxiliaries are equispaced, bracketing the main antenna. For the second case, the antennas are not equispaced. The third case is for two arbitrarily spaced auxiliary antennas and two narrowband (approximately zero bandwidth) jammers.

From (4.7), the main antenna voltage for the single jammer is

$$y_0(t) = n_0(t) + g_s j(t) \tag{7.39}$$

For equispaced, bracketing, omnidirectional, unit gain auxiliaries, the voltages are

$$\begin{aligned} y_1(t) &= n_1(t) + j(t + t_d)\exp(j\phi) \\ y_2(t) &= n_2(t) + j(t - t_d)\exp(-j\phi) \end{aligned} \tag{7.40}$$

where t_d denotes the respective reception times advance and delay with respect to that of the main antenna. The angle ϕ is a phase shift due to delay of the carrier. The covariance matrix and correlation vector are as defined by (5.49) through (5.53). We define the normalized correlation coefficient of $j(t)$ and $j(t \pm t_d)$ as ρ_m and that of $j(t + t_d)$ and $j(t - t_d)$ as ρ_a. Thus, the covariance matrix and correlation vector can be evaluated as

$$\begin{aligned} \mathbf{R} &= \begin{pmatrix} p_n + p_1 & \rho_a p_1 \exp(j2\phi) \\ \rho_a p_1 \exp(-j2\phi) & p_n + p_1 \end{pmatrix} \\ \mathbf{R}_0^{\mathrm{T}} &= g_s^* \rho_m p_1 [\exp(j\phi), \exp(-j\phi)] \end{aligned} \tag{7.41}$$

where, for notational convenience, p_{noise} is denoted as p_n.

The eigenvalues are the solutions of the quadratic equation:

$$(p_n + p_1 - d)^2 - (\rho_a p_1)^2 = 0 \tag{7.42}$$

so that the two eigenvalues are

$$\begin{aligned} d_1 &= p_n + p_1(1 + \rho_a) \\ d_2 &= p_n + p_1(1 - \rho_a) \end{aligned} \tag{7.43}$$

Note that, because ρ_a is close to unity, d_2 is very small compared to d_1. Thus, from (7.36), the additional residue power due to q_2 will converge slowly.

The non-unit-length eigenvectors can be found by assuming the first component is unity. Then, from (7.9), the second component of $\mathbf{F}_1$, f_{21}, is found from

$$(p_n + p_1 - d_1) + f_{21}\rho_a p_1 \exp(j2\phi) = 0$$

so that

$$f_{21} = \exp(-j2\phi) \tag{7.44}$$

The unit-length eigenvector thus is

$$\mathbf{F}_1^{\mathrm{T}} = \exp(-j\phi)[\exp(j\phi), \exp(-j\phi)]/\sqrt{2} \tag{7.45}$$

Note that $\mathbf{R}_0$ is colinear with $\mathbf{F}_1$ and, because $\mathbf{F}_1$ and $\mathbf{F}_2$ are orthogonal, $\mathbf{R}_0$ is orthogonal to $\mathbf{F}_2$ so that $q_2 = 0$. Thus, evaluating (7.31) gives

$$\begin{aligned} q_1 &= 2G_s\rho_m^2 p_1^2/[p_n + p_1(1 + \rho_a)] \\ q_2 &= 0 \end{aligned} \tag{7.46}$$

Note that, although d_2 is small so that the convergence time constant of this mode is very long, it is unimportant, as the power of this mode equals zero.

The two antennas are not equispaced about the main antenna for the second case. This geometry is probably more important than the first as prior analysis (Chapter 5) indicated that equispacing is undesirable when cancelling multiple jammers. Defining α as the ratio of the distances from the auxiliaries to the main antenna, (7.40) and (7.41) are modified to

$$\begin{aligned} y_1(t) &= n_1(t) + j(t + t_1)\exp(j\phi) \\ y_2(t) &= n_2(t) + j(t - t_2)\exp(-j\alpha\phi) \\ \mathbf{R} &= \begin{pmatrix} p_n + p_1 & \rho_a p_1 \exp[j(1+\alpha)\phi] \\ \rho_a p_1 \exp[-j(1+\alpha)\phi] & p_n + p_1 \end{pmatrix} \\ \mathbf{R}_0^{\mathrm{T}} &= g_s^* p_1[\rho_1 \exp(j\phi), \rho_2 \exp(-j\alpha\phi)] \end{aligned} \tag{7.47}$$

where ρ_1 and ρ_2 are the respective correlation coefficients for the main and auxiliary antennas. The eigenvalues for the perturbed covariance matrix are identical to those previously found. The eigenvector associated with d_1 evaluates to

$$\mathbf{F}_1^{\mathrm{T}} = \exp(-j\phi)[\exp(j\phi), \exp(-j\alpha\phi)]/\sqrt{2} \tag{7.48}$$

and, for this case, because $\rho_1 \neq \rho_2$, $\mathbf{R}_0$ is not colinear with $\mathbf{F}_1$. $\mathbf{F}_2$ can be found by the same procedure as used previously, or, as $\mathbf{F}_1$ and $\mathbf{F}_2$ are orthogonal, by inspection:

$$\mathbf{F}_2^{\mathrm{T}} = \exp(-j\phi)[\exp(j\phi), -\exp(-j\alpha\phi)]/\sqrt{2} \tag{7.49}$$

The modal powers q_1 and q_2 can then be evaluated as

$$\begin{aligned} q_1 &= G_s p_1^2(\rho_1 + \rho_2)^2/[2p_n + 2p_1(1 + \rho_a)] \\ q_2 &= G_s p_1^2(\rho_1 - \rho_2)^2/[2p_n + 2p_1(1 - \rho_a)] \end{aligned} \tag{7.50}$$

When the auxiliary-to-main-antenna spacings are identical, $\rho_1 = \rho_2$, and $q_2 = 0$ as found for the first case.

The difference between ρ_1 and ρ_2 depends on the degree of asymmetry of the auxiliary antennas and the radar bandwidth. To determine the effect of bandwidth on convergence time, note that the minimum residue power always exceeds the thermal-noise power. Thus, if q_2 is less than the thermal-noise power, slow convergence of this mode can have only a very minor effect on the overall convergence. Thus, a conservative requirement for the maximum bandwidth, for which the convergence time of the second mode is unimportant, is obtained from

$$q_2 \leq G_s p_1^2(\rho_1 - \rho_2)^2/2p_n \leq p_n \tag{7.51}$$

Assuming $G_s = 2$, an equivalent constraint is

$$(p_1/p_n)|\rho_1 - \rho_2| < 1 \tag{7.52}$$

Let us denote the distances between the main antenna and the two auxiliaries as d_a and d_b, so that the propagation time delays for a jammer angle of arrival, θ, and propagation velocity, c, are

$$\begin{aligned} t_1 &= d_a \sin\theta/c \\ t_2 &= d_b \sin\theta/c \end{aligned} \tag{7.53}$$

From (4.16b),

$$\rho_1 - \rho_2 = (\pi B \sin\theta/c)^2(d_a^2 - d_b^2)/6 \tag{7.54}$$

Assume $d_b = 0.9d_a$ and $d_a = \beta$ wavelengths. Then, these combine to obtain the constraint:

$$0.31(p_1/p_n)[\beta(B/f_0)\sin\theta]^2 < 1 \qquad (7.55)$$

As examples of typical parameters, let us consider

$$\begin{aligned} \beta &= 25 \\ B &= 1 \text{ MHz} \\ f_0 &= 3000 \text{ MHz} \\ \theta &= 45° \end{aligned}$$

Then, the convergence of the secondary mode is unimportant when the jammer-to-noise ratio is less than approximately 50 dB. Because the left-hand side of (7.55) varies as the square of bandwidth, increasing the bandwidth in the example to 10 MHz decreases the transition JNR to 30 dB.

The third case is for two statistically independent, narrowbandwidth jammers and two unequally spaced auxiliaries. The received voltages are

$$\begin{aligned} y_0 &= n_0 + g_1 j_1(t) + g_2 j_2(t) \\ y_1 &= n_1 + j_1(t)\exp(j\phi_1) + j_2(t)\exp(j\phi_2) \\ y_2 &= n_2 + j_1(t)\exp(-j\alpha\phi_1) + j_2(t)\exp(-j\alpha\phi_2) \end{aligned} \qquad (7.56)$$

Denote the nth column, mth row of the covariance matrix as r_{nm} and of the correlation vector as r_{0n}. Then

$$\begin{aligned} r_{11} &= r_{22} = p_n + p_1 + p_2 \\ r_{12} &= r_{21}^* = p\exp[j(1+\alpha)\phi_1] + p_2\exp[j(1+\alpha)\phi_2] \\ r_{01} &= g_1^* p_1 \exp(j\phi_1) + g_2^* p_2 \exp(j\phi_2) \\ r_{02} &= g_1^* p_1 \exp(-j\alpha\phi_1) + g_2^* p_2 \exp(-j\alpha\phi_2) \end{aligned} \qquad (7.57)$$

The eigenvalues can be evaluated as

$$\begin{aligned} d_{1,2} &= p_n + p_1 + p_2 \pm p_\phi \\ p_\phi &= \{p_1^2 + p_2^2 + 2p_1 p_2 \cos[(1+\alpha)(\phi_1 - \phi_2)]\}^{1/2} \end{aligned} \qquad (7.58)$$

When the first component of the eigenvector is unity, the second components, corresponding to d_1 and d_2, are

$$f = \pm p_\phi / \{p_1 \exp[j(1+\alpha)\phi_1] + p_2 \exp[j(1+\alpha)\phi_2]\} \tag{7.59}$$

Thus, we can see that deriving equations for the modal powers, q_1 and q_2, is algebraically tedious, even for this simple case. In addition, the resulting equations are functions of the variables g_1, g_2, p_1, p_2, ϕ_1, ϕ_2, and α and are difficult to interpret.

Some insight into the convergence behavior can be obtained by rewriting (7.56) in vector notation, as developed by (5.47) through (5.53). Thus,

$$\begin{aligned} \mathbf{Y} &= \begin{pmatrix} y_1 \\ y_2 \end{pmatrix} = \begin{pmatrix} n_1(t) \\ n_2(t) \end{pmatrix} + j_1(t)\mathbf{S}_1 + j_2(t)\mathbf{S}_2 \\ \mathbf{S}_1^{\mathbf{T}} &= [\exp(j\phi_1), \exp(-j\alpha\phi_1)] \\ \mathbf{S}_2^{\mathbf{T}} &= [\exp(j\phi_2), \exp(-j\alpha\phi_2)] \end{aligned} \tag{7.60}$$

so that

$$\begin{aligned} \mathbf{R}_0 &= g_1^* p_1 \mathbf{S}_1 + g_2^* p_2 \mathbf{S}_2 \\ \mathbf{R} &= p_n \mathbf{I} + p_1 \mathbf{S}_1 \mathbf{S}_1^{\mathbf{H}} + p_2 \mathbf{S}_2 \mathbf{S}_2^{\mathbf{H}} \end{aligned} \tag{7.61}$$

In general, as discussed above, finding the eigenvectors is tedious. However, a special case of interest is when the steering vectors, $\mathbf{S}_1$ and $\mathbf{S}_2$, are orthogonal. There are many geometries for which this occurs, as can be observed by using the notation of Figure 5.3(a). Orthogonality occurs for

$$\exp[j2\pi d_a (\sin\theta_2 - \sin\theta_1)/\lambda] + \exp[-j2\pi d_b(\sin\theta_2 - \sin\theta_1)/\lambda] = 0 \tag{7.62}$$

which is equivalent to

$$\exp[j2\pi(d_a + d_b)\sin\theta_2/\lambda] = \exp\left\{ j2\pi\left[i + \frac{1}{2} + (d_a + d_b)\sin\theta_1 \right] \Big/ \lambda \right\}$$

for any integer i, where λ is the wavelength. Thus, jammer angles of arrival related as

$$\theta_2(i) = \arcsin\left[\sin\theta_1 + \left(i + \frac{1}{2} \right) \lambda \Big/ (d_a + d_b) \right] \tag{7.63}$$

cause $\mathbf{S}_1$ and $\mathbf{S}_2$ to be orthogonal.

When $\mathbf{S}_1$ and $\mathbf{S}_2$ are orthogonal, they are each eigenvectors of $\mathbf{R}$. Although the current example is for a two-auxiliary-antenna system, the prior statement is true for any N-auxiliary-antenna system. To prove the eigenvector relationship, note that, as $\mathbf{S}_2$ and $\mathbf{S}_1$ are orthogonal:

$$\mathbf{RS}_1 = [p_n + p_1(\mathbf{S}_1^{\mathbf{H}}\mathbf{S}_1)]\mathbf{S}_1 \tag{7.64}$$

so that $\mathbf{S}_1$ is an eigenvector (non-unit-length) with eigenvalue equal to $p_n + p_1\mathbf{S}_1^{\mathbf{H}}\mathbf{S}_1$. Similarly, $\mathbf{S}_2$ is an eigenvector with eigenvalue $p_n + p_2\mathbf{S}_2^{\mathbf{H}}\mathbf{S}_2$. Note that the two eigenvalues are equal when the two jammer powers are equal.

The modal convergence powers q_1 and q_2 are also easily computed when $\mathbf{S}_1$ and $\mathbf{S}_2$ are orthogonal as q_1 and q_2 are only functions of the characteristics of jammers 1 and 2, respectively. Thus, the convergence characteristics are those that would occur for each jammer as if it were the only jammer.

7.2.5 Additional LMS Algorithm Convergence Examples

As indicated in the previous section, the LMS algorithm convergence characteristics depend upon many jammer geometric parameters. Evaluation of the performance *versus* different parameter values indicates that, in general, for an N-auxiliary-antenna system, the probability of any particular scenario of N jammers resulting in very slow convergence is greater than one-half. When the number of jammers is $N/2$ or less, the probability of a slow convergence is approximately zero. This section will show some examples of the investigations leading to this conclusion.

The convergence time also depends on the algorithm weight component's update gain, g. It must be chosen to ensure stability. From (7.25), the maximum allowable gain (MAG) value depends on the number and power levels of the jammers. A conservative value of g that ensures stability for all jammer power levels is obtained by noting that the maximum eigenvalue is always less than the trace of the covariance matrix, Np_{max}, where p_{max} is the power level that saturates the A/D converters used for each antenna. Let us define

$$g = g_{\text{LMS}}(2/Np_{\text{max}}) \tag{7.65}$$

Thus, from 7.24, stability is ensured when g_{LMS} is positive and less than unity. For the examples of this section, p_{max} is defined as 40 dB above thermal noise.

The examples are for a four-auxiliary-antenna system, and each of the antennas has an omnidirectional pattern. The jammer angles are chosen arbitrarily and the main antenna sidelobe gain for each jammer is selected as an independent random variable. The thermal-noise power is defined as 0 dB. The first example is for two narrowband jammers, each with JNR of 37 dB. The four calculated residue modal powers, associated eigenvalues, and convergence periods are tabulated in Table 7.1.

Table 7.1
Two-Jammer LMS Algorithm

Residue Modal Power (dB)	*Eigenvalue*	$g_{LMS} = 0.25$ *Convergence Time*
−85.9	1.0	—
−88.9	1.0	—
6.8	2673.2	86.5
36.6	37328.8	33.5

The powers for the first two modes should be $-\infty$ dB, and the finite values are caused by accuracy effects in the computing program. We can see that, although the two jammers are of equal power, the two large eigenvalues are unequal, indicating that the steering vectors are not orthogonal. The eigenvalues differ by more than an order of magnitude, and the two residue modal powers differ by about 30 dB. The convergence time for each mode is defined as the number of iterations required for the modal power to decay to the minimum residue power. Thus, from (7.30), the convergence time of each mode is the value of K, such that

$$q_n(1 - gd_n)^{2K} = p_{\min}$$

The value depends on the algorithm gain and the tabulated values are for $g_{\text{LMS}} = 0.25$. The scenario convergence time is defined as the maximum modal convergence time and equals 87 for this case.

Figure 7.1 gives the results of a simulaton of this scenario. Initially, the antenna voltages are due to thermal noise only. At the time index value of 100, the two equally powered jammers activate. Each jammer induces an average power level of 37 dB above thermal noise in each auxiliary antenna.

Figure 7.1 shows two curves. The smooth one is the residue power *versus* time, based on the mean-weight-vector analysis. From time point 0 to 100, when only thermal noise is present, the mean weight vector equals 0 and the theoretical residue power equals the thermal noise level of 0 dB. At time point 100, the residue power increases to approximately 40 dB, and then starts an exponential decay to its steady-state value. The optimum residue power after convergence equals 0.51 dB. The smooth curve shows the residue power approaching this power level because the mean weight vector converges to the optimum weight vector.

The jagged curve in the figure is the result of a random simulation. The voltage for each antenna is modeled as the sum of thermal noise and jammer induced voltages. The voltage contribution for each jammer is properly phase shifted to account

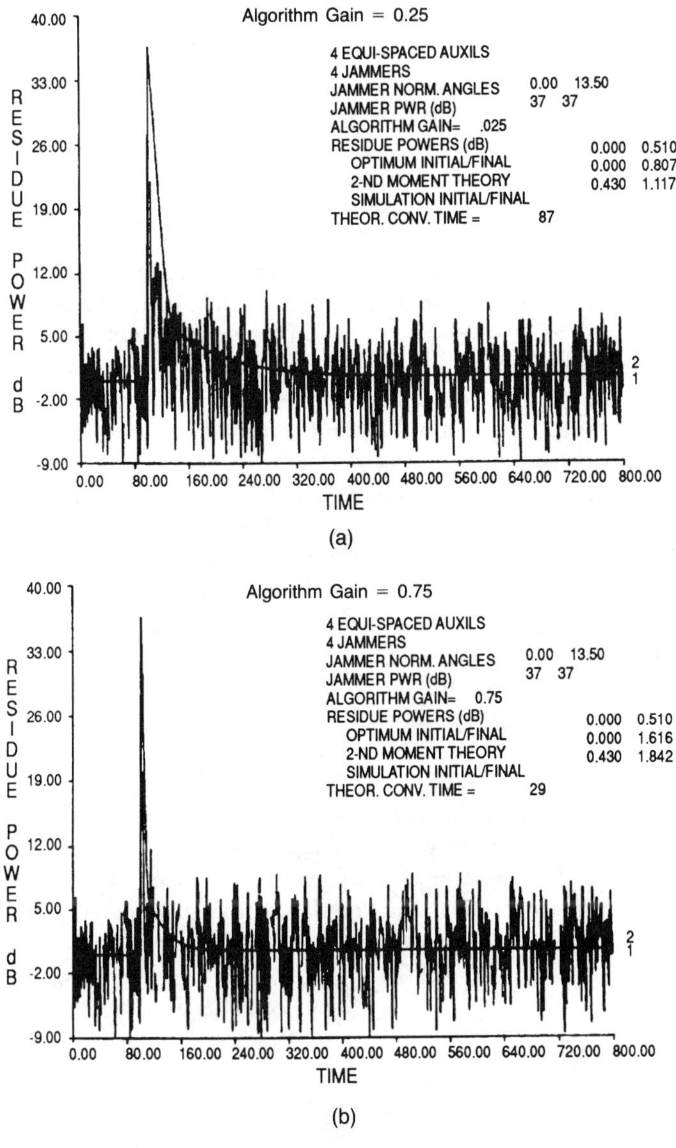

Figure 7.1 Two-jammer LMS algorithm convergence example.

for the physical distance between antennas. Thus, the simulation is narrowband in the sense that the jammer voltages are not modeled with time delay, but only as the phase shift resulting from the time delay. The residue power of the simulation for the prejammer and convergence condition is estimated by averaging 100 samples at the beginning and 350 samples at the end. These estimated power values are 0.43 and 1.117 dB, respectively. The residue power differs from the optimum weight's residue power because of the weight-noise effect. Section 7.4 develops the multiple-antenna LMS weight-noise residue power equations. Figure 7.1(b) is identical to Figure 7.1(a), except that the algorithmic gain has been increased by a factor of 3 to 0.75. We can see that the convergence time decreases and now equals 29. However, the steady-state residue power at convergence increases to 1.842 dB.

The next example is a four-jammer scenario of the same total jamming power as the previous example. The tabulated results for this case are shown in Table 7.2.

Note that, although there are four jammers, only two of the models contain power greater than thermal noise. Thus, the convergence time of the two low-power modes does not affect the scenario convergence time. Also note that the two unimportant modes are not associated with small eigenvalues.

The calculated convergence time is shown as 1101 samples. This is more than an order of magnitude greater than that for the two-jammer scenario. The reader should note that this example was not chosen to show long convergence times, but happens to have been the first four-jammer example simulated. Its results are typical and demonstrate the high probability of long convergence time when the number of jammers equals the number of auxiliaries.

The simulated and calculated convergence curves are shown in Figure 7.2(a) and (b) for algorithmic gains of 0.25 and 0.75, respectively. The convergence of the two modes is easily seen. The first mode of power, 36.6 dB, converges rapidly and when its power has decayed to less than that of the slowly converging second mode, only the second mode convergence rate can be seen. Note the close correspondence between the theoretical and simulation convergence characteristics.

Table 7.2
Four-Jammer LMS Algorithm

Residue Modal Power (dB)	*Eigenvalue*	$g_{LMS} = 0.25$ *Convergence Time*
−33.3	3.9	—
16.9	533.0	1101
−4.9	10144.6	—
36.6	29322.4	42.7

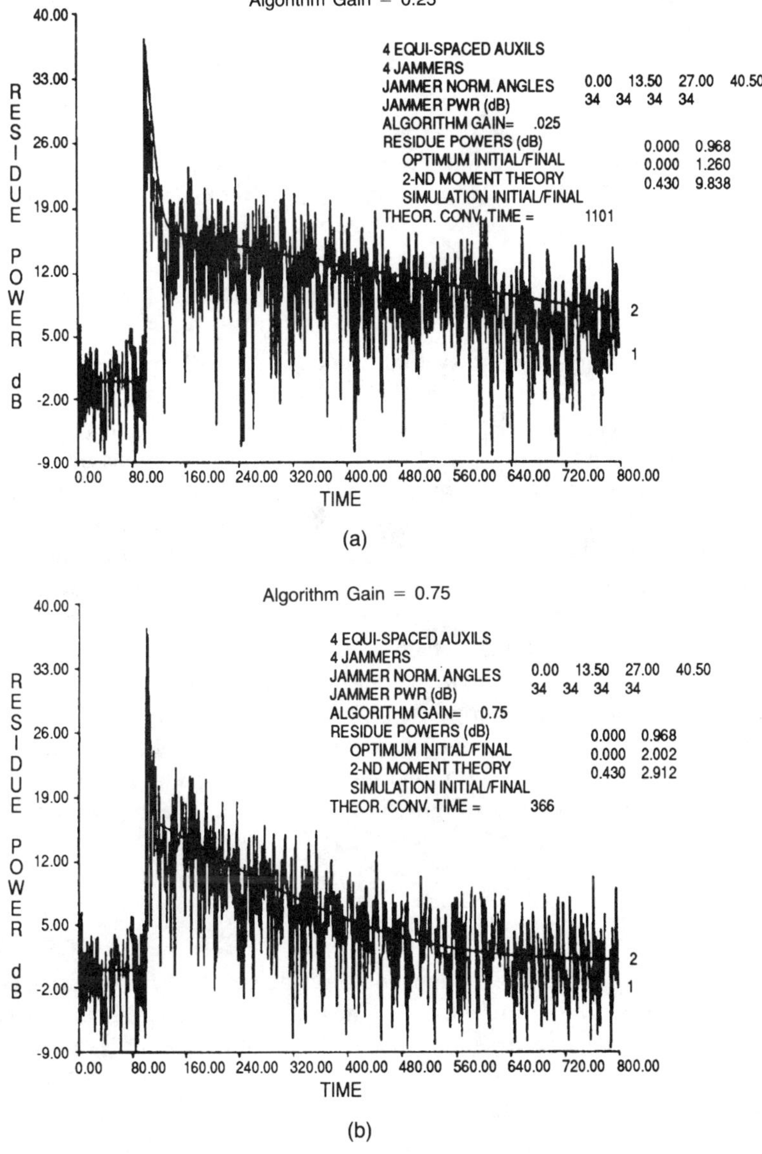

Figure 7.2 Four-jammer LMS algorithm convergence example.

7.3 NORMALIZED LMS ALGORITHM

As discussed in Chapter 6, for the single-auxiliary-antenna system, the significant feature of the NLMS algorithm is that the convergence time is independent of the power level of the auxiliary antenna. For the multiple-antenna system, there are several versions of the algorithm that have this property.

Generalizing (6.46), one version of the update equation for the nth-weight vector component is

$$w_n(k + 1) = w_n(k) + g_N r^*(k) y_n(k) / |y_n(k)|^2 \tag{7.66}$$

For the single-auxiliary-antenna system, the derivation of the average weight value, given by (6.47), is a single constant coefficient difference equation. As shown, the derived time constant is independent of the auxiliary power level. Obtaining the solution of the N coupled equations of (7.66) is more complicated. An alternative way of demonstrating that the convergence time is independent of the auxiliary antenna's power levels is by a mental experiment. Thus, imagine a particular realization of the set of measurements $\{y_n(k); n = 1, N; k = 1, K\}$. For an initial weight vector of zero, the initial output residue equals $y_0(1)$. For this residue voltage, the weight components are computed as a particular realization of $w_1(1)$ through $w_N(1)$. The residue voltage at time index 2 is then computed, and so forth. There will be some particular convergence time. Now, mentally modify each $y_n(k)$ by multiplying each by a constant β such that the modified measurement equals $\beta y_n(k)$. Multiplying by β is equivalent to changing the auxiliary power level from p_a to $\beta^2 p_a$. For the same initial weight vector of zero, the initial output residue equals $\beta y_0(1)$. For this residue voltage, using (7.66), the nth weight component value is computed as

$$\begin{aligned} w_n(1) &= w_n(0) + g_N [\beta y_0(1)]^* [\beta y_n(1)] / |\beta y_n(1)|^2 \\ &= g_N y_0^*(1) y_n(1) / |y_n(1)|^2 \end{aligned} \tag{7.67}$$

Note that each weight value is independent of the value of β. This independence is also obtained for all other time indices. Thus, all the weight values are independent of β, and therefore the convergence is independent of the power level of the auxiliary antennas.

An alternative update equation is

$$w_n(k + 1) = w_n(k) + g_N r^*(k) y_n(k) / \Sigma_m |y_m(k)|^2 \tag{7.68}$$

By the same mental experiment procedure, we can see that the weights computed by this method are also independent of the scaling constant, β. Thus, both versions of the algorithm have transient convergence characteristics that are independent of the auxiliary power level. Clearly, there are many similar algorithms that also have convergence times that are independent of the auxiliary antennas' power levels.

7.3.1 Preferred NLMS Algorithm Derivations

One method of deriving an indication that the second version of the NLMS algorithm is preferred to any other is by a more detailed examination of the reason for the convergence difficulty of the LMS algorithm. Combining (5.57) and (7.5) gives the average-weight-vector-update equation as

$$\mathbf{W}_a(k + 1) = \mathbf{W}_a(k) + g\mathbf{R}[\mathbf{W}_{\text{opt}} - \mathbf{W}_a(k)] \tag{7.69}$$

To interpret (7.69), consider that a "perfect" algorithm is one that from any initial weight vector produces the optimum weight vector in a single iteration. Alternately, a "good" algorithm is one that calculates a correction update that rapidly converges to the proper weight vector. For any specified weight vector, $\mathbf{W}_a(k)$, the direction to the desired weight vector is $\mathbf{W}_{\text{opt}} - \mathbf{W}_a(k)$. Note, however, that the update term in (7.69) is not in the desired direction, but instead is in the "wrong" direction $\mathbf{R}[\mathbf{W}_{\text{opt}} - \mathbf{W}_a(k)]$. We may conjecture that improved performance can be obtained by modifying the LMS weight-update equation to

$$\mathbf{W}(k + 1) = \mathbf{W}(k) + gr^*(k)\mathbf{R}^{-1}\mathbf{Y}(k) \tag{7.70}$$

Combining this with (6.32) and taking average values gives

$$\mathbf{W}_a(k + 1) = \mathbf{W}_a(k) + g[\mathbf{W}_{\text{opt}} - \mathbf{W}_a(k)] \tag{7.71}$$

Thus, the correction term is now in the correct direction. Also, if $g = 1$, the convergence always is done in a single iteration.

The difficulty of implementing this improvement is that the covariance matrix, and therefore its inverse, is not known so that (7.70) cannot be implemented. A method of estimating the inverse matrix thus is needed.

7.3.1.1 Direct Matrix Estimation Algorithm

One indication of the desirability of the second version of the NLMS algorithm can be obtained by first considering a totally different class of algorithm denoted as *direct matrix estimation*.

The several versions of the LMS algorithm are iterative and have small memory requirements. Specifically, at each time instant, the measurement vector $[y_0(k), \mathbf{Y}^{\mathrm{T}}(k)]^{\mathrm{T}}$

is used to iterate the estimated weight vector and then the measurement vector is discarded. An alternative procedure that requires significantly more memory is to use the vector measurements to estimate the covariance matrix and the correlation vector. For a batch of K vector measurements, the estimates are

$$\begin{aligned}\hat{\mathbf{R}} &= \Sigma_k \mathbf{Y}(k)\mathbf{Y}^{\mathbf{H}}(k)/K \\ \hat{\mathbf{R}}_0 &= \Sigma_k y_0^*(k)\mathbf{Y}(k)/K\end{aligned} \tag{7.72}$$

These estimates are hence used in (5.57) as approximations to the true matrices to obtain an estimate of the optimum vector weight as the solution of

$$\hat{\mathbf{R}}\hat{\mathbf{W}} = \hat{\mathbf{R}}_0 \tag{7.73}$$

In addition to the relatively large memory requirements, the solution for the weight vector is computationally intensive. A more detailed analysis of the algorithm's requirements and performance will be given in Section 7.6. Our primary current interest in (7.72) is that it leads to a method of estimating the inverse covariance matrix to obtain an iterative algorithm with better convergence characteristics than that of the LMS algorithm.

7.3.1.2 Generalized Matrix Inverse and NLMS

The matrix $\hat{\mathbf{R}}$ is nonsingular only when the number of vector measurements is equal to or greater than the dimension of $\mathbf{Y}(k)$. For singular $\hat{\mathbf{R}}$, there are an infinite number of solutions to (7.73). A unique solution can be defined as the particular $\hat{\mathbf{W}}$ of minimum length. The matrix that computes the minimum-length solution is defined as the generalized inverse. This matrix is defined as

$$\hat{\mathbf{W}} = \mathbf{R}^+\hat{\mathbf{R}}_0 \tag{7.74}$$

where $\mathbf{R}^+$ is the generalized inverse of $\hat{\mathbf{R}}$. When $K = 1$, we have

$$\mathbf{R}^+ = \mathbf{Y}(k)\mathbf{Y}^{\mathbf{H}}(k)/|\mathbf{Y}^{\mathbf{H}}(k)\mathbf{Y}(k)|^2 \tag{7.75}$$

The generalized inverse seems quite appropriate for the MSLC application. Specifically, as has been shown, a highly accurate, although approximate, explanation of the action of the optimum weight vector is that it is the vector that exactly cancels all jammers. For this condition, the residue power is due to the thermal-noise contribution from the main and auxiliary antennas. For equal noise power at each auxiliary antenna, the residue power equals the main antenna noise power plus the auxiliary antenna's noise power multiplied by the squared length of the weight vector. Thus, minimizing the length of the weight vector minimizes the residue power. Therefore, the minimum-length solution to (7.73) minimizes the residue power and the generalized inverse is appropriate.

Using $\mathbf{R}^+$ of (7.75) instead of $\mathbf{R}^{-1}$ in (7.70) gives

$$\mathbf{W}(k+1) = \mathbf{W}(k) + [g/\mathbf{Y}^{\mathbf{H}}(k)\mathbf{Y}(k)]r^*(k)\mathbf{Y}(k) \tag{7.76}$$

which is the vector generalizaton of (7.68).

7.3.2 Gradient Descent Equation Solutions and NLMS

Another indication of the desirability of the form of the NLMS algorithm given by (7.68) can be obtained by considering the gradient descent method of solving linear equations.

Linear matrix equations of the form of (7.73) arise in many applications. Their solution becomes increasingly difficult as the order of the matrix increases. One method of solution, termed gradient descent, is an iterative method of solution. Specifically, the gradient descent algorithm is applied to the case of solving for the vector **W** in

$$\mathbf{RW} = \mathbf{S} \tag{7.77}$$

when the square matrix **R** and the vector **S** are specified. One version of the update equation for iteratively computing **W** is

$$\mathbf{W}(k+1) = \mathbf{W}(k) + g_R[\mathbf{S} - \mathbf{RW}(k)] \tag{7.78}$$

The iterative procedure stops when the error, $\mathbf{S} - \mathbf{RW}(k)$, is sufficiently small. To emphasize the obvious, the matrices **R** and **S** are specified and unchanging. For the stochastic gradient descent algorithm, the update equation for the digitally implemented MSLC application is

$$\mathbf{W}(k+1) = \mathbf{W}(k) + g[y_0(k) - \mathbf{W}^{\mathbf{H}}(k)\mathbf{Y}(k)]^*\mathbf{Y}(k) \tag{7.79}$$

A contrast between (7.78) and (7.79) is that (7.78) only uses unchanging or "old" data, whereas (7.79) only uses "new" data. Therefore, a question arises as to whether improved performance would be attained by operating on the new data M times. Thus, consider

$$\begin{aligned} \mathbf{W}_1(k+1) &= \mathbf{W}_M(k) + g[y_0(k) - \mathbf{W}_M^{\mathbf{H}}(k)\mathbf{Y}(k)]^*\mathbf{Y}(k) \\ \mathbf{W}_m(k+1) &= \mathbf{W}_{m-1}(k+1) + g[y_0(k) - \mathbf{W}_{m-1}^{\mathbf{H}}(k+1)\mathbf{Y}(k)]^*\mathbf{Y}(k), \\ & \qquad m = 2, \ldots, M \end{aligned} \tag{7.80}$$

We can easily establish by simulation that this decreases convergence time and increasing M further decreases convergence time. We shall show below that, as M approaches infinity, the algorithm tends toward the NLMS algorithm.

7.3.2.1 Multiple-Pass Algorithm Analysis

The multiple-pass algorithm updates the weight vector M times for each data vector input. By suppressing the dependence on the time index k and rearranging, (7.80) can be rewritten as

$$\mathbf{W}_{m+1} = gy_0^*\mathbf{Y} + (\mathbf{I} - g\mathbf{Y}\mathbf{Y}^{\mathbf{H}})\mathbf{W}_m, \quad m = 1, 2, \ldots \tag{7.81}$$

where $\mathbf{W}_0$ is a specified boundary condition. The lack of time argument on y_0 and $\mathbf{Y}$ emphasizes the constancy of the values during the update. To solve this difference equation, assume a solution of the form:

$$\mathbf{W}_m = \alpha_1(m)\mathbf{Y} + \alpha_2(m)\mathbf{W}_0 \tag{7.82}$$

The boundary conditions on α_1 and α_2 are

$$\begin{aligned} \alpha_1(0) &= 0 \\ \alpha_2(0) &= 1 \end{aligned} \tag{7.83}$$

Substituting (7.83) into (7.81) gives the difference equations:

$$\begin{aligned} \alpha_1(m+1) &= gy_0^* - g\alpha_2(m)\mathbf{Y}^{\mathbf{H}}\mathbf{W}_0 + (1 - g\mathbf{Y}^{\mathbf{H}}\mathbf{Y})\alpha_1(m) \\ \alpha_2(m+1) &= \alpha_2(m) \end{aligned} \tag{7.84}$$

The second part of (7.84) indicates that α_2 must equal a constant. Then, the first part of (7.84) becomes a difference equation with constant coefficients. Thus, the solution is

$$\alpha_1(m) = \mathrm{c}(1 - g\mathbf{Y}^{\mathbf{H}}\mathbf{Y})^m + (y_0^* - \alpha_2\mathbf{Y}^{\mathbf{H}}\mathbf{W}_0)/\mathbf{Y}^{\mathbf{H}}\mathbf{Y} \tag{7.85}$$

where c is an arbitrary constant. Applying the boundary condition gives

$$\mathbf{W} = \mathbf{W}_0 + (y_0^* - \mathbf{Y}^{\mathbf{H}}\mathbf{W}_0)[1 - (1 - g\mathbf{Y}^{\mathbf{H}}\mathbf{Y})^m]/\mathbf{Y}^{\mathbf{H}}\mathbf{Y} \tag{7.86}$$

For sufficiently large M, the transient portion of the update equation is negligible. Reinserting the time dependence of $\mathbf{Y}$ and y_0 gives the update equation:

$$\mathbf{W}(k+1) = \mathbf{W}(k) + g_E[y_0^* - \mathbf{Y}^{\mathbf{H}}(k)\mathbf{W}(k)]/\mathbf{Y}^{\mathbf{H}}(k)\mathbf{Y}(k) \tag{7.87}$$

where g_E equals unity. More generally, the response characteristics of the algorithm can be modified by allowing g_E to be a value other than unity, and, as in 7.76, this is the vector generalization of 7.68. These are defined as the NLMS algorithm.

7.3.3 NLMS Algorithm Convergence Analysis

The weight-vector-update equation is obtained from (7.76) by combining with the definition of the residue voltage. Thus

$$\begin{aligned}\mathbf{W}(k+1) = \mathbf{W}(k) &+ g_N Y_0^*(k)\mathbf{Y}(k)/\mathbf{Y^H}(k)\mathbf{Y}(k) \\ &- g_N[\mathbf{Y}(k)\mathbf{Y^H}(k)/\mathbf{Y^H}(k)\mathbf{Y}(k)]\mathbf{W}(k)\end{aligned} \tag{7.88}$$

and, we should note that

$$\Sigma_m |y_m(k)|^2 = \mathbf{Y^H}(k)\mathbf{Y}(k) \tag{7.89}$$

By taking expectations, it follows that

$$\begin{aligned}\mathbf{W}_a(k+1) &= \mathbf{W}_a(k) + g_N \Omega_0 - g_N \Omega \mathbf{W}_a(k) \\ \Omega_0 &= E[y_0^*(k)\mathbf{Y}(k)/\mathbf{Y^H}(k)\mathbf{Y}(k)] \\ \Omega &= E[\mathbf{Y}(k)\mathbf{Y^H}(k)/\mathbf{Y^H}(k)\mathbf{Y}(k)]\end{aligned} \tag{7.90}$$

Note the great similarity between the NLMS algorithm average-weight-vector-update equation and that for the LMS algorithm, as given by (7.5). We can see that Ω_0 and Ω can be interpreted as the normalized correlation vector and normalized covariance matrix, respectively. By extrapolation, the transient response for the NLMS algorithm depends upon the eigenvalues of Ω and the inner products of the eigenvectors of Ω with the vector Ω_0. As shown below, the eigenvectors of the normalized covariance matrix Ω are identical to those of the covariance matrix $\mathbf{R}$. Also, the eigenvalues of Ω are expressible as a function of the eigenvalues of $\mathbf{R}$.

To find the eigenvalues and eigenvectors, define an N-dimensional, complex-valued, Gaussian distributed random vector $\mathbf{X}$ as

$$\begin{aligned}\mathbf{X} &= \mathbf{F^H}\mathbf{Y}(k) \\ \mathbf{Y}(k) &= \mathbf{F}\mathbf{X} \\ E(\mathbf{X}\mathbf{X^H}) &= \mathbf{F^H}\mathbf{R}\mathbf{F} = \mathbf{D}\end{aligned} \tag{7.91}$$

where, as defined in (7.17), $\mathbf{F}$ is the unitary matrix with columns that are the eigenvectors of $\mathbf{R}$, and $\mathbf{D}$ is a diagonal matrix of eigenvalues. Thus, each component of $\mathbf{X}$ is independently distributed, has zero mean, and the variance of the nth component is d_n, the nth eigenvalue of $\mathbf{R}$. Substituting (7.91) into (7.90) gives

$$\begin{aligned} \Omega &= \mathbf{F}\mathbf{D}'\mathbf{F}^{\mathbf{H}} \\ \mathbf{D}' &= E(\mathbf{X}\mathbf{X}^{\mathbf{H}}/\mathbf{X}^{\mathbf{H}}\mathbf{X}) \end{aligned} \tag{7.92}$$

We shall show that $\mathbf{D}'$ is a diagonal matrix.

The n, m component of $\mathbf{D}'$ is

$$(\mathbf{D}')_{nm} = E[x_n x_m^* / \Sigma_p |x_p|^2] \tag{7.93}$$

If we decompose x_n into its Rayleigh distributed magnitude, r_n, and its independently distributed phase angle, θ_n, then,

$$(\mathbf{D}')_{nm} = E[r_n r_m \exp[j(\theta_n - \theta_m)]/\Sigma_p r_p^2] \tag{7.94}$$

Because θ_n is uniformly distributed over 2π, $(\mathbf{D}')_{nm} = 0$ for $n \neq m$. Thus, $\mathbf{D}'$ is a diagonal matrix. From the definitions of (7.92), it follows that the eigenvectors of Ω are identical to the eigenvectors of $\mathbf{R}$ and the eigenvalues of Ω equal the diagonal entries of $\mathbf{D}'$. Let us denote these eigenvalues as d_n'. Then, the equivalent of (7.21) is

$$\mathbf{W}_a(k+1) = \mathbf{F}(\mathbf{I} - g_N\mathbf{D}')\mathbf{F}^{\mathbf{H}}\mathbf{W}_a(k) + g_N\Omega_0 \tag{7.95}$$

and the equivalent of (7.22) is

$$\mathbf{V}_a(k+1) = (\mathbf{I} - g_N\mathbf{D}')\mathbf{V}_a(k) + g_N\mathbf{F}^{\mathbf{H}}\Omega_0 \tag{7.96}$$

This is a set of N decoupled, first-order, constant coefficient difference equations. The solution for the nth component is

$$v_{an}(k+1) = [1 - (1 - g_N d_n')^k](\mathbf{F}_n^{\mathbf{H}}\Omega_0/d_n') \tag{7.97}$$

and the convergence is stable if

$$0 < g_N d_n' < 2 \tag{7.98}$$

From (7.92), note that

$$\mathrm{tr}(\mathbf{D}') = \mathrm{tr}[E(\mathbf{X}\mathbf{X}^{\mathbf{H}}/\mathbf{X}^{\mathbf{H}}\mathbf{X})] = 1 \tag{7.99}$$

Thus, because the order of taking the trace of a matrix and the expectation operation can be interchanged, the sum of the eigenvalues of Ω equals one under all conditions. Thus, each d_n' is less than one and a conservative convergence stability criterion, based upon a first moment analysis, is

$$0 < g_N < 2 \tag{7.100}$$

To compute (7.97), we need to determine the normalized correlation vector Ω_0. Therefore, consider the $N + 1$-dimensional extended snapshot vector $\mathbf{Y}_e$ defined in partitioned form as

$$\mathbf{Y}_e^{\mathbf{T}} = (y_0, \mathbf{Y}^{\mathbf{T}})^{\mathbf{T}} \tag{7.101}$$

The covariance matrix of $\mathbf{Y}_e$ in partitioned form is evaluated as

$$\mathbf{R}_e = E(\mathbf{Y}_e\mathbf{Y}_e^{\mathbf{H}}) = \begin{pmatrix} p_{\text{main}} & \mathbf{R}_0^{\mathbf{H}} \\ \mathbf{R}_0 & \mathbf{R} \end{pmatrix} \tag{7.102}$$

The inverse matrix can be found as

$$\begin{aligned} \mathbf{R}_e^{-1} &= \begin{pmatrix} 1 & -\mathbf{R}_0^{\mathbf{H}}\mathbf{R}^{-1} \\ -\mathbf{R}^{-1}\mathbf{R}_0 & \mathbf{R}^{-1}\mathbf{R}_0\mathbf{R}_0^{\mathbf{H}}\mathbf{R}^{-1} + p_{\min}\mathbf{R}^{-1} \end{pmatrix} / p_{\min} \\ p_{\min} &= (p_{\text{main}} - \mathbf{R}_0^{\mathbf{H}}\mathbf{R}^{-1}\mathbf{R}_0) \end{aligned} \tag{7.103}$$

The determinant of $\mathbf{R}_e$ is [1]:

$$\det(\mathbf{R}_e) = p_{\min} \det(\mathbf{R}) \tag{7.104}$$

Thus, the density function of $\mathbf{Y}_e$ is

$$\begin{aligned} p(\mathbf{Y}_e) &= \exp[(Q_1 + Q_2)/p_{\min}]/\pi^{N+1} \det(\mathbf{R}_e) \\ Q_1 &= \mathbf{Y}^{\mathbf{H}}(\mathbf{R}^{-1}\mathbf{R}_0\mathbf{R}_0^{\mathbf{H}}\mathbf{R}^{-1} + p_{\min}\mathbf{R}^{-1})\mathbf{Y} \\ Q_2 &= -y_0^* b \exp(j\phi) - y_0 b \exp(-j\phi) + |y_0|^2 \\ \mathbf{R}_0^{\mathbf{H}}\mathbf{R}^{-1}\mathbf{Y} &= b \exp(j\phi) \end{aligned} \tag{7.105}$$

The expectation defining Ω_0 can be formulated as the integral with respect to y_0 and then with respect to the vector $\mathbf{Y}$.

Expressing y_0 in terms of its magnitude r and phase θ gives

$$\Omega_0 = E(y_0^* \mathbf{Y}/\mathbf{Y}^{\mathbf{H}}\mathbf{Y})$$
$$= \int Q_3 \, \mathrm{d}\mathbf{Y} \left\{ \int_0^\infty \exp(-r^2/p_{\min}) \right.$$
$$\left. \int_{-\pi}^{\pi} r^2 \exp(-j\phi) \exp[2rb \cos(\theta - \phi)/p_{\min}]/\pi^{N+1} \det(R_e) \, \mathrm{d}\theta \, \mathrm{d}r \right\}$$
$$Q_3 = \mathbf{Y} \exp(-Q_1/p_{\min})/\mathbf{Y}^{\mathbf{H}}\mathbf{Y} \tag{7.106}$$

Integration on θ is done by using symmetry arguments and Gradshteyn and Ryzhik [2]: Thus,

$$\Omega_0 = 2 \exp(-j\phi) \int Q_3 \, \mathrm{d}\mathbf{Y} \int_0^\infty ir^2 J_1(-2ibr/p_{\min}) \exp(-r^2/p_{\min}) \, \mathrm{d}r \tag{7.107}$$

Using Gradshteyn and Ryzhik [2] and combining with (7.105) gives

$$\Omega_0 = \Omega \mathbf{R}^{-1}\mathbf{R}_0 \tag{7.108}$$

Expressing Ω and $\mathbf{R}$ in their eigenvalue decompositions gives

$$\Omega_0 = (\mathbf{F}\mathbf{D}'\mathbf{F}^{\mathbf{H}})(\mathbf{F}\mathbf{D}^{-1}\mathbf{F}^{\mathbf{H}})\mathbf{R}_0 = \mathbf{F}\mathbf{D}'\mathbf{D}^{-1}\mathbf{F}^{\mathbf{H}}\mathbf{R}_0 \tag{7.109}$$

Note that the product $\mathbf{D}'\mathbf{D}^{-1}$ is a diagonal matrix. Combining (7.109) with either (7.96) or (7.97) gives

$$v_{an}(k + 1) = [1 - (1 - g_N d_n')^k](\mathbf{F}_n^{\mathbf{H}}\mathbf{R}_0/d_n) \tag{7.110}$$

By comparing this to (7.23), we can see that the form of the convergence equations for the NLMS algorithm and the LMS algorithms are identical. The only difference is that the former depends on the product $g_N d_n'$, whereas the latter depends on the product of gd_n.

To develop a detailed comparison of the relative convergence properties of the LMS and NLMS algorithms, we need to derive expressions for the eigenvalues of Ω. Before deriving a general expression for the eigenvalues, let us consider some special-case examples. The first example is that of two narrowband jammers and two auxiliary antennas, as previously considered for the LMS algorithm evaluation.

In general, the values of d_n' can be obtained by noting from (7.94) that

$$d_n' = E(r_n^2/\Sigma_p r_p^2) \tag{7.111}$$

Let us define u_n as equal to r_n^2. Because the probability density function of **X** is Gaussian and the covariance matrix of **X** is diagonal, the components of **X** are statistically independent. Thus, the set of random variables $\{u_n\}$ is also statistically independent. The probability density function of u_n is exponential and is given in terms of the eigenvalues of **R** as

$$p_n(u) = \exp(-u/d_n)/d_n \tag{7.112}$$

and

$$d'_n = E(u_n/\Sigma_p u_p) \tag{7.113}$$

For the particular example, $N = 2$, so that

$$\begin{aligned} d'_1 &= E[u_1/(u_1 + u_2)] \\ d'_2 &= 1 - d'_1 \end{aligned} \tag{7.114}$$

and the second part of (7.114) follows from (7.99). From (7.112) and (7.114), we have

$$d'_1 = \int_0^\infty [u_1/(u_1 + u_2)] \exp[-(u_1/d_1 + u_2/d_2)](\mathrm{d}u_1\, \mathrm{d}u_2/d_1 d_2) \tag{7.115}$$

Changing to polar coordinates and recognizing that the integral, with respect to the magnitude, is essentially the γ function gives

$$d'_1 = \int_0^{\pi/2} [\cos\theta/(\cos\theta + \sin\theta)][\cos\theta/d_1 + \sin\theta/d_2]^{-2}(\mathrm{d}\theta/d_1 d_2) \tag{7.116}$$

The substitution $x = \tan\theta$ reduces (7.116) to

$$d'_1 = \int_0^\infty [\mathrm{d}x/(1 + x)(d_1^{-1} + d_2^{-1}x)^{-2}]/d_1 d_2 \tag{7.117}$$

Combining Gradshteyn and Ryzhik's equations 2.154 and 2.155 gives an expression for the indefinite integral:

$$\int dx/(a + bx)^n(\alpha + \beta x) = \left\{(\beta/\delta)^{n-1} \ln[(\alpha + \beta x)/(a + bx)] + \sum_{k=0}^{n-2} (\beta/\delta)^k/[(n - 1 - k)(a + bx)^{n-1-k}]\right\} \Big/ \delta \tag{7.118}$$

$$\delta = a\beta - \alpha b$$

For (7.117), n of (7.118) equals 2, so that

$$\begin{aligned} d_1' &= d_1 c_1\{[d_2 \ln(d_2/d_1)]/(d_2 - d_1) - 1\}/d_2 \\ c_1 &= d_2/(d_2 - d_1) \end{aligned} \tag{7.119}$$

When the two jammers' direction vectors are orthogonal, as shown by (7.64), the two eigenvalues are

$$\begin{aligned} d_1 &= p_n + Np_1 \\ d_2 &= p_n + Np_2 \end{aligned} \tag{7.120}$$

and, for this case, $N = 2$. From (7.34), the relative convergence times of the two modes equals the ratio of d_1/d_2 for the LMS algorithm and d_1'/d_2' for the NLMS algorithm. Consider the case where p_1 is 40 dB greater than thermal noise and p_2 is 20 dB greater than thermal noise. For the LMS algorithm, the transient-response time constant for the mode corresponding to the 20 dB power is 100 times greater than that for the 40 dB power.

Because the jammer powers are much greater than the noise power, an approximation for d_1' is

$$d_1' \approx [p_1 p_2/(p_2 - p_1)^2] \ln(p_2/p_1) - [p_1/(p_2 - p_1)] \tag{7.121}$$

and d_2' is given by (7.114). Then, d_1' and d_2' are evaluated as 0.9631 and 0.0369, respectively. The ratio of these eigenvalues is 26.11. Thus, the respective transient-response time constants of the two modes for the NLMS algorithm differ by a factor of only 26.11, instead of 100 as for the LMS algorithm.

As a second example, assume the jammer angles are the same as in the first example, but each jammer's power is reduced by a factor of 10. For the LMS algorithm, the convergence time of each mode increases by a factor of 10. However, as we can see from (7.121), the eigenvalues depend on ratios of the jammer powers so that the convergence time for the NLMS algorithm is the same as for the first example. This is the expected property of the NLMS algorithm. In general, its convergence properties are superior to those of the LMS algorithm because its convergence time does not depend on the absolute power levels of the jammers.

In addition to these examples of the improved convergence time attained by the NLMS algorithm, consider the case of two jammers of equal power, p, and arbitrary relative directional vectors. The total power received by each auxiliary antenna equals $2p$. Because the trace of the auxiliary antennas covariance matrix equals the sum of the two eigenvalues, neglecting thermal noise,

$$d_1 + d_2 = 4p$$

A parametric relation between the two eigenvalues is

$$\begin{aligned} d_1 &= 2p(1 + \beta) \\ d_2 &= 2p(1 - \beta) \end{aligned} \tag{7.122}$$

and β is a constant, the value of which depends on the relative directional vectors. For orthogonal direction vectors $\beta = 0$. For β varying between zero and one, the value of d_2 is less than that of d_1 and the smaller eigenvalue corresponds to the slow convergence mode. The relative convergence time constants of the slow and fast modes is a reasonable measure of the effects of jammer geometry on the convergence time.

We can see that as β approaches one, the convergence time of the slow mode becomes arbitrarily long. When the effects of thermal noise are considered, the smaller eigenvalue tends toward the thermal-noise power, rather than zero. The larger eigenvalue is approximately equal to $4p$. The ratio equals $4p/p_n$ and this is usually so large that approximating it by infinity is not unreasonable. Thus, the approximation of the eigenvalues using (7.122) should be satisfactory.

The resulting relative convergence times of the two modes for the LMS algorithm is

$$d_1/d_2 = (1 + \beta)/(1 - \beta)$$

Comparable relative convergence times for the NLMS algorithm can be obtained by combining (7.121) and (7.122) as a functon of β. The relative convergence times of the two modes for the two algorithms is plotted in Figure 7.3(a). We can see that, for either algorithm, the convergence time increases as β approaches one. However, the increase for the NLMS algorithm is considerably less than that of the LMS algorithm. The ratio of the values of the two curves is a measure of the decreased sensitivity of the convergence time constant of the NLMS algorithm as compared to

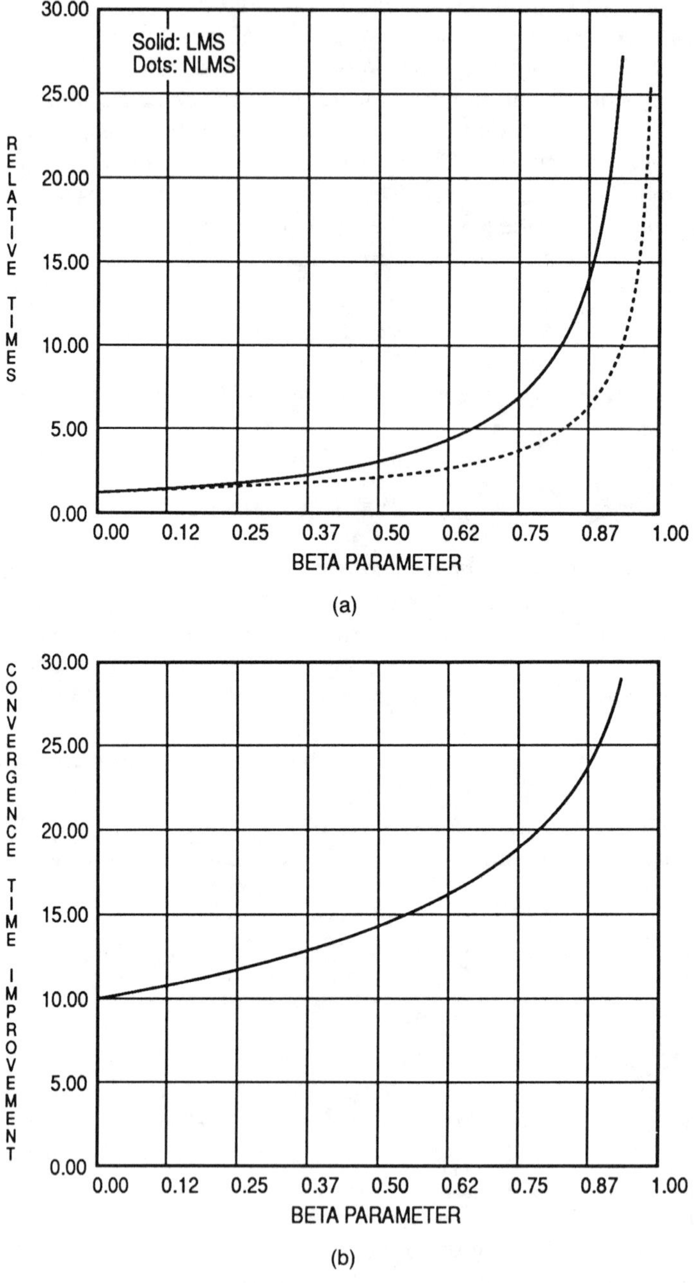

Figure 7.3 Geometric effect on LMS and NLMS convergence time constants.

the LMS algorithm for different jammer geometries. This ratio is plotted in Figure 7.3(b).

In summary, the primary feature of the NLMS algorithm compared to the LMS algorithm is that the convergence time constant of the former does not change when each jammer power is scaled by a common factor. This performance is superior to that of the LMS algorithm, the convergence time constant of which scales inversely as the power scaling factor. In addition, a secondary feature of the NLMS algorithm is that its convergence time constant is less sensitive to jammer geometry than is the LMS algorithm.

7.3.4 Multiple Antenna NLMS Convergence Analysis

The previous section derived a general expression for the eigenvalues of the normalized covariance matrix Ω in terms of the exponentially distributed random variables u_1 through u_N and evaluated it for the special case of two auxiliary antennas. This section will continue this analysis and derive an expression that can be evaluated for arbitrary N.

From (7.113), we have

$$d'_n = E\left(u_n \Big/ \sum_p u_p\right) = E[u_n/(u_n + t_n)]$$
$$t_n = \sum_{\substack{p=1 \\ p\neq n}}^{N} u_p \tag{7.123}$$

where u_n and t_n are statistically independent.

The statistical density function for t_n can be obtained via characteristic functions. The characteristic function of any u_p is

$$\phi_p(\mu) = E[\exp(j\mu u_p)] \tag{7.124}$$

and, by using (7.112), this can be evaluated as

$$\phi_p(\mu) = (1 - j\mu d_p)^{-1} \tag{7.125}$$

As t_n is the summation of independent random variables, its characteristic function is the product of the individual characteristic functions. Thus,

$$\phi_t(\mu) = \prod_{\substack{p=1 \\ p\neq n}}^{N} (1 - j\mu d_p)^{-1} \tag{7.126}$$

Assuming that the eigenvalues are distinct, this can be expanded by partial fractions as

$$\begin{aligned} \phi_t(\mu) &= \sum_{\substack{p=1 \\ p\neq n}}^{N} b_{pn}/(1 - j\mu d_p) \\ b_{pn} &= \prod_{\substack{j=1 \\ j\neq p,n}}^{N} d_p/(d_p - d_j) \end{aligned} \tag{7.127}$$

and, by comparison to (7.125), the statistical density function for t_n is

$$p_{tn}(t) = \sum_{\substack{p=1 \\ p\neq n}}^{N} b_{pn} \exp(-t/d_p)/d_p \tag{7.128}$$

The value of d_n' is then

$$d_n' = \sum_{p} b_{pn} \int_0^{\infty}\int_0^{\infty} [u_n/(u_n + t)] \exp[-(u_n/d_n + t/d_p)][\mathrm{d}u_n\, \mathrm{d}t/d_n d_p] \tag{7.129}$$

Comparing (7.129) and (7.115), we can see that the double integrals are identical in form. Thus,

$$\begin{aligned} d_n' &= d_n \sum_{\substack{p=1 \\ p\neq n}}^{N} c_p\{[d_p \ln(d_p/d_n)]/(d_p - d_n) - 1\}/d_p \\ c_p &= \prod_{\substack{j=1 \\ j\neq p}}^{N} [d_p/(d_p - d_j)] \end{aligned} \tag{7.130}$$

7.3.5 Additional NLMS Algorithm Convergence Examples

The examples discussed here are simulations of the same scenarios discussed for the LMS algorithm in Section 7.2.5.

The first case is the two-jammer scenario tabulated in Table 7.1. Table 7.3 includes the prior results as well as computed convergence data for the NLMS algorithm.

We can see that the convergence time of the NLMS algorithm is about four times less than that of the LMS algorithm.

Figure 7.4 is a simulation of the NLMS algorithm using the same set of random voltages used for the LMS algorithm convergence simulation of Figure 7.1. Again, the smooth curve is the residue power *versus* time as computed from the mean weight vector analysis. The jagged curve is for the random simulation. Comparing Figures 7.1 and 7.4 shows that the mean vector convergence time of the NLMS algorithm seems faster than that of the simulation. This divergence is probably due to the weight-noise effect. Note that, for the gain value of 0.25, the pre- and posttransient steady-state power values are 0.430 dB and 1.117 dB, respectively, for the LMS algorithm and are the larger values of 3.236 dB and 4.757 dB, respectively, for the NLMS algorithm.

The next example is the previously considered four-jammer scenario. The data include the LMS data of Table 7.2 as well as computed convergence data for the NLMS algorithm.

We can see that the convergence time for this scenario is about four times less than that for the LMS algorithm.

Figure 7.5 is a simulation of the NLMS algorithm convergence using the same set of random voltages as for the LMS algorithm convergence simulation of Figure 7.2. Comparing Figures 7.2 and 7.5 shows that the convergence time of the NLMS algorithm seems considerably faster for both the smooth computed curve and the jagged simulation curve. However, due to the weight-noise effect, the final converged residue power again is higher for the NLMS algorithm.

The sections to follow will derive equations for the weight noise associated with the two algorithms.

Table 7.3
Two-Jammer LMS and NLMS Algorithm

Residue Modal Power (dB)	*Eigenvalue*	$g_{LMS} = 0.25$ *Convergence Time*	$g_{NLMS} = 0.25$ *Convergence Time*
−85.9	1.0	—	—
−88.9	1.0	—	—
6.8	2673.2	86.5	20.1
36.6	37328.8	33.5	17.2

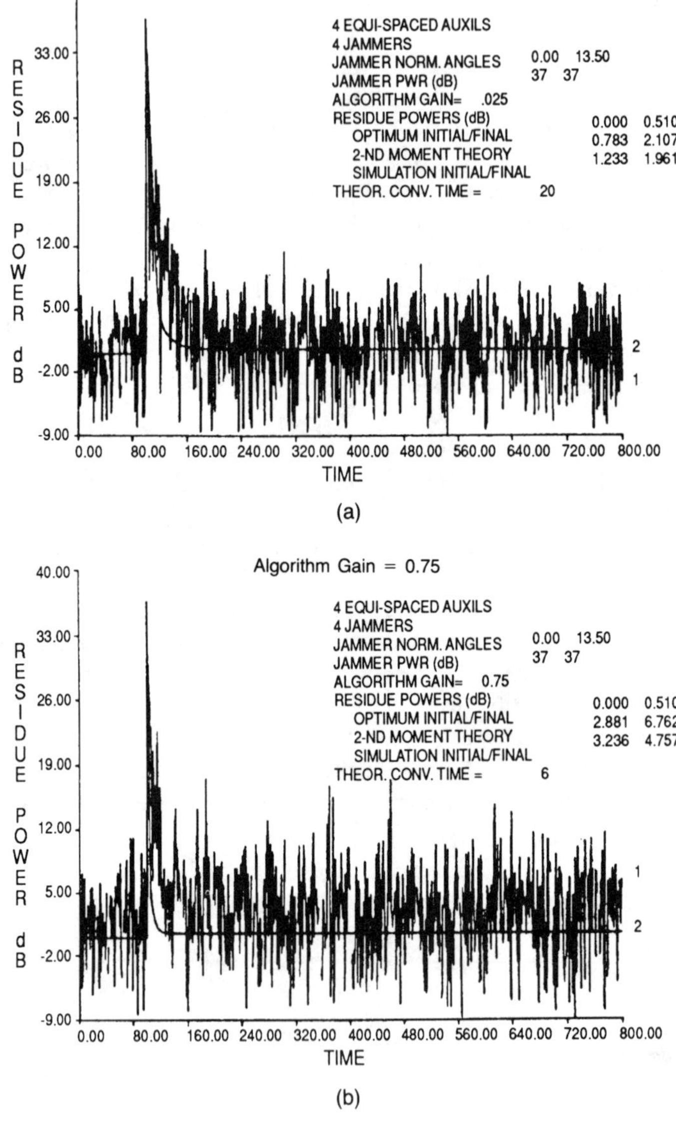

Figure 7.4 Two-jammer NLMS algorithm convergence example.

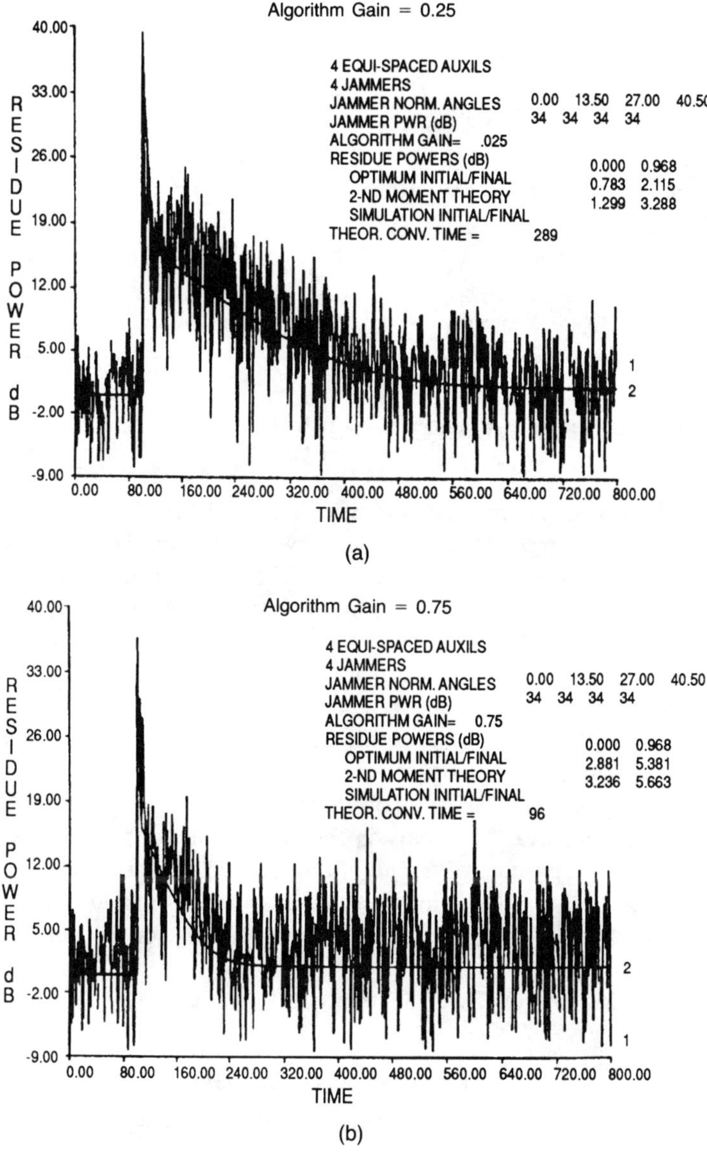

Figure 7.5 Four-jammer NLMS algorithm convergence example.

Table 7.4
Two-Jammer LMS and NLMS Algorithm

Residue Modal Power (dB)	*Eigenvalue*	$g_{LMS} = 0.25$ *Convergence Time*	$g_{NLMS} = 0.25$ *Convergence Time*
−33.3	3.9	—	—
16.9	533.0	1101	289
−4.9	10144.6	—	—
36.6	29322.4	43	23

7.4 LMS ALGORITHM WEIGHT NOISE

As previously indicated, the mean weight vector of the LMS algorithm for a multiple sidelobe canceller converges to the optimum weight vector. However, the steady-state residue power is greater than the minimum residue power achieved by the optimum weights due to weight noise. That is, because the algorithm forms the weight vector estimate by digital filtering of the antenna voltages, at convergence even though the observation time is infinite, the estimate effectively uses a finite data set. Therefore, the variance of the steady-state weights is nonzero. This variance is the weight noise, and it causes the steady-state residue power to be greater than that achieved by the optimum weights.

Section 6.4.1 analyzed the effect of weight noise for a single-auxiliary-antenna and single-jammer scenario. The analysis showed that the additional residue power is small only when the product of the algorithm gain and received auxiliary antenna power are small. As the product increases, the additional residue power increases without limit. As shown below, similar results are obtained for the multiple-auxiliary-antenna case.

By generalizing the results of Section 6.4.1, we have

$$\begin{aligned} w_n(k+1) &= w_n(k) + g_1 r^*(k) y_n(k) \\ r(k) &= y_0(k) - \mathbf{W}^{\mathbf{H}}(k)\mathbf{Y}(k) \end{aligned} \tag{7.131}$$

Again defining $r_0(k)$ as the residue when using the optimum weight gives

$$\begin{aligned} \mathbf{W}(k) &= \mathbf{W}_{\text{opt}} + \Delta(k) \\ r(k) &= r_0(k) - \Delta^{\mathbf{H}}(k)\mathbf{Y}(k) \end{aligned} \tag{7.132}$$

so that $\Delta(k)$ is the weight vector error. The vector equivalent of (6.55) is the difference equation:

$$\Delta(k+1) = \Delta(k) + gr_0^*(k)\mathbf{Y}(k) - g\mathbf{Y}(k)\mathbf{Y}^{\mathbf{H}}(k)\Delta(k) \tag{7.133}$$

As shown by (6.50), the additional residue power due to weight noise equals $\mathrm{tr}(\mathbf{R}\mathbf{R}_\Delta)$, and $\mathbf{R}_\Delta$ is the covariance matrix of the weight vector error at convergence. In general, the correlation matrix of $\Delta(k)$ is defined as

$$\mathbf{R}_\Delta(k) = E[\Delta(k)\Delta^{\mathbf{H}}(k)] \tag{7.134}$$

and $\mathbf{R}_\Delta(k)$ is time-dependent. By using (7.133), we obtain a difference equation for the correlation matrix of $\mathbf{R}_\Delta(k)$ as

$$\begin{aligned}\mathbf{R}_\Delta(k+1) = E[&\Delta(k) + gr_0^*(k)\mathbf{Y}(k) - g\mathbf{Y}(k)\mathbf{Y}^{\mathbf{H}}(k)\Delta(k)]\\ &[\Delta^{\mathbf{H}}(k) + gr_0(k)\mathbf{Y}^{\mathbf{H}}(k) - g\Delta^{\mathbf{H}}(k)\mathbf{Y}(k)\mathbf{Y}^{\mathbf{H}}(k)]\end{aligned} \tag{7.135}$$

As shown in Section 7.4.3, $\mathbf{R}_\Delta(k)$ converges to a constant matrix as k approaches infinity. Because, at convergence, the error vector has zero mean, the constant matrix equals the covariance matrix. Denoting $\mathbf{R}_\Delta(\infty)$ as simply $\mathbf{R}_\Delta$, we can also show that

$$\mathbf{R}_\Delta\mathbf{R} + \mathbf{R}\mathbf{R}_\Delta = g[p_{\min}\mathbf{R} + \mathbf{R}\mathbf{R}_\Delta\mathbf{R} + \mathbf{R}\,\mathrm{tr}(\mathbf{R}_\Delta\mathbf{R})] \tag{7.136}$$

Also, $\mathbf{R}$ and $\mathbf{R}_\Delta$ have the same eigenvectors. Therefore, the two matrices commute. Let us denote $\mathbf{R}_\Delta\mathbf{R}$ as $\mathbf{P}$. Then, (7.136) is equivalent to

$$(2\mathbf{I} - g\mathbf{R})\mathbf{P} = g[p_{\min} + \mathrm{tr}(\mathbf{P})]\mathbf{R}$$

and

$$\mathbf{P} = g[p_{\min} + \mathrm{tr}(\mathbf{P})][2\mathbf{R}^{-1} - g\mathbf{I}]^{-1} \tag{7.137}$$

where $\mathbf{I}$ is an identity matrix, as before. Taking the trace of (7.137) gives

$$\mathrm{tr}(\mathbf{P}) = \{gp_{\min}\,\mathrm{tr}[(2\mathbf{R}^{-1} - g\mathbf{I})^{-1}]\}/\{1 - g\,\mathrm{tr}[(2\mathbf{R}^{-1} - g\mathbf{I})^{-1}]\} \tag{7.138}$$

Note that the trace of a matrix equals the sum of its eigenvalues. Because the nth eigenvalue of $\mathbf{R}$ is denoted as d_n:

$$\mathrm{tr}[(2\mathbf{R}^{-1} - g\mathbf{I})^{-1}] = \sum_{n=1}^{N} [(2/d_n - g)^{-1}] = \sum_{n=1}^{N} d_n/(2 - gd_n) \quad (7.139)$$

The total residue power equals the minimum power plus the trace of $\mathbf{P}$. Thus,

$$p_{\text{res}} = p_{\text{min}}/[1 - g\Sigma d_n/(2 - gd_n)] \quad (7.140)$$

The steady-state residue power can be computed from (7.140) for any particular jammer scenario and algorithm gain setting. The remainder of the material in this section considers some special cases to obtain insight into the performance. We shall show that the single jammer is the most stressing scenario, as it results in the largest steady-state residue power. This outcome contrasts with the optimum weight performance. For optimum weights, the residue power is minimized for the single-jammer scenario.

We can see that the steady-state residue power depends on the gain value as well as the particular eigenvalues. By differentiation, we may easily show that the steady-state residue power increases monotonically with g. A convenient approach is to combine (7.65) and (7.140) to give

$$\begin{aligned} p_{\text{res}}/p_{\text{min}} &= [1 - \Sigma g_{\text{LMS}} e_n/(1 - g_{\text{LMS}} e_n)]^{-1} \\ \text{m}e_n &= d_n/Np_{\text{max}} \end{aligned} \quad (7.141)$$

Thus, e_n is a normalized eigenvalue. The sum of the normalized eigenvalues is unity when the jammer scenario is such that the total jammer power received by each auxiliary antenna equals the maximum design power level. Otherwise, the sum is less than unity.

7.4.1 Thermal-Noise Environment

For a thermal-noise-only environment, by symmetry, all eigenvalues are the same and equal to the thermal-noise value. Let us denote the ratio of p_M/p_{noise}, the dynamic range of the canceller, as α_D. Combining this with (7.142) and noting that $p_{\text{min}} = p_{\text{noise}}$ for this environment gives

$$p_{\text{res}} = p_{\text{noise}}[1 - g_{\text{LMS}}/(N\alpha_D)]/[1 - g_{\text{LMS}}(N + 1)/(N\alpha_D)]$$

and because g_{LMS} is less than one, and α_D is typically greater than 40 dB, this equals, to a very good approximation:

$$p_{\text{res}} = p_{\text{noise}}(1 + g_{\text{LMS}}/\alpha_D) \tag{7.142}$$

Thus, for a thermal-noise environment, the steady-state residue power differs from the theoretical minimum by less than 10^{-4} dB.

7.4.2 Jammer Environment

The single-jammer case with jammer power equal to the maximum design power level approximately corresponds to the situation where one value of e_n is unity and all others are zero. For this case,

$$p_{\text{res}}/p_{\text{min}} = (1 - g_{\text{LMS}})/(1 - 2g_{\text{LMS}}) \tag{7.143}$$

Because the steady-state residue power becomes infinite at $g_{\text{LMS}} = 1/2$, relation (7.143) implies that g_{LMS} must be less than 1/2. Note that this constraint on g_{LMS} is stronger than the requirement for stability of the mean-weight-vector transient response.

Another case of interest is that of two jammers. A useful approach for developing this case is initially considering the general multijammer case. Thus, for the mth jammer, denote the instantaneous voltage received in the first auxiliary antenna as $y_m(k)$. The voltage received in the nth antenna is a time-delayed version of $y_m(t)$. For a narrowband system, the time delay is well approximated as a phase shift. Then, the voltage received at the nth auxiliary antenna is $y_m(k)\exp(j\phi_{mn})$, where ϕ_{mn} is the phase shift. Let us define a unit-length vector $\mathbf{S}_m$, the nth entry of which is $\exp(j\phi_{mn})/\sqrt{N}$. The covariance matrix hence is

$$\mathbf{R} = p_n\mathbf{I} + N\Sigma_m p_m \mathbf{S}_m \mathbf{S}_m^{\mathbf{H}}$$

where p_n is the thermal-noise power of each auxiliary antenna and p_m is the average power of the mth jammer. Let us consider a special two-jammer case, where the jammer locations are such that $\mathbf{S}_1$ is perpendicular to $\mathbf{S}_2$. Thus, $\mathbf{S}_1$ and $\mathbf{S}_2$ are eigenvectors of $\mathbf{R}$, and the associated eigenvalues are $Np_1 + p_n$ and $Np_2 + p_n$. Assuming the sum of $p_1 + p_2 = p_{\text{max}}$, the normalized eigenvalues hence are approximately $e_1 \approx p_1/(p_1 + p_2)$; $e_2 = p_2/(p_1 + p_2)$; and all others are equal to zero. Note that $e_1 + e_2 = 1$. By substituting these relations into (7.142), we find that the minimum residue power occurs when $e_1 = e_2 = 1/2$. Thus,

$$p_{\text{res}}/p_{\text{min}} = (2 - g_{\text{LMS}})/(2 - 3g_{\text{LMS}}) \tag{7.144}$$

By comparing this result to (7.143), we can see that, for each value of g_{LMS}, the normalized residue power is less than that of the single-jammer case.

A more general two-jammer condition is where the vectors $\mathbf{S}_1$ and $\mathbf{S}_2$ are not eigenvectors. However, the result is still that the steady-state residue power is minimized when the two largest eigenvalues are equal. More generally, for J jammers and N auxiliary antennas with $J \leq N$, as shown in Section 7.4.2, for any fixed gain, the residue power is minimum when the eigenvalues are equal. This implies that the residue power is minimized when the number of jammers equals the number of auxiliary antennas. This result contrasts with the optimum weight residue power, which increases as the number of jammers increases. For N jammers, $e_n = 1/N$, for all n. Inserting this into (7.142) gives

$$(p_{res}/p_{min})_{min} = (N - g_{LMS})/[N - g_{LMS}(N + 1)] \tag{7.145}$$

Note that, for this case, the upper bound on the normalized gain is

$$g_{LMS} < N/(N + 1)$$

As this gain is greater than 1/2, the most restrictive scenario is the single-jammer case. Also note that the single-jammer, steady-state residue power is larger than the equal-eigenvalue case so that it represents a stressing case. Figure 7.6 is a graph of the minimum steady-state residue power normalized to the thermal-noise power *versus* the normalized gain parametric with N. Also note that (7.143) corresponds to (7.145) with N equal to unity. Thus, the curve for $N = 1$ also corresponds to the single-jammer case for any value of N. The normalized gain value should be set less than about 0.25 to keep it well removed from the knee of the curve.

7.4.3 Detailed Analysis of LMS Weight Noise

Expanding (7.135) and taking the expectation gives

$$\begin{aligned}\mathbf{R}_\Delta(k + 1) = \mathbf{R}_\Delta(k) + g\mathbf{R}_\Delta(k)\mathbf{R} + g^2 p_{min}\mathbf{R} - g\mathbf{R}\mathbf{R}_\Delta(k) \\ + g^2[\mathbf{R}\mathbf{R}_\Delta(k)\mathbf{R} + \mathbf{R}\,\mathrm{tr}(\mathbf{R}\mathbf{R}_\Delta(k))]\end{aligned} \tag{7.146}$$

In the above relation, we assume that $\mathbf{Y}(k)$ and $\Delta(k)$ are statistically independent because of the proper choice of the digitizer sampling frequency. Also, we use the relationship:

$$E[\mathbf{Y}(k)\mathbf{Y}^{\mathbf{H}}(k)\mathbf{R}_\Delta(k)\mathbf{Y}(k)\mathbf{Y}^{\mathbf{H}}(k)] = \mathbf{R}\mathbf{R}_\Delta(k)\mathbf{R} + \mathbf{R}\,\mathrm{tr}[\mathbf{R}\mathbf{R}_\Delta(k)] \tag{7.147}$$

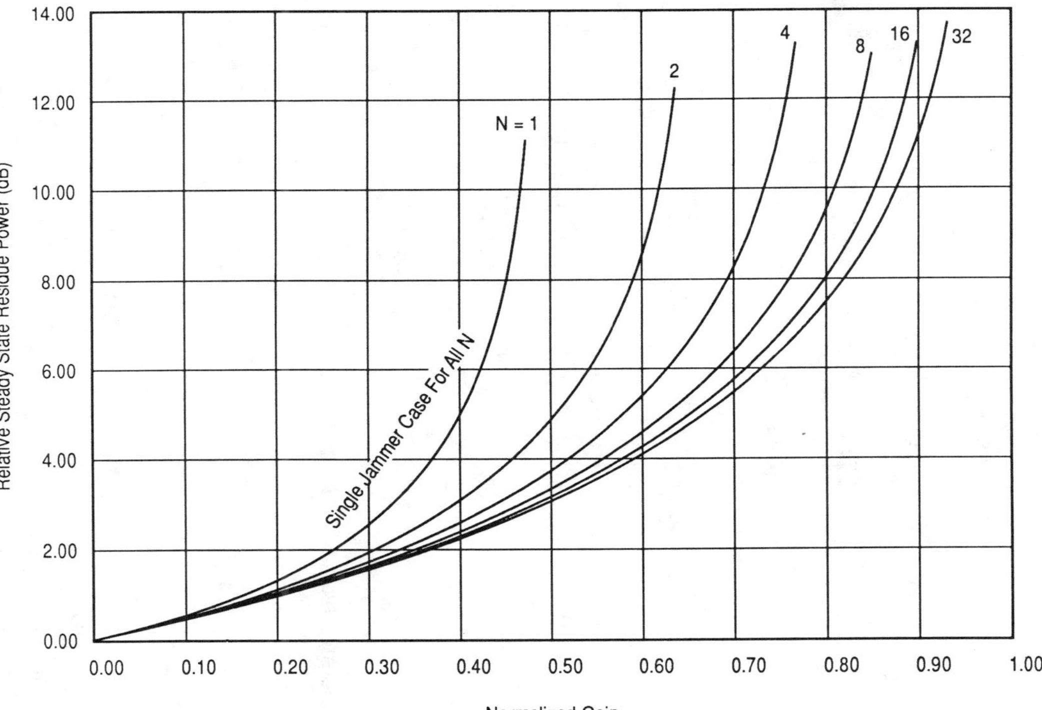

Figure 7.6 Relative steady-state residue.
Source: From R. Nitzberg, *Canceller Performance Degradation due to Estimation Noise*. Trans. IEEE, AES-17, No. 5, Sept 1981.

This theorem can be proved by using the transformation defined in (7.91). Thus, for arbitrary square matrix **A** (suppressing the time-dependence without loss of generality), gives

$$\begin{aligned} &E(\mathbf{YY^H AYY^H}) = \mathbf{F}[E(\mathbf{XX^H BXX^H})]\mathbf{F^H} = \mathbf{FQF^H} \\ &\mathbf{B} = \mathbf{F^H AF}; \quad \mathbf{Q} = E(\mathbf{XX^H BXX^H}) \end{aligned} \tag{7.148}$$

The ith row, jth column entry of **Q** therefore is

$$q_{ij} = \Sigma\Sigma E(x_i x_n^* x_m x_j^*) b_{nm} \tag{7.149}$$

The expectation can be evaluated by using a theorem of Reed [1]. For any zero-mean complex Gaussian random variables:

$$E(x_i x_n^* x_m x_j^*) = E(x_i x_n^*)E(x_m x_j^*) + E(x_i x_j^*)E(x_m x_n^*) \tag{7.150}$$

Because the $\{x_n\}$ values are statistically independent by the transformation of (7.91), the expectations of (7.150) are zero except for common subscripts. Thus,

$$\begin{aligned} &q_{ij} = d_i d_j b_{ij} + d_i[\delta(i-j)]\Sigma d_m b_{mm} \\ &\delta(i-j) = \begin{cases} 1; \, i = j \\ 0; \, i \neq j \end{cases} \end{aligned} \tag{7.151}$$

where $\{d_n\}$ denote the eigenvalues of **R**. Because

$$\Sigma d_m b_{mm} = \mathrm{tr}(\mathbf{DB})$$

we therefore obtain

$$\mathbf{Q} = \mathbf{DBD} + \mathbf{D}\,\mathrm{tr}(\mathbf{DB}) \tag{7.152}$$

Combining (7.152) and (7.148) results in

$$E(\mathbf{YY^H AYY^H}) = \mathbf{RAR} + \mathbf{R}\,\mathrm{tr}(\mathbf{RA}) \tag{7.153}$$

which proves (7.147).

Returning to (7.146), because the set of eigenvectors of $\mathbf{R}$ is a basis, there are constants $a_{mn}(k)$ such that

$$\mathbf{R}_\Delta(k)\mathbf{F}_n = \sum_{m=1}^{N} a_{mn}(k)\mathbf{F}_m \tag{7.154}$$

Therefore,

$$\mathbf{R}\mathbf{R}_\Delta(k)\mathbf{F}_n = \sum_{m=1}^{N} d_m a_{mn}(k)\mathbf{F}_m \tag{7.155}$$

Applying both sides of (7.146) to $\mathbf{F}_n$ and rearranging gives

$$\begin{aligned}&\Sigma_m[a_{mn}(k+1) - a_{mn}(k) + gd_n a_{mn}(k) + gd_m a_{mn}(k)\\ &\quad - g^2 d_n d_m a_{mn}(k)]\mathbf{F}_m = g^2[p_{\min} + \mathrm{tr}\mathbf{R}\mathbf{R}_\Delta(k)]d_n\mathbf{F}_n\end{aligned} \tag{7.156}$$

Note that the right-hand side of the above relation is proportional to $\mathbf{F}_n$. Thus, a difference equation for $a_{mn}(k)$ is

$$\begin{aligned}&a_{mn}(k+1) - (1 - gd_n - gd_m + g^2 d_n d_m)a_{mn}(k) = b(m)\\ &b(m) = 0; \quad m \neq n\\ &b(n) = g^2[p_{\min} + \mathrm{tr}\mathbf{R}\mathbf{R}_\Delta(k)]d_n\end{aligned} \tag{7.157}$$

For $m \neq n$, the difference equation is homogeneous with constant coefficients. The solution is

$$a_{mn}(k) = \mathrm{C}(1 - gd_n)^k(1 - gd_m)^k$$

where C is a constant. If g is chosen to satisfy the stability requirements, the limiting value of $a_{mn}(k)$ is zero. Thus, for sufficiently large k, (7.154) reduces to

$$\mathbf{R}_\Delta(k)\mathbf{F}_n = a_{nn}(k)\mathbf{F}_n \tag{7.158}$$

so that $\mathbf{F}_n$ is an eigenvector of $\mathbf{R}_\Delta(k)$ and $a_{nn}(k)$ is the associated eigenvalue. Therefore, for large k, we have

$$\mathrm{tr}\mathbf{R}\mathbf{R}_\Delta(k) = \sum_{n=1}^{N} d_n a_{nn}(k) \tag{7.159}$$

so that for $m = n$, (7.157) becomes

$$a_{nn}(k+1) - (1 - gd_n)^2 a_{nn}(k) = g^2\left[p_{\min} + \sum_{n=1}^{N} d_n a_{nn}(k)\right] d_n \tag{7.160}$$

This relation is a difference equation with constant coefficients. For sufficiently large k, the transient solution is zero. The particular solution is a constant. Thus, $\mathbf{R}_\Delta(k)$ converges to a constant matrix. Let us denote the limiting matrix as $\mathbf{R}_\Delta$. Then, taking the limit as k approaching infinity in (7.146) gives (7.136).

7.4.4 LMS Algorithm Weight Noise Dependence on Eigenvalue Distribution

A convenient approach is to consider the ratio of the residue power for optimum weights to the steady-state residue power. From (7.142), we have

$$p_{\min}/p_{\mathrm{res}} = 1 - \Sigma g_{\mathrm{LMS}} e_n/(1 - g_{\mathrm{LMS}} e_n) \tag{7.161}$$

The derivative with respect to g_{LMS} is

$$\mathrm{d}/\mathrm{d}g_{\mathrm{LMS}}(p_{\min}/p_{\mathrm{res}}) = -\Sigma e_n/(1 - g_{\mathrm{LMS}} e_n)^2 \tag{7.162}$$

Because the eigenvalues of a covariance matrix are always non-negative, the values of e_n are non-negative. Thus, the derivative in (7.162) is always negative so that p_{res} increases as g_{LMS} increases.

For J jammers, with $J \leq N$, the values of e_{J+1} through e_N are approximated as zero. The extrema of $p_{\min}/p_{\mathrm{res}}$ with respect to the e_n set, with the constraint that the sum of the e_n set is unity, can be investigated by substituting for e_J and then taking successive partials with respect to $e_1, e_2 \ldots, e_{J-1}$, which gives

$$\frac{\partial}{\partial e_k}(p_{\min}/p_{\mathrm{res}}) = g\{[1 - g(1 - e_1 - e_2, \ldots, e_{J-1})]^2 - (1 - ge_k)^2\} \tag{7.163}$$

for $e_k = 1 \ldots J - 1$. The only possible solution for the set of equations thus is e_k

$= 1/J$ for $k = 1$ to J. Thus, there is a unique extremum. Direct substitution easily verifies that this maximizes p_{min}/p_{res} (or minimizes p_{res}). Therefore,

$$(p_{res}/p_{min})_{min} = (J - g_{LMS})/[J - g_{LMS}(J + 1)] \qquad (7.164)$$

The dependence on J can be determined by differentiating this with respect to J, which gives

$$d/dJ[(p_{res}/p_{min})_{min}] = -g_{LMS}^2/[J - g_{LMS}(J + 1)]^2$$

Thus, the residue power decreases as J increases so that the smallest minimum is achieved for N jammers, this being the condition when all of the normalized eigenvalues of (7.161) are the same and equal to $1/N$.

7.5 NLMS ALGORITHM WEIGHT NOISE

By using a first moment analysis, we have shown in previous sections that the mean of the weight vector computed by the NLMS algorithm converges to the optimum weight vector faster than does the LMS algorithm. The present section shows that the superior convergence property is achieved at the cost of increased residue power after convergence, which occurs because of larger weight noise.

Similarly to (7.133), a difference equation for weight vector error is

$$\begin{aligned}\Delta(k + 1) = \Delta(k) + g_{NLMS} r_0^*(k)\mathbf{Y}(k)/\mathbf{Y}^{\mathbf{H}}(k)\mathbf{Y}(k) \\ - g_{NLMS}\mathbf{Y}(k)\mathbf{Y}^{\mathbf{H}}(k)\Delta(k)/\mathbf{Y}^{\mathbf{H}}(k)\mathbf{Y}(k)\end{aligned} \qquad (7.165)$$

and the analysis follows the procedure used in the Section 7.4. A difference equation for the time-dependent correlation matrix of $\mathbf{R}_\Delta(k)$ is of identical form to that of (7.135). As we show in Section 7.5.4, the time-dependent correlation matrix converges to a constant matrix. Denoting the covariance matrix at convergence as $\mathbf{R}_\Delta$, as shown in Section 7.5.4, it follows that

$$\begin{aligned}2\mathbf{R}_\Delta\Omega &= g_{NLMS}[\Omega\,\mathrm{tr}(\Omega R_\Delta) + \Omega\mathbf{R}_\Delta\mathbf{I} + p_{min}\Gamma] \\ \Omega &= E[\mathbf{Y}(k)\mathbf{Y}^{\mathbf{H}}(k)/\mathbf{Y}^{\mathbf{H}}(k)\mathbf{Y}(k)] \\ \Gamma &= E\{\mathbf{Y}(k)\mathbf{Y}^{\mathbf{H}}(k)/[\mathbf{Y}^{\mathbf{H}}(k)\mathbf{Y}(k)]^2\}\end{aligned} \qquad (7.166)$$

As shown in Sections 7.4.3 and 7.5.3, $\mathbf{R}$, $\mathbf{R}_\Delta$, Ω, and Γ all have the same eigenvectors. Thus, from (7.17), each has the matrix decomposition:

$$\begin{aligned}\mathbf{R} &= \mathbf{F}\mathbf{D}\mathbf{F}^{\mathbf{H}} \\ \mathbf{R}_\Delta &= \mathbf{F}\mathbf{D}_\Delta\mathbf{F}^{\mathbf{H}} \\ \Omega &= \mathbf{F}\mathbf{D}'\mathbf{F}^{\mathbf{H}} \\ \Gamma &= \mathbf{F}\mathbf{D}_g\mathbf{F}^{\mathbf{H}}\end{aligned} \tag{7.167}$$

where the respective eigenvalues are the elements of the diagonal matrices $\mathbf{D}$, $\mathbf{D}_\Delta$, $\mathbf{D}'$, and $\mathbf{D}_g$. From (6.50), the additional residue power due to weight noise equals the trace of the product $\mathbf{R}\mathbf{R}_\Delta$ so that the steady-state residue power is

$$p_{\text{res}} = p_{\min} + \operatorname{tr}(\mathbf{R}\mathbf{R}_\Delta) \tag{7.168}$$

Using (7.167) gives

$$\operatorname{tr}(\mathbf{R}\mathbf{R}_\Delta) = \operatorname{tr}(\mathbf{F}\mathbf{D}\mathbf{F}^{\mathbf{H}}\mathbf{F}\mathbf{D}_\Delta\mathbf{F}^{\mathbf{H}})$$

and, from (7.15), this reduces to

$$\begin{aligned}\operatorname{tr}(\mathbf{R}\mathbf{R}_\Delta) &= \operatorname{tr}(\mathbf{F}\mathbf{D}\mathbf{D}_\Delta\mathbf{F}^{\mathbf{H}}) = \operatorname{tr}(\mathbf{D}\mathbf{D}_\Delta\mathbf{F}^{\mathbf{H}}\mathbf{F}) \\ &= \operatorname{tr}(\mathbf{D}\mathbf{D}_\Delta) = \sum_{n=1}^{N} d_n d_{\Delta n}\end{aligned} \tag{7.169}$$

The sections to follow use (7.166) to evaluate the eigenvalues of $\mathbf{R}_\Delta$, and thus the steady-state residue power. The first case considered is the thermal-noise-only environment. As shown in Section 7.4.1, the steady-state residue power for this environment, when using the LMS algorithm, differs only slightly from the optimum weight's minimum residue power. Section 7.5.1 below shows that the steady-state residue power for the NLMS algorithm is significantly higher than the minimum residue power.

7.5.1 Thermal-Noise Environment

For the thermal-noise environment, $\mathbf{R}$ is proportional to the identity matrix. All eigenvalues are the same and equal to the thermal-noise power. From considerations of symmetry, Ω and Γ also each have equal eigenvalues. Because the eigenvectors of Ω and Γ are always the same as those of $\mathbf{R}$ and, for the thermal-noise environment, each has equal eigenvalues, the three matrices, $\mathbf{R}$, Ω, and Γ are proportional to the identity matrix. Note, from (7.166), that the trace of Ω always equals unity. Thus, for the thermal-noise environment and N auxiliary antennas, each eigenvalue equals $1/N$, so that

$$\Omega = \mathbf{I}/N$$

where $\mathbf{I}$ is the identity matrix.

Similarly, from (7.166), we have

$$\text{tr}(\Gamma) = E\left(\sum_{n=1}^{N} u_n\right)^{-1}$$

where, as in (7.112), the density function of each u_n is

$$p(u) = \exp(-u/p_{\text{noise}})/p_{\text{noise}} \tag{7.170}$$

Thus, $\text{tr}(\Gamma)$ is the N-fold integral:

$$\text{tr}(\Gamma) = \int_0^{\infty} \cdots \int_0^{\infty} \frac{\exp[-(u_1 + u_2 + \cdots + u_N)/p_{\text{noise}}]}{(u_1 + u_2 + \cdots + u_N)p_{\text{noise}}^N} \, du_1 \, du_2 \cdots du_N \tag{7.171}$$

From Gradshetyn and Ryshik [2], this reduces to

$$\text{tr}(\Gamma) = \int_0^{\infty} dx/(1 + xp_{\text{noise}})^N \tag{7.172}$$

This trace can be directly evaluated as $1/p_{\text{noise}}(N - 1)$, so that

$$\Gamma = \mathbf{I}/[p_{\text{noise}}N(N - 1)] \tag{7.173}$$

Substituting (7.167) and (7.172) and taking the trace of both sides gives

$$\text{tr}(\mathbf{R}_\Delta) = g_{\text{NLMS}}[N/(N - 1)](p_{\min}/p_{\text{noise}})/[2 - g_{\text{NLMS}}(1 + 1/N)] \tag{7.174}$$

so that each eigenvalue of $\mathbf{R}_\Delta = 1/N$ times the right-hand side of (7.174). Also note that, for a thermal-noise environment, $p_{\min} = p_{\text{noise}}$. From (7.168) and (7.174), the total steady-state residue power is

$$p_{\text{res}} = p_{\text{noise}}\{1 + g_{\text{NLMS}}[N/(N - 1)]/[2 - g_{\text{NLMS}}(1 + 1/N)]\} \tag{7.175}$$

For large N and g_{NLMS} equal to unity, the steady-state residue power is evaluated as 3 dB greater than the thermal-noise power. For large N and $g_{\mathrm{NLMS}} = 1/4$, the steady-state residue power is evaluated as 0.58 dB greater than the thermal-noise power. Thus, the NLMS weight-noise effect is substantially greater than that of the LMS for the thermal-noise environment.

7.5.2 Jammers Environment

As defined by (7.167), the nth eigenvalues of $\mathbf{R}_\Delta$, Ω, and Γ are denoted as $d_{\Delta n}$, d'_n, and d_{gn}, respectively. As all of these matrices have the same eigenvectors, the first relation of (7.166) can be manipulated to

$$d_{\Delta n} = [g_{\mathrm{NLMS}}/(2 - g_{\mathrm{NLMS}} d'_n)](p_{\min} d_{gn}/d'_n + \mathbf{F}_\mathbf{A}^\mathbf{T}\mathbf{F}_\Delta) \tag{7.176}$$

where $\mathbf{F}_\mathbf{A}$ and $\mathbf{F}_\Delta$ are vectors, the nth components of which are d'_n and $d_{\Delta n}$, respectively. This generalizes to the vector relationship:

$$\mathbf{F}_\Delta = g_{\mathrm{NLMS}}(p_{\min}\mathbf{F}_\mathbf{c} + \mathbf{F}_\mathbf{B}\mathbf{F}_\mathbf{A}^\mathbf{T}\mathbf{F}_\Delta) \tag{7.177}$$

where the superscript $\mathbf{T}$ denotes the transpose. The nth components of the vectors $\mathbf{F}_\mathbf{B}$ and $\mathbf{F}_\mathbf{C}$ are $(2 - g_{\mathrm{NLMS}} d'_n)^{-1}$ and $d_{gn}/[d'_n(2 - g_{\mathrm{NLMS}} d'_n)]$, respectively. This relation can be solved for $\mathbf{F}_\Delta$ as

$$\mathbf{F}_\Delta = g_{\mathrm{NLMS}} p_{\min}(\mathbf{I} - g_{\mathrm{NLMS}}\mathbf{F}_\mathbf{B}\mathbf{F}_\mathbf{A}^\mathbf{T})^{-1}\mathbf{F}_\mathbf{C} \tag{7.178}$$

This result can be simplified by using the matrix inversion lemma. Define $\mathbf{X}$ and $\mathbf{Y}$ as arbitrary vectors. We can verify, by multiplying the right-hand side of (7.179) by $\mathbf{I} - \mathbf{X}\mathbf{Y}^\mathbf{H}$ and obtaining the identity matrix, that

$$(\mathbf{I} - \mathbf{X}\mathbf{Y}^\mathbf{H})^{-1} = \mathbf{I} + \mathbf{X}\mathbf{Y}^\mathbf{H}/(1 + \mathbf{Y}^\mathbf{H}\mathbf{X}) \tag{7.179}$$

Combining (7.178) and (7.179) gives

$$\mathbf{F}_\Delta = g_{\mathrm{NLMS}} p_{\min}\{\mathbf{F}_\mathbf{C} + [g_{\mathrm{NLMS}}\mathbf{F}_\mathbf{A}^\mathbf{T}\mathbf{F}_\mathbf{C}/(1 + g_{\mathrm{NLMS}}\mathbf{F}_\mathbf{A}^\mathbf{T}\mathbf{F}_\mathbf{B})]\mathbf{F}_\mathbf{B}\} \tag{7.180}$$

Thus, the values of $d_{\Delta n}$ can be computed from (7.180) when the values of d'_n and d_{bn} are known.

The values of d_n' are given in terms of the eigenvalues of $\mathbf{R}$ by (7.130). The values of d_{bn} are derived in Section 7.5.3 as

$$d_{bn} = \sum_{\substack{p=1 \\ p \neq n}}^{N} (c_p/d_p)[1 - d_n \ln(d_p/d_n)/(d_p - d_n)] \tag{7.181}$$

with c_p as defined in (7.130).

7.5.3 Comparison of NLMS and LMS Convergence Characteristics

Figures 7.1 (Section 7.2.5) and 7.2 (Section 7.3) show the convergence of the LMS algorithm to a two- and four-jammer scenario. Each figure gives two curves. The smooth one is the theoretical convergence curve computed from the equations, based on an eigenvalue decomposition of the time-dependent mean-weight vector. The jagged curve is due to a Monte Carlo random simulation. Each curve also tabulates residue powers. The tabulations show the initial (prior to the jammer being turned on) and final (after convergence) residue powers for optimum weights, the LMS algorithm's weights (based on the weight noise equations derived in Section 7.4), and the actual random simulation values. Figures 7.7 and 7.8 are similar to 7.1 and 7.2, except that they are for the NLMS algorithm. The relative convergence characteristics are tabulated in Table 7.5. We can see that, for both algorithms, the convergence time increases and the residue power decreases as the algorithm gain decreases. The data indicate that the performance of the two algorithms is similar when the total jammer power is at the design maximum. Table 7.6 shows the comparative performance when the powers of each jammer are decreased by 20 dB. The NLMS algorithm, at a gain of 0.1, converges much more rapidly and its residue power is approximately the same as that of the LMS algorithm. Thus, for the two-jammer scenario, the NLMS algorithm is preferred.

Tables 7.7 and 7.8 provide the comparative performance for the full-power and −20 dB power four-jammer scenarios. Note that the convergence times for the LMS algorithm exceed the total simulation time. Thus, the final residue power for the simulation in Figure 7.8 is inaccurate. Similar to the two-jammer scenarios, none of the algorithms produce clearly superior performance for the full-power case. However, for the decreased power case (see Table 7.8), the NLMS algorithm of gain 0.1 again seems superior to the others considered. In addition, by comparing the entries of Tables 7.7 and 7.8, we can see that, for the decreased power case, the relative convergence time of the LMS algorithm increases, whereas that of the NLMS algorithm *decreases*. This fact is another indication of the superiority of the NLMS algorithm.

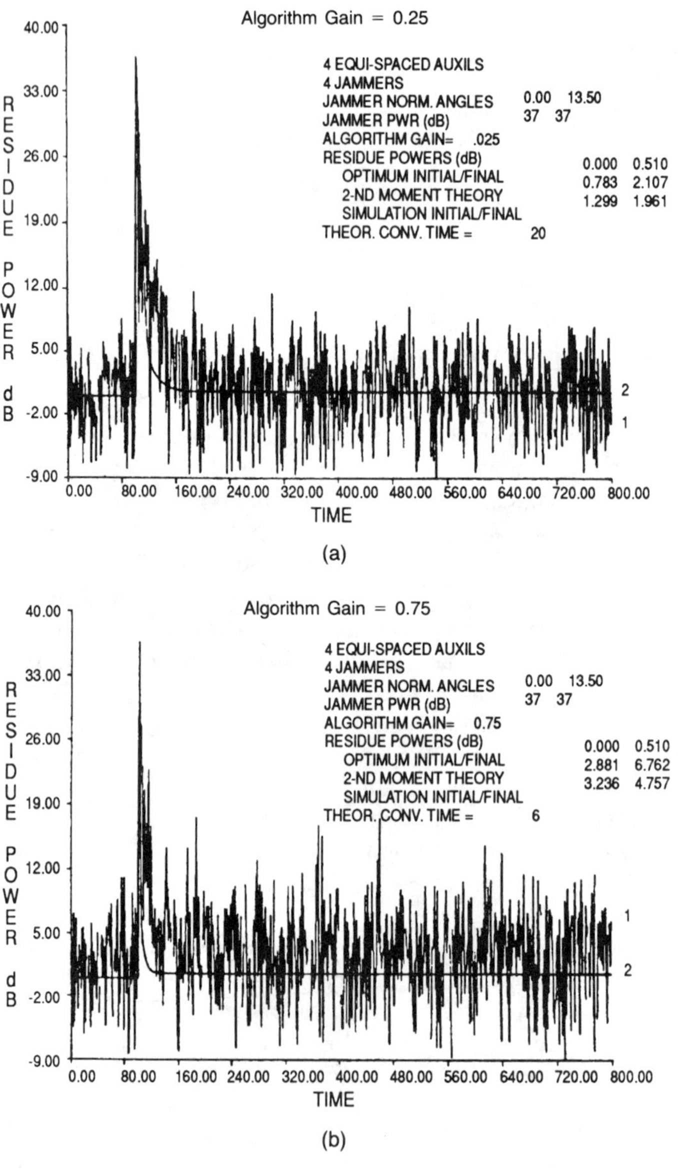

Figure 7.7 Two-jammer NLMS algorithm convergence example.

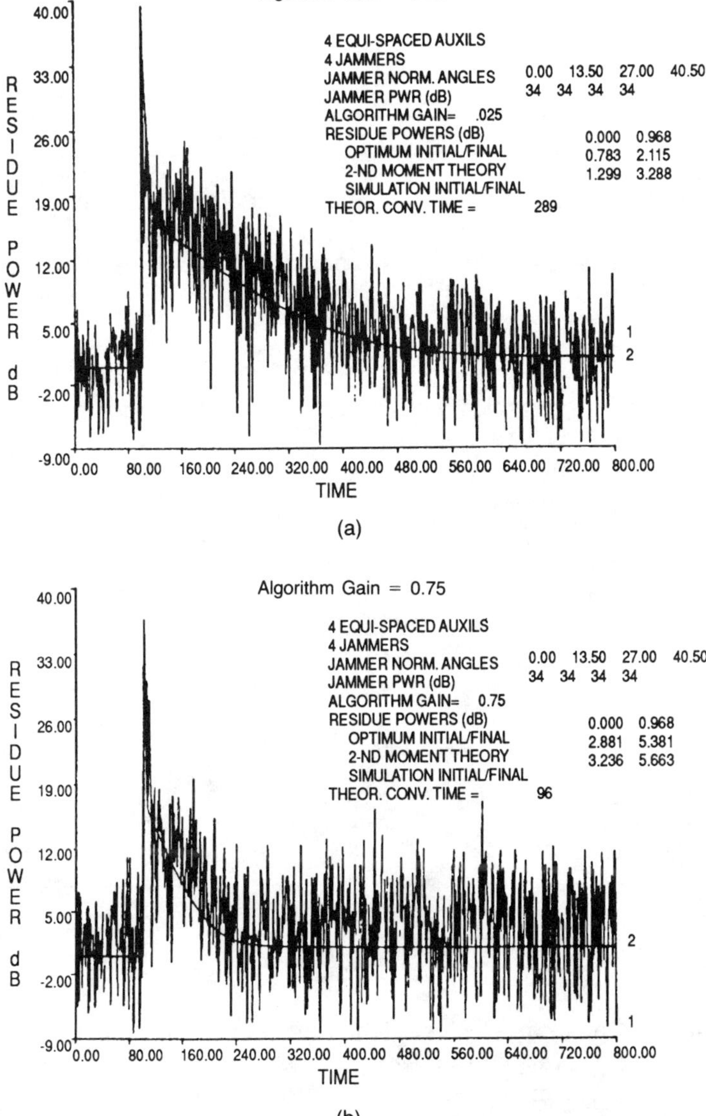

Figure 7.8 Four-jammer NLMS convergence example.

Table 7.5
Initial and Final Residue Power (dB) and Convergence Time for Two-Jammer, Full-Power Scenario

	Optimum Weights: 0.00/0.51	
Algorithm	*Second Moment Theory*	*Simulation*
LMS; Gain = 0.25	0.000/0.807/87	0.430/1.117
LMS; Gain = 0.75	0.000/1.616/29	0.430/1.842
NLMS; Gain = 0.10	0.298/1.120/51	0.705/1.268
NLMS; Gain = 0.25	0.783/2.107/20	1.299/1.961
NLMS; Gain = 0.50	2.881/6.762/16	3.236/4.757

Table 7.6
Initial and Final Residue Power (dB) and Convergence Time for Two-Jammer, −20 dB Power Scenario

	Optimum Weights: 0.00/0.510	
Algorithm	*Second Moment Theory*	*Simulation*
LMS; Gain = 0.25	0.00/0.537/256	0.430/1.382
LMS; Gain = 0.75	0.00/0.594/84	0.430/0.966
NLMS; Gain = 0.10	0.298/1.112/34	0.705/1.253
NLMS; Gain = 0.25	0.783/2.088/12	1.299/1.922
NLMS; Gain = 0.75	2.881/6.712/3	3.236/4.461

7.5.4 Detailed Analysis of NLMS Weight Noise

The first analysis shows that the correlation matrix converges to a constant matrix. The derivation method is quite similar to that used in Section 7.4.3. From (7.165), an update equation for the correlation matrix of the time-dependent weight-vector error is given by

$$\begin{aligned}\mathbf{R}_\Delta(k+1) = \mathbf{R}_\Delta(k) &- g_{\mathrm{NLMS}}[\mathbf{R}(k)\Omega + \Omega\mathbf{R}_\Delta(k)]\\ &+ g_{\mathrm{NLMS}}^2\{p_{\min}\Gamma + E[\mathbf{Y}(k)\mathbf{Y}^{\mathbf{H}}(k)\Delta(k)\Delta^{\mathbf{H}}(k)\mathbf{Y}(k)\mathbf{Y}^{\mathbf{H}}(k)/(\mathbf{Y}^{\mathbf{H}}(k)\mathbf{Y}(k))^2]\}\end{aligned}$$

The expectation can be evaluated by using (7.153), giving

Table 7.7
Initial and Final Residue Power (dB) and Convergence Time for Four-Jammer Full-Power Scenario

Algorithm	*Second Moment Theory*	*Simulation*
LMS; Gain = 0.25	0.000/1.260/1101	0.430/—
LMS; Gain = 0.75	0.000/2.002/366	0.430/2.2912
NLMS; Gain = 0.10	0.298/1.404/723	0.705/—
NLMS; Gain = 0.25	0.783/2.115/289	1.299/3.288
NLMS; Gain = 0.75	2.881/5.381/96	3.236/5.663

Table 7.8
Initial and Final Residue Power (dB) and Convergence Time for Four-Jammer −20 dB Power Scenario

	Optimum Weights: 0.000/0.915	
Algorithm	*Second Moment Theory*	*Simulation*
LMS; Gain = 0.25	0.00/0.917/1966	0.430/—
LMS; Gain = 0.75	0.00/0.923/625	0.430/—
NLMS; Gain = 0.10	0.298/1.341/27	0.705/1.526
NLMS; Gain = 0.25	0.783/2.037/10	1.299/2.054
NLMS; Gain = 0.75	2.881/5.244/3	3.236/4.211

$$\begin{aligned}\mathbf{R}_\Delta(k+1) = \mathbf{R}_\Delta(k) &- g_{\text{NLMS}}[\mathbf{R}(k)\Omega + \Omega\mathbf{R}_\Delta(k)] \\ &+ g_{\text{NLMS}}^2\{p_{\min}\Gamma + \Omega\mathbf{R}_\Delta(k)\Omega + \Omega\,\text{tr}[\Omega\mathbf{R}_\Delta(k)]\}\end{aligned} \tag{7.182}$$

Because the set of eigenvectors of **R** is a basis function, there are constants $b_{mn}(k)$, such that

$$R_\Delta(k)\mathbf{F}_n = \sum_{m=1}^{N} b_{mn}\mathbf{F}_m = \mathbf{F}\mathbf{B}_n(k)$$

where **F** is the square matrix, the *m*th column of which is the *m*th eigenvector of **R**, and the mth component of the vector $\mathbf{B}_n(k)$ is $b_{mn}(k)$. Let us define the square matrix $\mathbf{B}(k)$ such that its nth column is $\mathbf{B}_n(k)$. Then,

$$\mathbf{F}^{\mathbf{H}}\mathbf{R}_{\Delta}(k)\mathbf{F} = \mathbf{B}(k) \tag{7.183}$$

We shall premultiply and postmultiply (7.182) by $\mathbf{F}^{\mathbf{H}}$ and $\mathbf{F}$, respectively. Hence, substituting (7.183) gives

$$\begin{aligned}\mathbf{B}(k+1) = \mathbf{B}(k) &- g_{\mathrm{NLMS}}[\mathbf{B}(k)\mathbf{D}' + \mathbf{D}'\mathbf{B}(k)] \\ &+ g_{\mathrm{NLMS}}^2\{p_{\min}\mathbf{D}_g + \mathbf{D}' \operatorname{tr}[\Omega\mathbf{R}_{\Delta}(k)] + \mathbf{D}'\mathbf{B}(k)\mathbf{D}'\}\end{aligned} \tag{7.184}$$

For the nondiagonal terms of $\mathbf{B}(k)$, the above equation reduces to

$$b_{nm}(k+1) = (1 - g_{\mathrm{NLMS}}d_n')(1 - g_{\mathrm{NLMS}}d_m')b_{nm}(k)$$

This expression is a homogeneous constant-coefficient difference equation, which converges to zero when g_{NLMS} is properly chosen to ensure convergence of the mean-weight vector. Thus, $\mathbf{B}(k)$ converges to a diagonal matrix and, from (7.183), the eigenvectors of $\mathbf{R}$ equal the eigenvectors of $\mathbf{R}_{\Delta}(k)$.

For the main diagonal terms of $\mathbf{B}(k)$, (7.184) reduces to a nonhomogeneous constant coefficient difference equation. When g_{NLMS} is properly chosen to ensure convergence of the mean-weight vector, each $b_{nn}(k)$ converges to a constant. Thus, $\mathbf{R}_{\Delta}(k)$ converges to a constant matrix, the eigenvectors of which are the same as those of $\mathbf{R}$, and therefore of Ω as well.

Let us denote the converged matrix as $\mathbf{R}_{\Delta}$. From (7.182), we thus have

$$2\mathbf{R}_{\Delta}\Omega = g_{\mathrm{NLMS}}[p_{\min}\Gamma + \Omega\mathbf{R}_{\Delta}\Omega + \Omega \operatorname{tr}(\Omega\mathbf{R}_{\Delta})]$$

and this completes the derivaton of (7.166).

7.5.4.1 Eigenvalue Derivation

The eigenvalues of Ω, denoted as $\{d_n'\}$ are derived in Section 7.3.4 for the case of nonequal eigenvalues of $\mathbf{R}$. The eigenvalues of Γ, denoted as $\{d_{bn}\}$ are derived in this subsection.

The transformation of (7.91) can be used to establish that

$$\mathbf{D}_g = E[\mathbf{X}\mathbf{X}^{\mathbf{H}}/(\mathbf{X}^{\mathbf{H}}\mathbf{X})^2]$$

Using a procedure like that of (7.123) and (7.129) gives

$$d_{bn} = E[u_n/(u_n + t_n)^2]$$

$$= \sum_p b_{pn} \int\int_0^\infty [u_n/(u_n + t)^2] \exp[-(u_n/d_n + t/d_p)][du_n \, dt/d_n d_p]$$

Changing to polar coordinates, as used in progressing from (7.115), again gives a γ function when integrating on the magnitude. The equivalent of (7.117) is

$$d_{bn} = \int_0^\infty [dx/(1 + x)^2(d_n^{-1} + d_p^{-1}x)^{-2}]/d_n d_p$$

Using (7.118) results in (7.181).

7.6 SIDELOBE BLANKING

The multiple sidelobe canceller works well for rejecting noise jammers received by the main antenna's sidelobes, but does not reject false-target repeater jammers received by the main antenna's sidelobes. The difference in performance for the two jammer classes is because the algorithms used to estimate the optimum MSLC weights work well with high-duty-factor waveforms, but poorly for low-duty-factor waveforms. As an example, for a single-false-target jammer environment, the weights developed by the LMS or other algorithms relax toward the quiescent value of zero during the period between jammer pulses. For multiple-false-target jammer environments, not only do the weights relax toward zero between received pulses, but the weights also tend toward different values during each received pulse, as successive received pulses can be from different jammers. Thus, an alternate ECCM technique is needed to counter false-target repeater jammers.

A conceptual block diagram of the fundamental sidelobe blanker is shown in Figure 7.9. The depiction of the two antenna patterns indicates that the gain of the auxiliary antenna exceeds the sidelobe gain of the main antenna. The sidelobe blanker's action decides whether a pulse is received by the main lobe of the main antenna (a true target return) or the sidelobes of the main antenna (a false target generated by a jammer) by comparing the relative magnitudes of the voltages received by the two antennas. In the absence of thermal noise, the magnitude of the auxiliary antenna is always greater than that of the main antenna for a false target, and *vice versa* for a true target. As shown in the figure, the comparator circuit gates off when a false target is detected. Otherwise, the gate is in the *on* mode.

The next subsection contains an analysis of the performance of the circuit of Figure 7.9. In addition, we include a discussion of a more practical implementation.

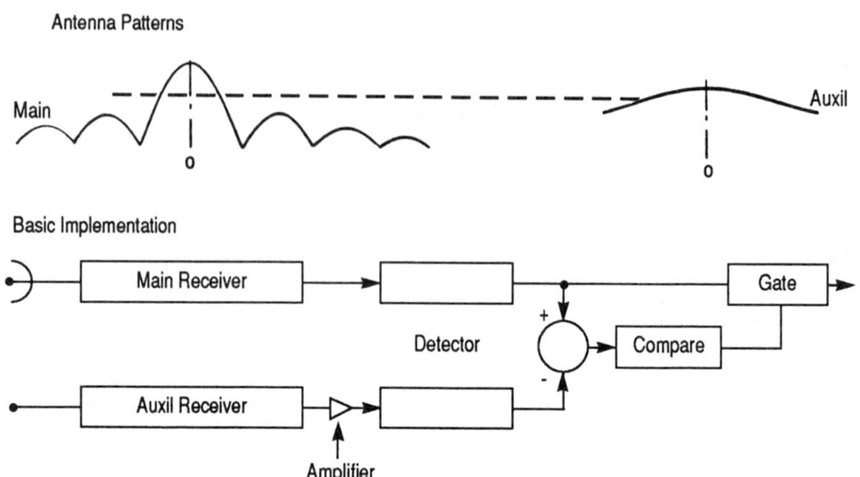

Figure 7.9 Conceptual block diagram of fundamental sidelobe blanker.

Two major problems of this circuit are that it performs poorly in either a noise-jamming environment or a clutter environment. The noise-jamming environment problem is that a sidelobe noise jammer causes the blanker to gate off continuously. To prevent this, it is necessary to incorporate the MSLC technique in both the main and blanker auxiliary channels prior to the blanking comparator. The clutter environment problem is that the clutter can mask the presence of a sophisticated false-target jammer until after doppler processing and pulse compression circuitry. To prevent this, the design must include doppler processing and pulse compression circuitry in both the main and auxiliary channels.

7.6.1 Fundamental Sidelobe Blanker Analysis

The predetector main-channel voltage of the fundamental sidelobe blanker, y_0, shown in Figure 7.9, is defined as

$$y_0 = n_0 + s \tag{7.185}$$

where n_0 depicts the thermal noise and s is the amplitude of a pulse return due to either a true or false target. For convenience and without loss of generality, we assume that the variance of $n_0 = 1$. The predetector auxiliary channel voltage, y_1, is

$$y_1 = g(n_1 + ks) \tag{7.186}$$

where g is the voltage gain of the amplifier prior to the detector, n_1 is the unit-variance thermal-noise of the channel (statistically independent of n_0), and k is the ratio of the voltage gain of the auxiliary antenna to that of the main antenna in the direction of the pulse's angle of arrival. For a nonfluctuating pulse and Gaussian distributed noise, the density functions of the outputs of the envelope detectors, r_0 and r_1, corresponding to the inputs y_0 and y_1, are

$$p(r_0) = r_0 \exp[-(r_0^2 + s^2)/2]I_0(sr_0)$$
$$p(r_1) = (r_1/g^2) \exp[-(r_1^2 + s^2g^2k^2)/2g^2]I_0(ksr_1/g) \quad (7.187)$$

where $I_0(\cdot)$ is the modified Bessel function of the first kind and zero order.

A target present decision is made when r_1 is less than r_0, and r_0 exceeds a threshold T. The value of T is, as usual, determined by the allowable false-alarm probability. The false-alarm probability is obtained by setting $s = 0$ in (7.187). Thus,

$$p_{FD} = \int_r^\infty r_0 \exp(-r_0^2/2) \int_0^{r_1} [(r_1/g^2) \exp(-r_1^2/2g^2)dr_1]dr_0$$

$$p_{FD} = \exp(-T^2/2) - \exp(-T^2G/2)/G \quad (7.188)$$
$$G = 1 + 1/g^2$$

The probability of detection apparently cannot be expressed in closed form. After T is determined from (7.188), a convenient method of determining the probability of detection is by digital computer simulation. Specifically, methods of generating Gaussian distributed complex random numbers are used to emulate the noises, y_0 and y_1. The remainder of the blanker circuit voltages are then computed and compared. The experiment is repeated as many times as necessary to obtain a statistically valid estimation of the value of the detection probability. Specifically, for each experiment, the probability that a target present declaration is made equals p_D. The probability that a target absent decision is made is $1 - p_D$. The experiment is repeated N times. The estimate of p_D equals the number of target present declarations divided by the total number of experiments. The technique, because of its obvious connection to games of chance, is called Monte Carlo. The probability density function of the ensemble of outcomes is the binomial density. The mean value of the estimate of p_D is the true value. The variance of the estimate, $V(p_D)$, is

$$V(p_D) = p_D(1 - p_D)/N$$

A method of determining the required number of experiments is to specify that the estimate should be within 1% of the true value. As the estimate almost always will fall within the range of plus or minus two standard deviations bracketing the mean, this implies

$$p_D + 2[V(P_D)]^{1/2} \leq 1.01 p_D$$

$$N \geq (4 \times 10^4)(1 - p_D)/p_D$$

At $p_D = 1/2$, about 40,000 experiments are required.

The analytical results are shown in Figure 7.10 for two cases. For the right-hand side, $k = -23$ dB. This corresponds to a true target, as the pulse amplitude received by the main beam exceeds that of the auxiliary antenna by 23 dB. Note that when $g = 0$, the sidelobe blanker is not activated. The blanker is only activated for nonzero values of g. We can see that the probability of target detection decreases as g increases. The left-hand side of the figure is for $k = +5$ dB. This corresponds to the case where the pulse amplitude received by the auxiliary antenna exceeds that received by the main antenna. Thus, the pulse is received by the sidelobes of the

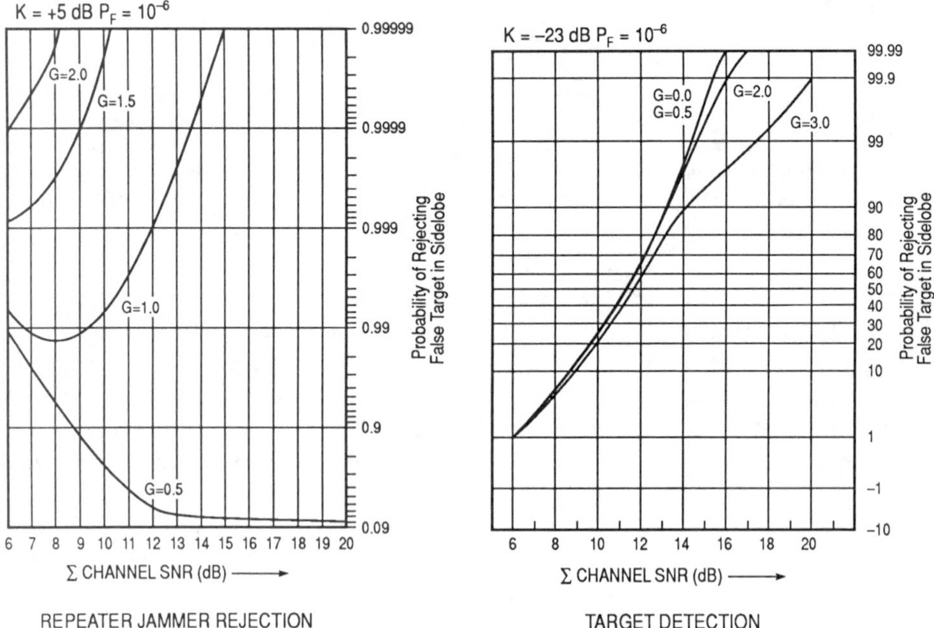

Figure 7.10 Sidelobe blanker performance.

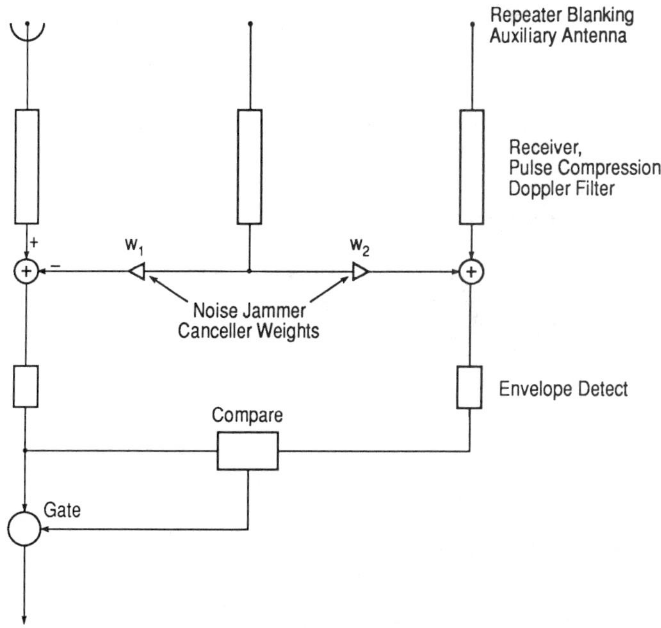

Figure 7.11 Combined noise-jammer–repeater-jammer ECCM.

main antenna and corresponds to a repeater jammer. We can see that the probability of rejecting a jammer increases as g increases. Thus, this is a conflict between the two requirements of detecting true targets with high probability and rejecting false target jammers with high probability, and a trade-off is required.

A combined sidelobe blanker and canceller is depicted in Figure 7.11, where two auxiliary antennas are shown. One is used for a blanker channel; the other is used as a noise-jammer cancellation channel for both the main antenna and blanker channel. A pulse compression network and doppler processor (delay-line canceller or doppler filter bank) are shown for the main and blanker channel prior to the blanking comparison point. The noise canceller's subtraction point is also shown prior to the blanking comparison point. For best performance, the adaptive-weight computer of the noise canceller should obtain data that are not corrupted by clutter or false targets. One way of obtaining this data is to use a time period prior to pulse transmission to sample the noise-jamming environment and compute the adaptive weights. These weights are then held for the duration of the doppler processing period. Thus, the assumption is that the noise-jamming environment does not change after the adaptive weights are computed. This assumption is grossly incorrect if the jammers modulate their characteristics. The assumption is slightly incorrect if the radar beam is continuously scanned during the doppler processing period, as the sidelobe antenna gains in the jammers' directions will change.

REFERENCES

[1] A.D. Whalen, *Detection of Signals in Noise,* New York, Academic Press, 1971, p. 367.

[2] I.S. Gradshteyn and I.M. Ryzhik, *Table of Integrals, Series and Products,* New York, Academic Press, 1965.

[3] I.S. Reed, "On a Moment Theorem for Complex Gaussian Processes," *IRE Trans. Info. Theory,* April 1962, pp. 194–195.

Chapter 8
Output Signal-to-Interference-plus-Noise (SINR) Maximization Criterion

8.1 INTRODUCTION

The basic MSLC optimization criterion is residue power minimization. The effect is that the optimum adaptive weights are chosen without any constraint concerning the gain of the combined main and auxiliary antenna system in the direction of the target. For systems with only a few low-gain auxiliary antennas and a high-gain main antenna, the perturbation gain in the target direction due to the auxiliary antennas is usually small compared to that of the main antenna. This criterion is satisfactory. However, when the number of adaptive weights is equal to a large fraction of the main antenna's gain, significantly better target detection performance can be attained by use of a criterion that determines the optimum weights as a function of both the combined antennas' target gain and the residue power. Signal-to-interference-plus-noise (SINR) maximization is the preferred criterion. In addition to the adaptive antenna application, as discussed later, this criterion is appropriate for anticlutter processing by adaptive doppler processors.

8.2 SINR MAXIMIZATION CRITERION DERIVATION

The output residue voltage is given as

$$r(t) = y_0(t) - \Sigma w_n^* y_n(t) = y_0(t) - \mathbf{W}^{\mathbf{H}}\mathbf{Y}(t) \tag{8.1}$$

We can conveniently generalize this equation to

$$r(t) = \Sigma g_n^* y_n(t) = \mathbf{G}^{\mathbf{H}}\mathbf{Z} \tag{8.2}$$

where **G** and **Z** are $N + 1$ = dimensional vectors with components g_n and $y_n(t)$, respectively. The main difference between (8.1) and (8.2) is that the former constrains the gain applied to the main antenna to be unity, whereas the latter can be any value. In addition, to avoid confusion with the notation used for the residue power minimization criterion, the adaptive gain values are denoted as g_n, rather than w_n.

In the absence of clutter and noise, the output target voltage is

$$r_s = \Sigma g_n^* s_n = \mathbf{G^H S} \tag{8.3}$$

where s_n is the target voltage for the nth antenna element and **S** is a vector, the nth component of which equals s_n. In the absence of a target, the residue power is

$$p_{\text{res}} = E|\mathbf{G^H Z}|^2 = \mathbf{G^H R_Z G} \tag{8.4}$$

The SINR is defined as the ratio of the squared magnitude of r_s to p_{res}. Thus,

$$\text{SINR} = |\mathbf{G^H S}|^2/\mathbf{G^H R_Z G} \tag{8.5}$$

The optimum weight vector for this criterion is that which maximizes (8.5). The following derives the optimum weight vector by two complementary procedures.

8.2.1 Lagrangian Multiplier Derivation

One technique is based on noting that, if the weight vector is a particular choice, denoted as G_1, the SINR value is the same value for all weight vectors that are scalar (nonzero) multiples of G_1. Then, the maximization problem can be recast to constrain the set of weight vectors to only weight vectors having an inner product with **S** as a specified constant (e.g., unity). Hence, the weight vector of the constrained set that minimizes the residue power maximizes the SINR. Minimization of a function subject to a constraint can be done by using Lagrangian multipliers. Thus, consider minimizing the function:

$$q_{\text{res}} = p_{\text{res}} - \alpha(\mathbf{G^H S} - 1) - \alpha^*(\mathbf{S^H G} - 1) \tag{8.6}$$

where α is the Lagrangian multiplier. Equating the partial derivative with respect to $\mathbf{G^H} = 0$ gives

$$\mathbf{R_Z G}_{\text{opt}} = \alpha \mathbf{S} \tag{8.7}$$

Because, as previously noted, the output SINR is unchanged when the weight vector is multiplied by a scale factor, the optimum weight vector is given by (8.7) for any value of α, despite the fact that (8.7) was derived under the assumption that α is a variable in the Lagrangian technique.

Note that the optimum weight vector for SINR maximization, as given by (8.7), is very similar in form to that for residue minimization, given by (5.57). However, the definitions of the symbols and the equations are different.

8.3 COVARIANCE MATRIX DECOMPOSITION DERIVATION

An alternative method of deriving the optimum weight is based on special decomposition properties of the covariance matrix. Recognizing special decomposition properties also will be useful to simplify implementation of equipment.

8.3.1 Square Root Matrix

The covariance matrix, $\mathbf{R_Z}$, is defined as

$$\mathbf{R_Z} = E(\mathbf{Z}\mathbf{Z}^{\mathbf{H}}) \tag{8.8}$$

Consider the quadratic form, defined as

$$\mathbf{X}^{\mathbf{H}}\mathbf{R_Z}\mathbf{X} = E(\mathbf{X}^{\mathbf{H}}\mathbf{Z}\mathbf{Z}^{\mathbf{H}}\mathbf{X}) = E|\mathbf{X}^{\mathbf{H}}\mathbf{Z}|^2 \tag{8.9}$$

for any nonzero vector, $\mathbf{X}$. The last entry of (8.9) indicates that the quadratic form of a covariance matrix is always a nonzero real number. In addition, as $\mathbf{Z}$ always includes thermal noise, the quadratic form cannot equal zero, but is always positive. This matrix property is defined as *positive-definite*.

The eigenvector decomposition of a Hermitian matrix is given by (7.17) as

$$\mathbf{R} = \mathbf{F}\mathbf{D}\mathbf{F}^{\mathbf{H}}$$

where the columns of $\mathbf{F}$ are unit-length eigenvectors and $\mathbf{D}$ is a diagonal matrix with eigenvalues on the main diagonal. From (8.9) and (7.17), we have

$$\mathbf{X}^{\mathbf{H}}\mathbf{R_Z}\mathbf{X} = \mathbf{X}^{\mathbf{H}}\mathbf{F}\mathbf{D}\mathbf{F}^{\mathbf{H}}\mathbf{X} = \mathbf{P}^{\mathbf{H}}\mathbf{D}\mathbf{P} = \Sigma d_n|p_n|^2 > 0 \tag{8.10}$$

where, by definition, $\mathbf{P}$ is a vector. The strictly positive sign of (8.10) is because the matrix is positive-definite. Because (8.10) is true for any $\mathbf{X}$, the summation inequality of (8.10) is true for any p_n set, and this implies that each eigenvalue is strictly positive. Thus, the eigenvalues of a positive-definite matrix are all positive. Therefore, there is a diagonal matrix with real, positive entries equal to the square root of the eigenvalues. We denote this matrix as $\mathbf{D}^{1/2}$. Similarly, there is a diagonal matrix with real, positive entries equal to the inverse square root of the eigenvalues. We denote this matrix as $\mathbf{D}^{-1/2}$. Let us consider the matrices:

$$\begin{aligned} \mathbf{Q}_1 &= \mathbf{F}\mathbf{D}^{1/2}\mathbf{F}^{\mathbf{H}} \\ \mathbf{Q}_2 &= \mathbf{F}\mathbf{D}^{-1/2}\mathbf{F}^{\mathbf{H}} \end{aligned} \tag{8.11}$$

By construction, the above expressions are positive-definite Hermitian matrices. As $\mathbf{F}^{\mathbf{H}}\mathbf{F}$ and $\mathbf{F}\mathbf{F}^{\mathbf{H}}$ equals the identity, the product of $\mathbf{Q}_1\mathbf{Q}_2$ equals the identity, so that $\mathbf{Q}_2$ is the inverse of $\mathbf{Q}_1$. Also,

$$\mathbf{Q}_1\mathbf{Q}_1^{\mathbf{H}} = \mathbf{F}\mathbf{D}\mathbf{F}^{\mathbf{H}} = \mathbf{R}_{\mathbf{Z}} \tag{8.12}$$

so that $\mathbf{Q}_1$ can be defined as $\sqrt{\mathbf{R}_{\mathbf{Z}}}$ and this square root matrix has the same eigenvectors as $\mathbf{R}_{\mathbf{Z}}$. There are other matrices that have a square root property, but their eigenvectors are not necessarily the same as that of the original matrix.

As $\mathbf{Q}_1$ is a reasonable definition of a square root matrix and $\mathbf{Q}_2$ is the inverse of $\mathbf{Q}_1$, $\mathbf{Q}_2$ can be defined as the inverse square root matrix. Note that this is a consistent definition, because

$$\mathbf{Q}_2\mathbf{Q}_2^{\mathbf{H}} = \mathbf{F}\mathbf{D}^{-1}\mathbf{F}^{\mathbf{H}} = \mathbf{R}_{\mathbf{Z}}^{-1} \tag{8.13}$$

8.3.2 SINR Maximization via Square Root Matrix

Let us denote the derived square root and inverse square root matrices as $\mathbf{R}_{\mathbf{Z}}^{1/2}$ and $\mathbf{R}_{\mathbf{Z}}^{-1/2}$, respectively. Note that

$$\begin{aligned} \mathbf{R}_{\mathbf{Z}}^{1/2}\mathbf{R}_{\mathbf{Z}}^{-1/2} &= \mathbf{I} \\ \mathbf{R}_{\mathbf{Z}}^{1/2}\mathbf{R}_{\mathbf{Z}}^{1/2} &= \mathbf{R}_{\mathbf{Z}} \end{aligned} \tag{8.14}$$

and each is Hermitian. Combining (8.5) and (8.14) gives

$$\mathrm{SINR} = |\mathbf{G}^{\mathbf{H}}\mathbf{R}_{\mathbf{Z}}\mathbf{R}_{\mathbf{Z}}^{-1/2}\mathbf{S}|^2\mathbf{G}^{\mathbf{H}}\mathbf{R}_{\mathbf{Z}}^{1/2}\mathbf{R}_{\mathbf{Z}}^{1/2}\mathbf{G} \tag{8.15}$$

Let us define the vectors:

$$\mathbf{V} = \mathbf{R}_Z^{1/2}\mathbf{G}$$
$$\mathbf{S}_1 = \mathbf{R}_Z^{-1/2}\mathbf{S} \tag{8.16}$$

Then,

$$\text{SINR} = |\mathbf{V}^H\mathbf{S}_1|^2/\mathbf{V}^H\mathbf{V} \tag{8.17}$$

The Schwartz inequality on the inner product of vectors is that for arbitrary vectors **X** and **Y**:

$$|\mathbf{X}^H\mathbf{Y}|^2 \leq (\mathbf{X}^H\mathbf{X})(\mathbf{Y}^H\mathbf{Y}) \tag{8.18}$$

and equality is attained only when **X** is a scalar multiple of **Y**. Combining (8.17) and (8.18) gives

$$(\text{SINR})_{max} = \mathbf{S}_1^H\mathbf{S}_1 = \mathbf{S}^H\mathbf{R}_Z^{-1}\mathbf{S}$$
$$\mathbf{V}_{opt} = \mathbf{S}_1 \tag{8.19}$$

Clearly, (8.7) and (8.19) agree in the equation for the optimum weight vector. In addition, (8.19) gives an equation for the maximum output SINR.

8.4 PERFORMANCE EXAMPLE

Section 5.5.1.2 derived the covariance matrix for the auxiliary antennas with MSLC. The equations for the covariance matrix, $\mathbf{R}_Z$, for the SINR maximization criterion are identical in form, and are given by (5.49) and (5.50).

An example of the attained performance is given by Figure 8.1. The antenna is a uniformly spaced 16-element line array, and each element is omnidirectional. For any set of antenna weights the array pattern is computed using equation 3.2. The pattern is periodic and the horizontal axis range equals the period. When the thermal noise is statistically independent on each element and there is no jamming, the covariance matrix is proportional to the identity matrix. The optimum weight vector then equals the steering vector, and the resulting pattern is as shown on the left side of the figure. The adapted pattern on the right side is for the case of three jammers. Each jammer has a power such that the JNR at each array element equals 50 dB. The jammer angular locations are depicted by the arrows. We can see that the adapted pattern has deep nulls at these angles.

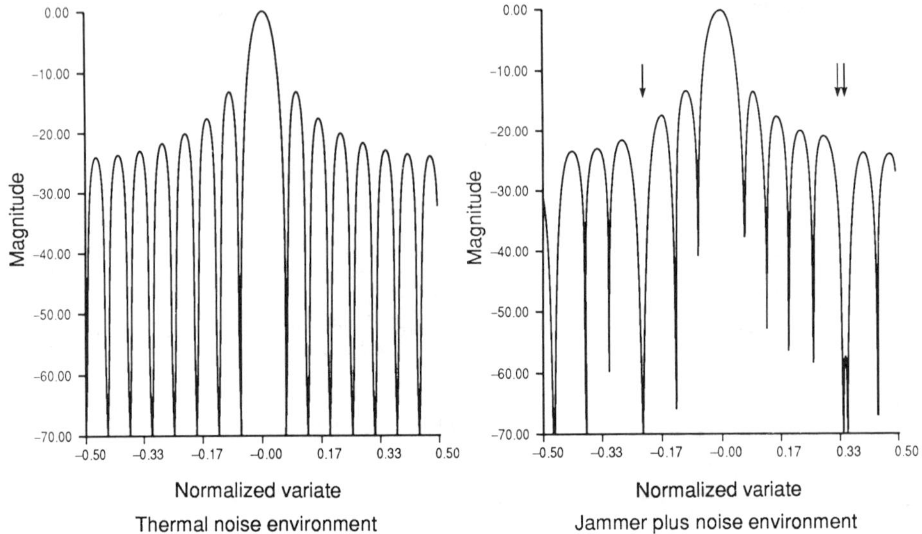

Figure 8.1 Optimum adapted array.

8.5 SIMILARITY OF THE RESIDUE MINIMIZATION AND SINR MAXIMIZATION CRITERIA

For the previous example, all elements of the array are identical. Another interesting case is similar to the MSLC architecture, where one of the array elements is actually the main-beam antenna and the other array elements are the MSLC auxiliary antennas. For N auxiliary antennas, the vector $\mathbf{Z}$ of (8.2) is of dimension $N + 1$. The first component corresponds to the main antenna's output; the others correspond to the auxiliary antennas' outputs. In partitioned form, the vector $\mathbf{Z}$ is

$$\mathbf{Z} = \begin{pmatrix} y_0 \\ \mathbf{Y} \end{pmatrix}$$

where y_0 denotes the main antenna output and $\mathbf{Y}$ denotes the vector of auxiliary outputs. Then, using the notation of Section 5.5:

$$\mathbf{R_Z} = E(\mathbf{Z}\mathbf{Z}^{\mathbf{H}}) = \begin{pmatrix} p_{\text{main}} & \mathbf{R}_0^{\mathbf{H}} \\ \mathbf{R}_0 & \mathbf{R} \end{pmatrix}$$

where $\mathbf{R}$ is the covariance matrix of the auxiliaries, $\mathbf{R}_0$ is the vector of cross-correlations of the auxiliaries and main antenna, and p_{main} is the power of the main antenna due to the thermal noise and jammers. Then, from (8.7), the optimum SINR maximization criterion weight vector, in partitioned form, is the solution of

$$\begin{pmatrix} p_{\text{main}} & \mathbf{R}_0^{\mathbf{H}} \\ \mathbf{R}_0 & \mathbf{R} \end{pmatrix} \begin{pmatrix} g_0 \\ \mathbf{G}_N \end{pmatrix} = \begin{pmatrix} s_0 \\ \mathbf{S}_N \end{pmatrix}$$

where s_0 denotes the target voltage received by the main antenna and $\mathbf{S}_N$ is the vector of target inputs received by the array of auxiliary antennas. For high-gain main antennas and omnidirectional auxiliary antennas, the magnitude of each component of $\mathbf{S}_N$ is much less than that of s_0. If $\mathbf{S}_{\text{N}}$ is approximated as zero, the above gives

$$g_0 p_{\text{main}} + \mathbf{R}_0^{\mathbf{H}} \mathbf{G}_{\text{N}} = s_0$$

$$g_0 \mathbf{R}_0 + \mathbf{R}\mathbf{G}_{\text{N}} = 0$$

In the MSLC notation, the weight applied to the main antenna equals unity and $\mathbf{G}_{\text{N}}$ equals the negative of the MSLC optimum weight vector $\mathbf{W}_{\text{opt}}$. Thus, the SINR maximization criterion is identical to the residue minimization criterion when the target voltage induced in the auxiliary antennas is negligible, compared to that induced in the main antenna.

8.6 PERFORMANCE EXAMPLES ANALYSIS

The maximum output SINR is given by (8.20). This section evaluates it for several example scenarios to gain insight into performance and the variation of performance with parameters such as radar bandwidth.

All examples are for an M-element phased array, and not an MSLC system. Thus, the target voltage vector, $\mathbf{S}$, for a target spatial angle, θ, is of the form:

$$\begin{aligned} \mathbf{S} &= s\mathbf{S}_\theta \\ \mathbf{S}_\theta^{\mathbf{T}} &= (\mathrm{e}^{j\phi_1}, \quad \mathrm{e}^{j\phi_2}, \ldots, \quad \mathrm{e}^{j\phi_M}) \end{aligned} \tag{8.20}$$

where s is the common complex-valued target voltage and the set $\{\phi_n\}$ denotes element phase shifts. Note that, as the factor s is common, there is an implicit assumption that the elements are identical and mutual coupling is neglected. The vector $\mathbf{S}_\theta$ is termed the target steering vector.

8.6.1 Thermal-Noise Environment

The interference solely consists of zero-mean thermal noise. A convenient assumption is a gain control that adjusts the elements' noise powers to equal unity. Then, various important performance ratios, such as SNR or JNR, simply equal the signal energy or the jammer power.

The covariance matrix equals the identity matrix. The per-element SINR equals $|s|^2$. From (8.7), the optimum array weighting vector is a scalar multiple of $\mathbf{S}$, and we conveniently define $\mathbf{G}_{opt}$ as equal to $\mathbf{S}_\theta$. Clearly, the optimum vector steers the array to the target direction using uniform weighting. As discussed in Section 3.2.2, this effect maximizes the array gain. From (8.19), the maximum output SINR = $\mathbf{S}^\mathbf{H}\mathbf{S}$ and, from (8.20), this equals $|s|^2M$. Thus, the ratio of output-to-input SINR for a thermal-noise environment equals M, and this factor is the well known *array power gain*.

8.6.2 Single Narrowband Jammer

The presence of a jammer modifies the covariance matrix to equal the sum of the covariance matrix due to noise and a jammer-induced matrix. For narrowband jamming, the jammer portion of the covariance matrix is of the same form as that of Section 5.6. The jammer steering vector is of the form of (8.20). Thus,

$$\mathbf{R_Z} = \mathbf{I} + p\mathbf{S}_1\mathbf{S}_1^\mathbf{H} \tag{8.21}$$

and, from (5.62), we have

$$\mathbf{R_Z}^{-1} = \mathbf{I} - (p/1 + pM)\mathbf{S}_1\mathbf{S}_1^\mathbf{H} \tag{8.22}$$

so that the optimum weight vector is

$$\mathbf{G}_{opt} = \mathbf{S}_\theta - [(\mathbf{S}_1^\mathbf{H}\mathbf{S}_\theta)p/(1 + pM)]\mathbf{S}_1 \tag{8.23}$$

The two terms of the optimum weight vector can be interpreted as forming two beams. One beam, due to $\mathbf{S}_\theta$, forms a beam directed to the target. The second beam, due to $\mathbf{S}_1$, forms a beam directed to the jammer. The combination of the two beams creates the adapted beam with low gain (approximately zero) at the jammer angle.

The output SINR can be evaluated by combining (8.18) and (8.22), giving

$$(\text{SINR})_{max} = \mathbf{S}^\mathbf{H}\mathbf{S} - (p/(1 + pM)|\mathbf{S}_1^\mathbf{H}\mathbf{S}|^2) \tag{8.24}$$

Note that $|\mathbf{S}_1^\mathbf{H}\mathbf{S}|^2$ is the power gain of the unadapted antenna in the direction of the jammer. As this value is much less than the peak unadapted antenna gain, $\mathbf{S}^\mathbf{H}\mathbf{S}$, (8.24) indicates that after cancellation the output SINR is almost equal to its value in the absence of jamming.

8.6.3 Additional Environments

Section 5.6 evaluated the performance of the MSLC using the residue minimization criterion for several simple example environments. Because, as discussed in Section 8.5, the performance for the residue minimization criterion is often close to that of the SINR maximization criterion, the additional examples of Section 5.6 are pertinent. As a specific example, the attained $(\text{SINR})_{\max}$ is expected to degrade as the radar system bandwidth increases. The degradation is expected to depend on the number of jammers and adaptable weights. When the number of adaptable weights significantly exceeds the number of jammers, the bandwidth degradation is expected to decrease substantially.

8.7 ADAPTIVE ARRAY ARCHITECTURES

The previous discussion of the SINR maximization criterion, as applied to adaptive arrays, was in terms of each array element being adaptively weighted. This is termed *full adaptivity*. Often, the array consists of a large number of elements, and full adaptivity is prohibitively expensive. In addition, prior results of the performance evaluation of the MSLC indicate that excellent jammer cancellation can be achieved when the number of adaptable elements equals twice the number of jammers. For many scenarios, the numerical value of twice the number of jammers is much less than the number of array elements. Thus, not only is full adaptivity prohibitively expensive, its performance may not be significantly better than that achievable with a much smaller number of adaptable weights. Use of a number of adaptable weights that are much less than the number of array elements is termed *partial adaptivity*.

Several partially adaptive architectures are described below. At present, there is no preferred architecture. We conjecture that the relative performance depends on the interference scenario.

8.7.1 Element-Based Partial Adaptivity

A conceptual block diagram of the element-based, partially adaptive array is given by Figure 8.2(a). As shown, each element of the array has nominal steering weights. These consist of a phase shift for steering the direction of the peak of the beam and tapered magnitudes for sidelobe-level control. In addition, some of the elements are adaptively weighted. For those elements, the total element weighting is the sum of the nominal weights plus the adaptive weights.

The architecture is much like that of the original multiple sidelobe canceller. Figure 8.2(b) is Figure 8.2(a) as redrawn to emphasize the main-beam–auxiliary-array configuration. Thus, the nominal weights configure a main beam and the adaptive weights configure an adaptive beam.

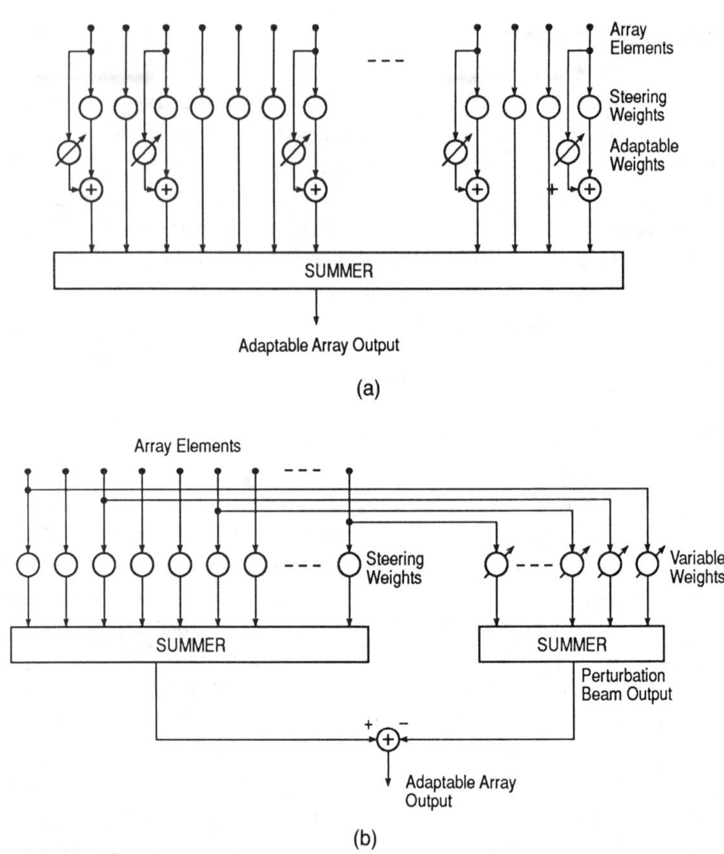

Figure 8.2 Element-based partially adaptive array.
(a) Conceptual block diagram of element-based partially adaptive array.
(b) Main-beam-perturbation beam decomposition.

The performance of this architecture is like that of the conventional MSLC with some significant differences. One similarity is that the technique can be configured by using either the residue minimization criterion or SINR maximization criterion. A second similarity is that the implementation can be by using any of the previously discussed algorithms, such as LMS, NLMS, or direct matrix estimation.

A significant architectural complication is deciding which elements should be selected for adaptability. Good methods for element selection have not yet been developed for large numbers of adaptable elements. The preferred selection apparently depends on the interference scenario and is very sensitive to errors of the nominal steering weights. As a particular example, the results of an evaluation of the capability of element-based partial adaptivity for cancellation of spatially distributed sidelobe clutter is discussed below.

Figure 8.3 shows the receiving pattern for a 128-element, equispaced, linear array steered to broadside with errorless −60 dB Taylor weighting. Figure 8.4 shows the adapted pattern operating with spatially distributed clutter of 70 dB CNR and eight adaptable elements. The adaptively weighted elements are elements 1, 2, 4, 8, 121, 125, 127, and 128, clustered toward the array edges. We can see that the pattern has a broad "null" over the clutter region. The clutter cancellation equals 50 dB.

Figure 8.5 shows the main-beam receiving pattern when the Taylor weighting has small errors. The errors cause the sidelobe level to approximate −60 dB everywhere. An important aspect of the errors is believed to be that the sidelobe pattern is noiselike, rather than the smoothly varying pattern of Figure 8.3. Figure 8.6 shows the adapted pattern for the errored-sidelobe case. The cancellation performance has severely degraded, and is now only 3 dB.

Other adaptive element set choices have been found to exhibit very different behavior. As an example, an alternative choice is to have elements clustered toward the middle of the array, rather than toward the edges. For this choice, the errorless Taylor weighting clutter cancellation was found to equal about 40 dB, rather than the edge-clustered 50 dB, but, with errors, the clutter cancellation is still approximately 40 dB, instead of the edge-clustered 3 dB.

There are two versions of the element-based architecture. Figure 8.2 is termed "reused" or "common" elements because the array elements are used both to form the main beam and to serve as auxiliary elements. The other uses separate or not common elements, as the auxiliary elements are not used for the main beam. The latter is more like the architecture used for the multiple sidelobe canceller configured with reflector antennas. The former minimizes implementation costs for an array. For some jammer environments, the attained performance is drastically different for these two architectures. The mathematical explanation for the difference in performance is the differing correlation properties due to the *quiescent noise* (background noise plus receiver noise). For the former architecture, the outputs of the main and auxiliary channels are partially correlated. For the latter, the outputs are statistically independent. Note that, although the former is correlated, the correlation coefficient is relatively low. The correlation coefficient is on the order of the reciprocal of the square root of the number of array elements. Quite surprisingly this small difference of correlation coefficient values causes an enormous performance difference.

Both element-based architectural configurations attain approximately the same jammer cancellation. The major performance difference for these two configurations is that the average sidelobe level of the adapted pattern for the not-common-element configuration is approximately the same as that of the unadapted pattern, but, for

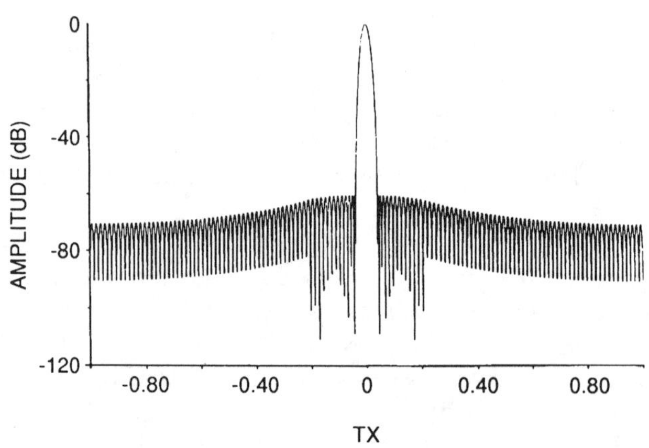

Figure 8.3 Errorless Taylor weighting.
Source: From R. Nitzberg, *OTH Radar Aurora Clutter Rejection When Adapting a Fraction of the Array Elements*. 1976 EASCON, Washington, D.C., Sept 1976.

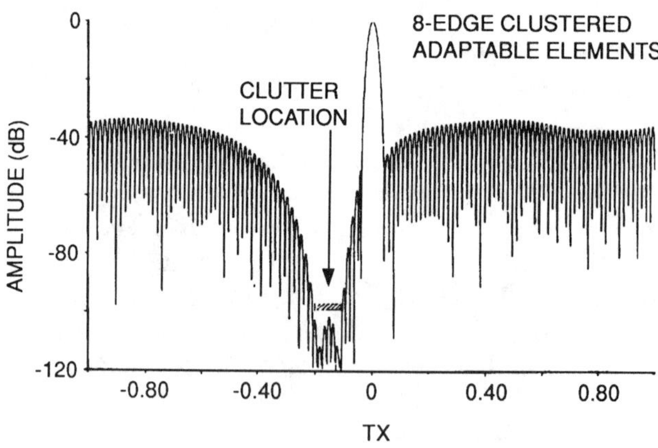

Figure 8.4 Adapted antenna pattern for errorless Taylor weighting.
Source: From R. Nitzberg, *OTH Radar Aurora Clutter Rejection When Adapting a Fraction of the Array Elements*. 1976 EASCON, Washington, D.C., Sept 1976.

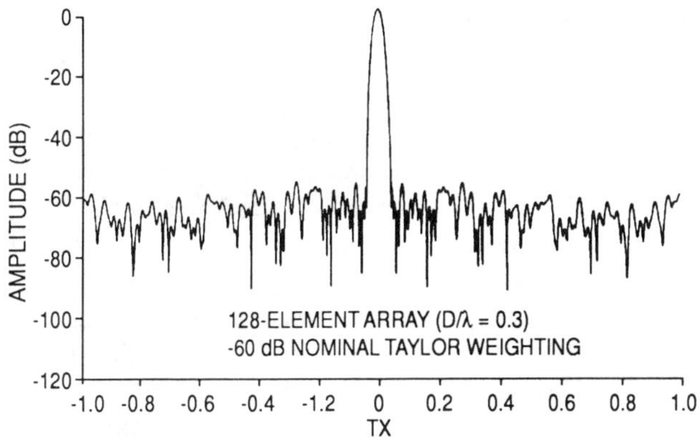

Figure 8.5 Main-beam pattern for small-error Taylor weighting.

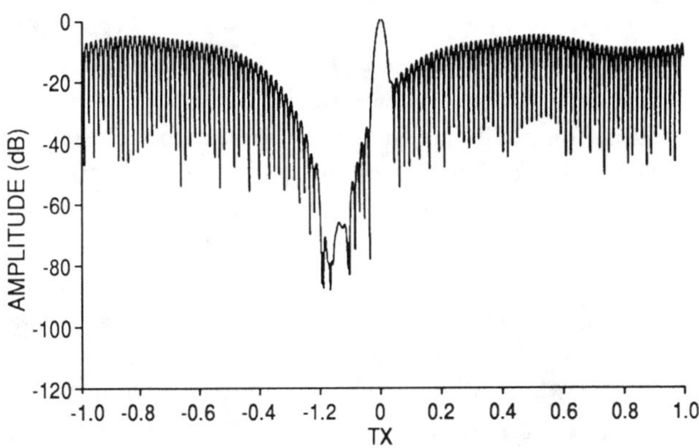

Figure 8.6 Adaptive antenna pattern for errored Taylor weighting.
Source: From R. Nitzberg, *OTH Radar Aurora Clutter Rejection When Adapting a Fraction of the Array Elements*. 1976 EASCON, Washington, D.C., Sept 1976.

the common-element configuration, the average sidelobe level of the adapted pattern is sometimes tens of decibels higher than that of the unadapted pattern.

8.7.2 Beam-Based Architecture

The beam-based architecture uses all of the array elements to form several beams. One of the beams is the "main" beam, steered to the target. The others are auxiliary beams. For the residue minimization criterion, the outputs of the latter beams are adaptively weighted and subtracted from the main beam's output. For the SIR maximization criterion, the outputs of all beams are adaptively weighted and summed. A significant issue for this architecture pertains to the steering directions of the auxiliary beams. One technique is to steer the beams in the directions of jammers. This requires knowledge of the number of jammers and their locations. Note that this version of this architecture is potentially more complex than element-based adaptivity. Alternatively, to simplify implementation, the auxiliary beams could be steered in arbitrary directions.

The relative performance of these two beam-based architectures has not yet been completely analyzed. Preliminary analysis indicates that the relative performance depends on the array's size and the radar bandwidth. For small bandwidths, the residue power decreases when the antenna gain of the auxiliary antenna is large. Therefore, for small bandwidths, steering the beam to the jammer angle should tend to improve performance. For large radar bandwidth, the residue power is relatively independent of the gain of the auxiliary antenna so that steering the beam to the jammer direction has little advantage. In fact, some cases have been observed where better performance was obtained by steering the beam away from the jammer. A possible explanation for this performance difference is due to the apparent spatial width of a wideband jammer. The jammer is in the sidelobe region of the main beam. If an auxiliary beam is steered far from the jammer angle, the jammer is in the sidelobe region of all beams. The shapes of the main and auxiliary beams are similar at the jammer angle, and cancellation over an angular region can occur "easily." However, the null-to-null width of the main-beam region of the auxiliary beam and the sidelobe region of an antenna pattern are quite different. If an auxiliary beam is steered to the jammer angle, the shapes of the two beams are not similar at the jammer angle, and cancelling over an angular region is difficult.

A third architecture is based on adaptively combining subarrays. Conceptually, the number of elements of each subarray may be different, and the subarrays may overlap such that a particular element can be used in more than one subarray. Thus, for a specified array, the number of different subarray configurations can be very large. This architecture has not been extensively analyzed because of the large number of possible subarray configurations.

8.8 ALGORITHMS

There are several potential methods of implementing the SIR maximization criterion. The previously described LMS algorithm is an iterative method of computing an estimate of the optimum weights. The direct matrix estimation algorithm, discussed briefly in Subsection 7.3.1.1, is also a candidate algorithm.

8.8.1 LMS Algorithm

By using the notation of (8.2) and (8.3), the update equation for the nth adaptive weight is

$$g_n(k+1) = g_n(k) + gs_n - gz_n(k)\mathbf{Z}^{\mathbf{H}}(k)\mathbf{G}(k) \tag{8.25}$$

and the vector generalization is

$$\mathbf{G}(k+1) = \mathbf{G}(k) - g\mathbf{Z}(k)\mathbf{Z}^{\mathbf{H}}(k)\mathbf{G}(k) + g\mathbf{S} \tag{8.26}$$

The mean weight vector of the algorithm converges to the optimum weight vector. This result can be shown by taking the average value of (8.26). Let us denote the average value of $\mathbf{G}(k)$ as $\mathbf{G}_a(k)$. Then,

$$\mathbf{G}_a(k+1) = \mathbf{G}_a(k) - g\mathbf{R}_{\mathbf{Z}}\mathbf{G}_a(k) + g\mathbf{S} \tag{8.27}$$

which is similar to (7.5). Extrapolating from this analysis indicates that the average weight vector converges to the optimum weight vector with convergence time constants that depend on the eigenvalues of the covariance matrix.

8.8.2 Direct Matrix Estimation or Sample Matrix Inversion (SMI) Algorithm

As indicated previously, the performance characteristics of the LMS and NLMS algorithms are scenario-dependent. Thus, their convergence time and weight noise effects depend on the specific jammer environment. A significant feature of the direct matrix estimation algorithm is that its performance characteristics are insensitive to the environment. They depend only upon the number of adaptable weights (the size of the covariance matrix) and the number of vector samples used to estimate the weight vector.

An important, although potentially confusing, aspect of this algorithm is that its descriptors are misnomers. If the covariance matrix is available, either actual or estimated, to solve for the weights via a matrix inversion procedure is computationally undesirable. Techniques with superior computational properties, such as Cholesky decomposition, are well known and will be discussed in later sections. In addition, the initial source of data usually is the observed voltages at the output ports of the antennas. These voltages can be used to estimate the covariance matrix as the name "direct matrix estimation" implies. Hence, Cholesky decomposition techniques could be applied to the sample covariance matrix to solve for the estimated weights. However, as discussed later, sometimes this also is computationally undesirable. The preferred computational procedure is the QR decomposition, discussed in Subsection 8.8.4.4.

Although the name *sample matrix inversion* (SMI) is an erroneous descriptor, it is the usual terminology and will be used to denote all versions of the technique. The SMI technique will be discussed in terms of the SIR maximization criterion. Thus, from (8.7), the optimum weight is estimated as $\hat{\mathbf{G}}$ by using

$$\hat{\mathbf{R}}\hat{\mathbf{G}} = \mathbf{S}$$

where $\hat{\mathbf{R}}$ is an estimate of the covariance matrix. As shown in Subsection 7.3.1.1, the method is applicable to the MSLC residue power minimization criterion architecture. One way of applying the SMI results to the MSLC case is with the associated material of Section 8.5.

An additional important aspect of the SMI algorithm is that, although it is a reasonable technique, it has no known statistical optimality properties. Thus, there could be other techniques with superior performance.

There being no known statistical optimality property occurs for several reasons. One is that the technique is used both when the statistical description of the interference is zero-mean Gaussian and when it is not. In the latter case, optimal properties cannot be known, as the technique is derived independently of the statistics of the interference. For the former case, the derivation is based on maximum-likelihood (ML) estimation of the covariance matrix and the estimated weight vector thus is calculated from (7.73). Note that this procedure is *not* the maximum-likelihood estimate of the optimum weight vector. Further, even if it were the maximum-likelihood estimator of the optimum weight vector, there would be no known optimality properties of maximum-likelihood estimation, except in an asymptotic sense.

8.8.3 Maximum Likelihood Estimation of a Covariance Matrix for Gaussian Statistics

The probability density function for a multidimensional Gaussian variate, X, is given by

$$p(X) = c_1 \exp[-c(\mathbf{X} - \mathbf{U})^{\mathbf{H}}\mathbf{R}^{-1}(\mathbf{X} - \mathbf{U})]/[\det(\mathbf{R})]^c \qquad (8.28)$$

where $\mathbf{R}$ is the covariance matrix and $\mathbf{U}$ is the mean vector. For the case of interest, the mean vector always equals zero. The equation is applicable for both real and complex variables. The values of c_1 and c depend on whether the variates are real or complex. The value of c_1 is associated with roots of 2π. As it is a scalar multiplier, the estimator is independent of c_1. The value of $c = 1$ when the variates are complex valued and $c = 1/2$ when the variates are real-valued. The estimator is also independent of this value.

The derivation of the covariance-matrix ML estimator depends on whether the covariance matrix has special symmetries. Thus, all covariance matrices are Hermitian. The number of real values required to specify this matrix equals N^2. Note that specifying the main diagonal requires N real values. The remainder of the matrix consists of $N^2 - N$ complex values. Because of the Hermitian symmetry, only $(N^2 - N)/2$ are independent. As each complex value is specified by two real values, the total number of real values required to specify a Hermitian matrix equals N^2.

For some applications, the covariance matrix has additional symmetry so that fewer than N^2 real values are required to specify the matrix. An example is when the antenna is a line array of identical, equispaced elements without mutual coupling. Then, the entries of the covariance matrix depend only on the difference between the row and column numbers. Thus, r_{nm} depends only on $n - m$ so that it is usually denoted as $r(n - m)$. Thus, the entries are the same on all diagonals. This is called a Hermitian-Toeplitz matrix. An example of a fourth-order Hermitian-Toeplitz matrix is

$$\mathbf{R} = \begin{pmatrix} r_0 & r_1^* & r_2^* & r_3^* \\ r_1 & r_0 & r_1^* & r_2^* \\ r_2 & r_1 & r_0 & r_1^* \\ r_3 & r_2 & r_1 & r_0 \end{pmatrix}$$

Note that the entries of the first column specify the total matrix. The number of real values required to specify the matrix equals $2N - 1$.

A third example also is for a line array. If the elements are symmetrically spaced with respect to the phase center (so that the kth element and the $N + 1 - k$th element are the same distance from the array phase center and straddle it), the covariance matrix is termed *persymmetric*. As mutual coupling should also be symmetric, the persymmetry effect would occur, even in the presence of mutual coupling. An example of a fourth-order Hermitian-persymmetric matrix is

$$\mathbf{R} = \begin{pmatrix} a & \alpha^* & \beta^* & \gamma^* \\ \alpha & b & \delta^* & \beta^* \\ \beta & \delta & b & \alpha^* \\ \gamma & \beta & \alpha & a \end{pmatrix}$$

Note that

$$r_{nm} = r_{N-m+1,\ N-n+1} \tag{8.29}$$

Thus, as an example, $r_{21} = r_{43}$. Therefore, the matrix has an additional symmetry with respect to the other main diagonal. Note that a Hermitian-Toeplitz matrix is also Hermitian-persymmetric.

An important analytical property of persymmetric matrices is that its inverse is also persymmetric. Thus, the inverse of a Hermitian-Toeplitz matrix is persymmetric, but the inverse is generally not Toeplitz. There are no known special symmetries to the inverse of a Hermitian-Toeplitz matrix.

8.8.3.1 Hermitian Covariance Matrix Estimation

The density function for K statistically independent vector observations is the K-fold product of the density function given by (8.28). The ML estimator is most easily obtained by the logarithm of the density function. Neglecting constants and noting that $[\det(\mathbf{R})]^{-1} = \det(\mathbf{R}^{-1})$, we have

$$\log(p) = cK \log[\det(\mathbf{R}^{-1})] - c\Sigma_k \mathbf{X}_k^{\mathbf{H}} \mathbf{R}^{-1} \mathbf{X}_k \tag{8.30}$$

This may be put into a more useful form by manipulations involving the trace of a matrix. Note that, when **A**, **B**, and **C** are matrices such that all indicated products are defined, we obtain

$$\mathrm{tr}(\mathbf{ABC}) = \mathrm{tr}(\mathbf{BCA})$$

and a scalar equals the trace of a scalar. Then,

$$\mathrm{tr}[\Sigma_k \mathbf{X}_k^{\mathbf{H}} \mathbf{R}^{-1} \mathbf{X}_k] = \mathrm{tr}[\Sigma_k \mathbf{R}^{-1} \mathbf{X}_k \mathbf{X}_k^{\mathbf{H}}] = \mathrm{tr}\{\mathbf{R}^{-1}[K(\Sigma_k \mathbf{X}_k \mathbf{X}_k^{\mathbf{H}})/K]\}$$

The term $(\Sigma_k \mathbf{X}_k \mathbf{X}_k^{\mathbf{H}})/K$ is the *sample matrix,* denoted as **S**. Then, (8.30) reduces to

$$\log(p) = cK\{\log[\det(\mathbf{R}^{-1})] - \mathrm{tr}(\mathbf{R}^{-1}\mathbf{S})\} \tag{8.31}$$

Note that

$$\mathrm{tr}(\mathbf{R}^{-1}\mathbf{S}) = \Sigma_i \Sigma_k r^{ik} s_{ki} \tag{8.32}$$

where r^{ik} and s_{ik} are the ith row, kth column entries of $\mathbf{R}^{-1}$ and **S**, respectively.

The ML estimator can be determined by computing the partial derivatives of $\log(p)$ with respect to the elements of $\mathbf{R}^{-1}$. From (8.32), the partial derivative of $\mathrm{tr}(\mathbf{R}^{-1}\mathbf{S})$ is

$$\partial \mathrm{tr}(\mathbf{R}^{-1}\mathbf{S})/\partial r^{ik} = s_{ki} \tag{8.33}$$

The determinant can be evaluated by a cofactor expansion. Hence,

$$\partial \det(\mathbf{R}^{-1})/\partial r^{ik} = \mathrm{cof}(r^{ik})$$

where $\mathrm{cof}(\cdot)$ denotes cofactor, and we thus have

$$\partial \log[\det(\mathbf{R}^{-1})]/\partial r^{ik} = r_{ki}$$

Therefore,

$$\partial \log(p)/\partial r^{ik} = cK[r_{ki} - s_{ki}] \tag{8.34}$$

Equating this partial derivative to zero results in the classical ML estimator of a Hermitian covariance matrix:

$$\begin{aligned} \hat{r}_{ki} &= s_{ki} \\ \hat{\mathbf{R}} &= \mathbf{S} = \Sigma_k \mathbf{X}_k \mathbf{X}_k^{\mathbf{H}}/K \end{aligned} \tag{8.35}$$

8.8.3.2 *Persymmetric Covariance Matrix Estimation*

Note that the step from (8.34) to (8.35) did not explicitly consider the Hermitian aspect of the covariance matrix. Thus, r_{ki} and r_{ik} are a complex conjugate pair, and the estimate must include this constraint. However, as s_{ik} and s_{ki} are a complex conjugate pair, (8.35) follows from (8.34). In general, the commonality of elements cannot be ignored. To account explicitly for the symmetry of (8.29) it is necessary to derive an estimator for the persymmetric covariance matrix.

By using the chain rule of differentiation, if the entries of $\mathbf{R}^{-1}$ are functions of the variable t, we have

$$\partial \mathbf{R}^{-1}/\partial t = \Sigma_i \Sigma_k (\partial \mathbf{R}^{-1}/\partial r^{ik})(\partial r^{ik}/\partial t)$$

When several entries of $\mathbf{R}^{-1}$ are identical, denoted as r, we thus have

$$\partial \mathbf{R}^{-1}/\partial r = \Sigma_M \partial \mathbf{R}^{-1}/\partial r^{ik}$$

where M is the set:

$$M = \{i,k | r^{ik} = r\}$$

Combining (8.29) and (8.34) gives

$$r_{ki} = (s_{ki} + s_{N+1-i,\ N+1-k})/2 \tag{8.36}$$

An alternative view of this estimator can be obtained by a symmetry consideration. Specifically, note that, for a Hermitian-persymmetric covariance matrix, the order of the observations can be conjugated and permuted without changing the covariance matrix. Thus, the vectors $\mathbf{X}$ and $\mathbf{Y}$:

$$\begin{aligned} \mathbf{X}^{\mathbf{T}} &= (x_1, x_2, x_3, \ldots, x_{N-1}, x_N) \\ \mathbf{Y}^{\mathbf{T}} &= (x_N^*, x_{N-1}^*, \ldots, x_3^*, x_2^*, x_1^*) \end{aligned} \tag{8.37}$$

both have the same covariance matrix. Thus, the covariance matrix estimator that is the equivalent of (8.35) is

$$\mathbf{R}_{\mathbf{Y}} = \Sigma_k \mathbf{Y}_k \mathbf{Y}_k^{\mathbf{H}}/K \tag{8.38}$$

Combining both matrices gives the ML estimator of a persymmetric matrix, $\mathbf{R}_p$, as

$$\hat{\mathbf{R}}_p = \Sigma_k(\mathbf{X}_k \mathbf{X}_k^{\mathbf{H}} + \mathbf{Y}_k \mathbf{Y}_k^{\mathbf{H}})/K \tag{8.39}$$

which is the matrix form of (8.36).

8.8.3.3 Hermitian-Toeplitz Covariance Matrix Estimation

The derivation of this estimator cannot proceed from the previous derivations because the Toeplitz symmetry does not explicitly appear in the matrix inverse. In addition, the resulting equation is not very useful, as it is highly nonlinear and extraordinarily difficult to solve. We can show that the estimator, $\hat{\mathbf{R}}_T$, is the solution of

$$\mathrm{tr}[(\hat{\mathbf{R}}_T^{-1}\hat{\mathbf{R}}_p\hat{\mathbf{R}}_T^{-1} - \hat{\mathbf{R}}_T^{-1})\mathbf{Q}] = 0 \tag{8.40}$$

where $\hat{\mathbf{R}}_p$ is as given by (8.39) and $\mathbf{Q}$ is any positive-definite Hermitian-Toeplitz matrix.

8.8.4 Computational Procedures

Different methods of computing the optimum or estimated optimum weights are used, depending on the particular problem. One case of interest is evaluating the theoretical performance of the adaptive processor. For this case, the covariance matrix, or, equivalently, the jammer parameters are known *a priori*. Another case is computing the estimated optimum weights from observed data. There are several methods of handling this second case. One is to form the sample matrix and to use the techniques of the first case. Alternatively, we can use the **QR** decomposition, discussed in Subsection 8.4.4.4. As shown, the **QR** decomposition technique is applicable when the covariance matrix estimator is the sample matrix of (8.35). However, the **QR** decomposition is not applicable when the covariance matrix estimator is modified to incorporate persymmetric or Toeplitz symmetry.

8.8.4.1 Optimum Weight Computation

To evaluate the optimum weight performance of an adaptive processor is usually desirable, as this is the unattainable upper bound for the performance of an implemented adaptive processor. The optimum weights of an adaptive processor are given by a matrix equation:

$$\mathbf{RG} = \mathbf{S} \tag{8.41}$$

A formal solution for $\mathbf{G}$ is

$$\mathbf{G} = \mathbf{R}^{-1}\mathbf{S} \tag{8.42}$$

Thus, a direct method of evaluating $\mathbf{G}$ is to invert $\mathbf{R}$ and then multiply the inverse by $\mathbf{S}$. The standard computing literature indicates that, because of the constraints imposed by the finite-word-length arithmetic of digital computers, this direct method is not preferred. Alternative evaluation methods are faster and more accurate. The particular method that is recommended depends on the specifics of the matrix $\mathbf{R}$.

For adaptive processors, the matrix is always Hermitian, and this case is considered below.

Triangular Decomposition of Hermitian Matrices

The matrix is a covariance matrix. The *i-j*th entry, r_{ij}, is given by

$$r_{ij} = E(y_i y_j^*)$$

The set y_i represents complex envelopes of particular voltages. For some processing techniques, the values are array element voltages; for other techniques, they are voltages at various taps of a delay line, *et cetera*.

An important property of a covariance matrix is that

$$r_{ji} = E(y_j y_i^*) = E(y_i y_j^*)^* = r_{ij}^*$$

Thus, there is a form of symmetry inherent in these matrices. For reasons that will become clear later, a convenient approach is to decompose **R** into the product:

$$\mathbf{R} = \mathbf{T}\mathbf{T}^{\mathrm{H}} \tag{8.43}$$

where **T** is a lower triangular matrix. The prior notation has denoted **R** as an $N + 1 \times N + 1$ matrix. For convenience, $N + 1$ is denoted as M. Thus,

$$\mathbf{T} = \begin{pmatrix} t_{11} & 0 & 0 & \cdots & 0 & 0 \\ t_{21} & t_{22} & 0 & \cdots & 0 & 0 \\ t_{31} & t_{32} & t_{33} & \cdots & 0 & 0 \\ \vdots & & & & & \\ t_{M-1,1} & t_{M-1,2} & t_{M-1,3} & \cdots & t_{M-1,M-1} & 0 \\ t_{M1} & t_{M2} & t_{M3} & \cdots & t_{M,M-1} & t_{MM} \end{pmatrix} \tag{8.44}$$

Then,

$$\begin{pmatrix} r_{11} & r_{12} & \cdots & r_{1M} \\ r_{21} & & & \\ \vdots & & & \\ r_{M1} & & \cdots & r_{MM} \end{pmatrix} = \begin{pmatrix} t_{11} & 0 & \cdots & 0 \\ t_{21} & t_{22} & \cdots & 0 \\ \vdots & & & \vdots \\ t_{M1} & \cdots & & t_{MM} \end{pmatrix} \begin{pmatrix} t_{11} & t_{21}^* & \cdots & t_{M1}^* \\ 0 & t_{22} & & \\ \vdots & \vdots & & \\ 0 & 0 & \cdots & t_{MM} \end{pmatrix} \tag{8.45}$$

Note that multiplying the first row of $\mathbf{T}$ by the first column of $\mathbf{T}^{\mathrm{H}}$ gives

$$r_{11} = t_{11}^2$$

so that

$$t_{11} = r_{11}^{1/2}$$

Multiplying the first row of $\mathbf{T}$ by the second column of $\mathbf{T}^{\mathrm{H}}$ gives

$$r_{12} = t_{11}t_{21}^*$$

so that

$$t_{21} = r_{12}^*/t_{11}$$

and note that the value of t_{11} is known. In general, therefore,

$$t_{k1} = r_{1k}^*/t_{11}; \quad k = 1, 2, \ldots, M$$

so that the first column of $\mathbf{T}$ can be calculated.

Now, consider multiplying the second row of $\mathbf{T}$ by the second column of $\mathbf{T}^{\mathrm{H}}$. This gives

$$|t_{21}|^2 + t_{22}^2 = r_{22}$$

so that

$$t_{22} = [r_{22} - |r_{12}|^2/r_{11}]^{1/2}$$

Multiplying the second row of $\mathbf{T}$ by the third column of $\mathbf{T}^{\mathrm{H}}$ gives

$$t_{21}t_{31}^* + t_{22}t_{32}^* = r_{23}$$

Because the only unknown in the above equation is t_{32}, its value may be determined. In general, therefore,

$$t_{k2} = (r_{2k} - t_{21}t_{k1}^*)^*/t_{22}; \quad k = 2, 3, \ldots, M$$

so that the second column of $\mathbf{T}$ can be calculated. This procedure may be continued.

Each new element of **T** can be expressed only in terms of previously determined elements of **T**.

8.8.4.2 *Solution of Equations with Triangular Matrices*

Because **R** is decomposed into triangular matrices, (8.41) may be rewritten as

$$\mathbf{T}\mathbf{T}^{\mathrm{H}}\mathbf{G} = \mathbf{S} \tag{8.46}$$

Let us define the column vector **P** as

$$\mathbf{P} = \mathbf{T}^{\mathrm{H}}\mathbf{G} \tag{8.47}$$

so that (8.46) becomes

$$\mathbf{TP} = \mathbf{S}$$

Stated in words, the problem is to find **P** when **T** and **S** are known. Writing these matrices in expanded form:

$$\begin{pmatrix} t_{11} & 0 & 0 & \cdots & 0 \\ t_{21} & t_{22} & 0 & \cdots & 0 \\ t_{31} & t_{32} & t_{33} & \cdots & 0 \\ \vdots & & & & \\ t_{M1} & t_{M2} & \cdots & & t_{MM} \end{pmatrix} \begin{pmatrix} p_1 \\ p_2 \\ \vdots \\ p_M \end{pmatrix} = \begin{pmatrix} s_1 \\ s_2 \\ \vdots \\ s_M \end{pmatrix}$$

Multiplying the first row gives

$$t_{11}p_1 = s_1$$

so that the value of p_1 can be calculated. Multiplying the second row gives

$$t_{21}p_1 + t_{22}p_2 = s_2$$

and the value of p_2 can be calculated. The general equation to find the nth entry of **P** is

$$p_n = (s_n - \Sigma t_{ni}p_i)/t_{nn}; \quad n = 1, 2, 3, \ldots, M$$

After **P** is determined, **G** can be computed by using back-substitution in (8.47). Writing the matrices of (8.47) in expanded form:

$$\begin{pmatrix} t_{11} & t_{21}^* & t_{31}^* & \cdots & t_{M1}^* \\ 0 & t_{22} & t_{32}^* & & t_{M2}^* \\ 0 & 0 & t_{33} & & \\ \vdots & \vdots & \vdots & & \vdots \\ 0 & 0 & 0 & & t_{M,M-1}^* \\ 0 & 0 & 0 & & t_{MM} \end{pmatrix} \begin{pmatrix} g_1 \\ g_2 \\ \vdots \\ g_M \end{pmatrix} = \begin{pmatrix} p_1 \\ p_2 \\ p_2 \\ \vdots \\ p_M \end{pmatrix}$$

Multiplying through the Mth row gives

$$t_{MM}g_M = p_M$$

thus determining the Mth component of $\mathbf{G}$. Multiplying through the $M - 1$th row gives

$$t_{M-1,M-1}g_{M-1} + t_{M,M-1}^*g_M = p_{M-1}$$

and the $M - 1$th component of $\mathbf{G}$ is determined. In general, back-substitution finds all the components.

8.8.4.3 *Data Matrix Computational Techniques*

As previously indicated, the covariance matrix usually is unknown, and an estimated optimum weight vector can be obtained by first estimating the covariance matrix. A common estimation technique is to use data from K time samples. As indicated in Subsection 8.8.3.1, when the probability density function of the set of K independent observation vectors $\{\mathbf{Z}_k : k = 1, K\}$ is zero-mean complex Gaussian and the covariance matrix has no symmetries other than Hermitian, the maximum-likelihood estimates of the covariance matrix components are

$$\hat{\mathbf{r}}_{nm} = \Sigma z_n(k) z_m^*(k)/K \tag{8.48}$$

Denote the matrix estimate as $\hat{\mathbf{R}}_\mathbf{Z}$. Hence, the estimate of the optimum weight vector $\hat{\mathbf{G}}$ is

$$\hat{\mathbf{R}}_\mathbf{Z}\hat{\mathbf{G}} = \mathbf{S}$$

A convenient way to write (8.48) is in expanded matrix notation. Note that $\mathbf{Z}_k$ is the M-dimensional vector of data obtained from an antenna at time t_k. The nth component of $\mathbf{Z}_k = z_n(k)$. Then,

$$\hat{\mathbf{R}}_{\mathbf{Z}} = \Sigma \mathbf{Z}_k \mathbf{Z}_k^{\mathbf{H}}/K \tag{8.49}$$

This equation may be further simplified by defining the $M \times K$ matrix, $\mathbf{Z}$, in partitioned form, the kth column of which equals $\mathbf{Z}_k$. Thus,

$$\mathbf{Z} = (\mathbf{Z}_1, \mathbf{Z}_2, \ldots, \mathbf{Z}_K) \tag{8.50}$$

Then,

$$\hat{\mathbf{R}}_{\mathbf{Z}} = \mathbf{Z}\mathbf{Z}^{\mathbf{H}}/K \tag{8.51}$$

so that the estimated weight vector is the solution of

$$\mathbf{Z}\mathbf{Z}^{\mathbf{H}}\hat{\mathbf{G}} = \mathbf{S}K = \mathbf{S}_1 \tag{8.52}$$

where, for convenience, $\mathbf{S}_1$ is defined as the product of the $\mathbf{S}$ matrix and the scalar $\mathbf{K}$. The indicated vector products of (8.52) can be performed in order to solve for $\hat{\mathbf{G}}$. However, considerable computational simplification can be achieved by avoiding the multiplication and instead making use of the **QR** decomposition theorem.

QR Decomposition

Note that $\mathbf{Z}^{\mathbf{H}}$ is a rectangular matrix with more rows than columns. In addition, with probability equal to 1, its rank equals the number of columns. We can show that such matrices can be decomposed as*

$$\mathbf{Z}^{\mathbf{H}} = \mathbf{Q}\mathbf{R} \tag{8.53}$$

where $\mathbf{Q}$ is a $K \times K$ unitary matrix and $\mathbf{R}$ is a $K \times N$ trapezoidal matrix. Specifically, the first N rows of $\mathbf{R}$ equal a nonsingular upper triangular matrix, and the remaining rows all equal to zero. Thus,

$$\mathbf{R} = \begin{pmatrix} \mathbf{R}_1 \\ \mathbf{0} \end{pmatrix} \tag{8.54}$$

where $\mathbf{R}_1$ is an $N \times N$ upper triangular matrix. From (8.53) and (8.54),

*Unfortunately, standard notation uses the same symbol **R** to denote covariance matrices and the second matrix of (8.53). The traditional symbology will be used, as the proper symbol interpretation always is expected to be clear from the context.

$$\mathbf{ZZ^H} = \mathbf{R_1^H R_1} \tag{8.55}$$

Thus, (8.52) becomes

$$\mathbf{R_1^H R_1 \hat{G}} = \mathbf{S_1} \tag{8.56}$$

Note that this is identical in form to the equation that results from the Cholesky decomposition of the covariance matrix, as given by (8.46). Note that this equation indicates that to compute the estimate of the covariance matrix is unnecessary, as the **QR** decomposition works directly on the data matrix. In addition, computing the data matrix is undesirable, as the number of bits required to compute its entries without error is more than twice that of the data matrix. The computer literature indicates that, when using finite bit arithmetic, to solve for $\mathbf{\hat{G}}$ from (8.56) is much more accurate than forming the sample matrix and then using Choleksy decomposition.

QR Decomposition Applied to the Residue Power Minimization Criterion

Denote the vector of observed voltages from the auxiliary antennas at time t_k as $\mathbf{Y}_k$. The observed voltage at the output of the main antenna is $y_0(k)$. The maximum likelihood estimates of the auxiliary antenna's covariance matrix, $\mathbf{\hat{R}}$, and the correlation vector $\mathbf{\hat{R}}_0$ is*

$$\begin{aligned} \mathbf{\hat{R}} &= \Sigma \mathbf{Y}_k \mathbf{Y}_k^{\mathbf{H}}/K \\ \mathbf{\hat{R}}_0 &= \Sigma y_0^*(k) \mathbf{Y}_k/K \end{aligned} \tag{8.57}$$

Then,

$$\mathbf{\hat{R}\hat{W}} = \mathbf{\hat{R}}_0$$

These equations may be simplified by defining the $N \times K$ matrix, similar to that of (8.50). Thus, let us define $\mathbf{Y}$ in partitioned form as one in which its kth column equals $\mathbf{Y}_k$. Thus,

$$\mathbf{Y} = (\mathbf{Y}_1, \mathbf{Y}_2, \ldots, \mathbf{Y}_K) \tag{8.58}$$

A vector containing the K components of the main antenna output voltage is defined as

*Note that the covariance matrix of the auxiliaries denoted by **R** is *not* the covariance matrix for the SIR maximization criterion analysis, denoted by $\mathbf{R_Z}$, of the previous section.

$$\mathbf{Y}_M^{\mathbf{T}} = (y_0(1), y_0(2), \ldots, y_0(K)) \tag{8.59}$$

Then,

$$\begin{aligned} \hat{\mathbf{R}} &= \mathbf{Y}\mathbf{Y}^{\mathbf{H}}/K \\ \hat{\mathbf{R}}_0 &= \mathbf{Y}\mathbf{Y}_M^*/K \end{aligned} \tag{8.60}$$

so that the estimated weight vector is the solution of

$$\mathbf{Y}\mathbf{Y}^{\mathbf{H}}\hat{\mathbf{W}} = \mathbf{Y}\mathbf{Y}_M^* \tag{8.61}$$

The **QR** decomposition is also useful to solve this equation. Similarly to (8.53), the data matrix $\mathbf{Y}^{\mathbf{H}}$ can be decomposed as

$$\mathbf{Y}^{\mathbf{H}} = \mathbf{Q}\mathbf{R}$$

Q and **R** have the same orthogonal and trapezoidal forms as previously discussed. Then, (8.61) reduces to

$$\mathbf{R}_1^{\mathbf{H}}\mathbf{R}_1\hat{\mathbf{W}} = \mathbf{R}^{\mathbf{H}}\mathbf{Q}^{\mathbf{H}}\mathbf{Y}_M^*$$

Let us partition **Q** as

$$\mathbf{Q} = (\mathbf{Q}_1, \mathbf{Q}_2) \tag{8.62}$$

where $\mathbf{Q}_1$ is as the rectangular matrix equal to the first N columns of **Q**. Then,

$$\begin{aligned} \mathbf{R}^{\mathbf{H}}\mathbf{Q}^{\mathbf{H}} &= \mathbf{R}_1^{\mathbf{H}}\mathbf{Q}_1^{\mathbf{H}} \\ \mathbf{R}_1^{\mathbf{H}}\mathbf{R}_1\hat{\mathbf{W}} &= \mathbf{R}_1^{\mathbf{H}}\mathbf{Q}_1^{\mathbf{H}}\mathbf{Y}_M^* \end{aligned}$$

and, because $\mathbf{R}_1$ is nonsingular:

$$\mathbf{R}_1\hat{\mathbf{W}} = \mathbf{Q}_1^{\mathbf{H}}\mathbf{Y}_M^* \tag{8.63}$$

As $\mathbf{R}_1$ is a lower triangular matrix, to solve for $\hat{\mathbf{W}}$ is easy. Note, again, that this equation indicates that to compute the estimate of the covariance matrix is unnecessary, as the **QR** decomposition works directly on the data matrix.

8.8.5 LMS Formulation for Non-Gaussian Statistics

The preceding used a derivation based upon maximum likelihood matrix estimation under the assumption of Gaussian statistics. The same result for the residue power minimization criterion can be obtained for non-Gaussian statistics via the least mean square (LMS) formulation. Specifically, the vector of residue voltages, $\mathbf{E}_0$, is

$$\mathbf{E}_0 = \mathbf{Y}_M - \mathbf{Y}^T\mathbf{W}^*$$

where the kth component of E_0 equals $e_0(k)$ and the other vectors have been previously defined. The LMS error, ε, is

$$\varepsilon = \mathbf{E}_0^H\mathbf{E}_0 = (\mathbf{Y}_M^H - \mathbf{W}^T\mathbf{Y}^*)(\mathbf{Y}_M - \mathbf{Y}^T\mathbf{W}^*)$$

To find the weight vector that minimizes ε, differentiate with respect to $\mathbf{W}$. This leads to the equation

$$\mathbf{Y}\mathbf{Y}^H\mathbf{W} = \mathbf{Y}\mathbf{Y}_M^*$$

and this is the identical result obtained for the Gaussian statistics case.

8.8.6 Implementation Issues

The preceding section covered the use of the **QR** decomposition, but did not indicate how to obtain the decomposition. The literature discusses several methods, known as

1. Householder's transformation;
2. Givens;
3. Gram-Schmidt;
4. Modified Gram-Schmidt.

As an example of the procedure, Householder's transformation will be considered in detail. The literature states that the best accuracy is obtained by using Householder's transformation. However, often, hardware constraints, such as systolic array implementation requirements, cause one of the others to be preferable.

8.8.6.1 Householder's Transformation

Householder's transformation uses a succession of unitary transformations to transform the columns of a matrix sequentially into new columns, the components of which below the main diagonal are zero. Specifically, given a rectangular matrix $\mathbf{C}$, with columns $\mathbf{C}_1$ through $\mathbf{C}_N$, there is a square unitary matrix $\mathbf{Q}_1$, such that

$$\mathbf{Q}_1\mathbf{C} = \mathbf{C}^{(1)} \tag{8.64}$$

and the components of the first column of $\mathbf{C}^{(1)}$, except for the first, equal zero. Then, there is a square unitary matrix $\mathbf{Q}_2$, such that

$$\mathbf{Q}_2\mathbf{Q}_1\mathbf{C} = \mathbf{C}^{(2)} \tag{8.65}$$

The first column of $\mathbf{C}^{(2)}$ is identical to the first column of $\mathbf{C}^{(1)}$ and only the first two components of the second column of $\mathbf{C}^{(2)}$ are nonzero. This is continued until the N columns of $\mathbf{C}$ have been transformed into an upper trapezoidal matrix.

Householder's transformation defines $\mathbf{Q}_1$ in the form:

$$\mathbf{Q}_1 = \mathbf{I} - \mathbf{V}\mathbf{V}^{\mathbf{H}}/p \tag{8.66}$$

where $\mathbf{I}$ is the identity matrix, $\mathbf{V}$ is a column vector, and p is a scalar. Define the vector $\mathbf{E}_1$ as

$$\mathbf{E}_1^{\mathbf{T}} = (1, 0, 0, \ldots, 0)$$

Denote the kth column of $\mathbf{C}$ as $\mathbf{C}_k$. The first column of $\mathbf{C}^{(1)}$, $\mathbf{C}_1^{(1)}$, should be

$$\mathbf{Q}_1\mathbf{C}_1 = \mathbf{C}_1^{(1)} = -\sigma\mathbf{E}_1 \tag{8.67}$$

where σ is a scalar and the negative sign is used for convenience. If we assume $\mathbf{V}$ can be expressed as

$$\mathbf{V} = \mathbf{C}_1 + \sigma\mathbf{E}_1 \tag{8.68}$$

then,

$$\mathbf{Q}_1\mathbf{C}_1 = \mathbf{C}_1(1 - \mathbf{V}^{\mathbf{H}}\mathbf{C}_1/p) - (\mathbf{V}^{\mathbf{H}}\mathbf{C}_1/p)\sigma\mathbf{E}_1 \tag{8.69}$$

Note that, if $\mathbf{V}^{\mathbf{H}}\mathbf{C}_1/p$ equals unity, (8.67) is satisfied. Thus,

$$p = \mathbf{C}_1^{\mathbf{H}}\mathbf{C}_1 + \sigma^* c_{11} \tag{8.70}$$

where c_{11} is the first component of $\mathbf{C}_1$.

The value of σ is determined from the requirement that $\mathbf{Q}_1$ is unitary. Combining (8.66) and (8.68) gives

$$\mathbf{Q}_1\mathbf{Q}_1^{\mathbf{H}} = \mathbf{I} + \mathbf{V}\mathbf{V}^{\mathbf{H}}[\mathbf{V}^{\mathbf{H}}\mathbf{V}/|p|^2 - p^{-1} - (p^*)^{-1}] \tag{8.71}$$

For this expression to equal the identity matrix, the bracketed term equals zero, so that p must be real. Because, in (8.70), $\mathbf{C}_1^{\mathbf{H}}/\mathbf{C}_1$ is always real, this requires that $\sigma^* c_{11}$

be real. If c_{11} is zero, the phase of σ is arbitrary. If c_{11} is nonzero, the phase of σ equals the phase of c_{11}. From (8.67), $|\sigma|^2 = \mathbf{C}_1^H\mathbf{C}_1$ so that the complex value of σ required to make p real can be determined. The value of p is then calculated from (8.70). Therefore, by construction, a unitary transform has been found, which causes all components of $\mathbf{C}_1^{(1)}$, except the first, to be zero.

To find the $\mathbf{Q}_2$ value that transforms the second column of $\mathbf{C}^{(1)}$ into the desired form, consider the partitioned matrix:

$$\mathbf{Q}_2 = \begin{pmatrix} 1 & 0 \\ 0 & \mathbf{P}_2 \end{pmatrix}$$

where $\mathbf{P}_2$ is a square matrix of order $\mathbf{Q}_2 - 1$. Note that $\mathbf{Q}_2$ operating on $\mathbf{C}^{(1)}$ produces $\mathbf{C}^{(2)}$ such that the first row and column of $\mathbf{C}^{(2)}$ is identical to the first row and column of $\mathbf{C}^{(1)}$. Thus, $\mathbf{P}_2$ operating on $\mathbf{C}^{(1)}$ can only modify the second column with respect to the terms on and below the second row. Therefore, the technique used to derive the value of $\mathbf{Q}_1$ from the first column of $\mathbf{C}$ can be used to find a unitary $\mathbf{P}_2$ from the one-component-shortened second column of $\mathbf{C}^{(1)}$. For an N-column matrix, this technique is iterated N times to obtain the desired trapezoidal form.

8.8.7 SMI Performance Analysis

The performance analysis of the LMS and NLMS algorithms evaluated the convergence time of the algorithms and the additional residue power due to weight noise effects after convergence. These performance criteria are pertinent, as the estimated weight vector is random and does not converge to the optimum weight vector. Only the mean value of the estimated weight vector converges to the optimum weight vector. The SMI performance analysis differs from that of the iterative algorithms for several reasons. One reason is that the estimated weight vector converges with probability one to the optimum weight vector. Thus, given a sufficient number of data vectors (i.e., large enough K in (8.51)), the weight-noise effect becomes arbitrarily small. The convergence time is defined as the value of K such that the performance loss compared to that for the optimum weights is small (e.g., 3 dB; 1 dB).

Another difference between SMI and the iterative algorithms is the dependence of performance on environmental conditions. As previously shown, the convergence time and weight-noise effects of the LMS and NLMS algorithms strongly depend on the number, powers, and spatial locations of the jammers. As shown below, the convergence time of the SMI algorithm is independent of the jammer environment. Specifically, the convergence time is shown to be the same for any covariance matrix.

The output SINR for arbitrary weight vector $\mathbf{G}$ is defined by (8.5). For the optimum weight vector, the maximum output SINR is given by (8.19). The ratio of

(8.5) and (8.19) is thus the SINR loss when using an arbitrary weight vector. The loss η, thus is

$$\eta = |\mathbf{G}^{\mathbf{H}}\mathbf{S}|^2/[(\mathbf{S}^{\mathbf{H}}\mathbf{R}_{\mathbf{Z}}^{-1}\mathbf{S})(\mathbf{G}^{\mathbf{H}}\mathbf{R}_{\mathbf{Z}}\mathbf{G})] \tag{8.62}$$

Note that η is a real value bounded between zero and one. When $\mathbf{G}$ is computed via the SMI algorithm, $\mathbf{G}$ is a random vector, and η is a random variable. The goal of this section is to derive the probability density function of η. This will be done in two steps. The first step is to show that η is independent of $\mathbf{R}_{\mathbf{Z}}$. The second step is to derive the density function of η for any $\mathbf{R}_{\mathbf{Z}}$, as the derived density function is appropriate for all $\mathbf{R}_{\mathbf{Z}}$. The chosen $\mathbf{R}_{\mathbf{Z}}$ should be one that simplifies the analysis. This part of the analysis will take $\mathbf{R}_{\mathbf{Z}}$ as equal to the identity matrix.

8.8.7.1 Loss Invariance with Covariance Matrices

The observed kth data vector is $\mathbf{Z}_k$. Let us define the vector $\mathbf{X}_k$ and a transformed steering vector $\mathbf{S}_x$ as

$$\begin{aligned} \mathbf{X}_k &= \mathbf{U}^{\mathbf{H}}\mathbf{R}_{\mathbf{Z}}^{-1/2}\mathbf{Z}_k; \quad \mathbf{Z}_k = \mathbf{R}_{\mathbf{Z}}^{1/2}\mathbf{U}\mathbf{X}_k \\ \mathbf{S}_x &= \mathbf{U}^{\mathbf{H}}\mathbf{R}_{\mathbf{Z}}^{-1/2}\mathbf{S}; \qquad \mathbf{S} = \mathbf{R}_{\mathbf{Z}}^{1/2}\mathbf{U}\mathbf{S}_x \end{aligned} \tag{8.73}$$

where $\mathbf{R}_Z^{-1/2}$ is the Hermitian inverse square root matrix, defined in Section 8.3.1 by (8.11) and $\mathbf{U}$ is an as yet unspecified unitary matrix. The covariance matrix of $\mathbf{X}_k$ is the identity matrix because

$$\begin{aligned} E(\mathbf{X}_k\mathbf{X}_k^{\mathbf{H}}) &= E(\mathbf{U}^{\mathbf{H}}\mathbf{R}_{\mathbf{Z}}^{-1/2}\mathbf{Z}_k\mathbf{Z}_k^{\mathbf{H}}\mathbf{R}_{\mathbf{Z}}^{-1/2}\mathbf{U}) \\ &= \mathbf{U}^{\mathbf{H}}\mathbf{R}_{\mathbf{Z}}^{-1/2}\mathbf{R}_{\mathbf{Z}}\mathbf{R}_{\mathbf{Z}}^{-1/2}\mathbf{U} = \mathbf{I} \end{aligned} \tag{8.74}$$

where E is the expectation. Because $\mathbf{U}$ is unitary, using (8.73) in (8.49) gives

$$\hat{\mathbf{R}}_{\mathbf{Z}} = (\Sigma\mathbf{Z}_k\mathbf{Z}_k^{\mathbf{H}})/K = \mathbf{R}_{\mathbf{Z}}^{1/2}\mathbf{U}(\Sigma\mathbf{X}_k\mathbf{X}_k^{\mathbf{H}}/K)\mathbf{U}^{\mathbf{H}}\mathbf{R}_{\mathbf{Z}}^{1/2} \tag{8.75}$$

Let us denote $(\Sigma\mathbf{X}_k\mathbf{X}_k^{\mathbf{H}}/K)$ as $\hat{\mathbf{I}}$. Then,

$$\begin{aligned} \hat{\mathbf{R}}_{\mathbf{Z}}^{-1} &= \mathbf{R}_{\mathbf{Z}}^{-1/2}\mathbf{U}\hat{\mathbf{I}}^{-1}\mathbf{U}^{\mathbf{H}}\mathbf{R}_{\mathbf{Z}}^{-1/2} \\ \mathbf{G} &= \hat{\mathbf{R}}_{\mathbf{Z}}^{-1}\mathbf{S} = \hat{\mathbf{R}}_{\mathbf{Z}}^{-1}\mathbf{R}_{\mathbf{Z}}^{1/2}\mathbf{U}\mathbf{S}_x = \mathbf{R}_{\mathbf{Z}}^{-1/2}\mathbf{U}\hat{\mathbf{I}}^{-1}\mathbf{S}_x \\ \mathbf{G}^{\mathbf{H}}\mathbf{S} &= \mathbf{S}_x^{\mathbf{H}}\hat{\mathbf{I}}^{-H}\mathbf{U}^{\mathbf{H}}\mathbf{R}_{\mathbf{Z}}^{-1/2}\mathbf{R}_{\mathbf{Z}}^{1/2}\mathbf{U}\mathbf{S}_x = \mathbf{S}_x^{\mathbf{H}}\hat{\mathbf{I}}^{-\mathbf{H}}\mathbf{S}_x = \mathbf{S}_x^{\mathbf{H}}\hat{\mathbf{I}}^{-1}\mathbf{S}_x \end{aligned} \tag{8.76}$$

Similarly,

$$\begin{aligned} \mathbf{S}^{\mathbf{H}}\mathbf{R}_{\mathbf{Z}}^{-1}\mathbf{S} &= \mathbf{S}_x^{\mathbf{H}}\mathbf{U}^{\mathbf{H}}\mathbf{R}_{\mathbf{Z}}^{1/2}\mathbf{R}_{\mathbf{Z}}^{-1}\mathbf{R}_{\mathbf{Z}}^{1/2}\mathbf{U}\mathbf{S}_x = \mathbf{S}_x^{\mathbf{H}}\mathbf{S}_x \\ \mathbf{G}^{\mathbf{H}}\hat{\mathbf{R}}_{\mathbf{Z}}\mathbf{G} &= \mathbf{S}_x^{\mathbf{H}}\hat{\mathbf{I}}^{-2}\mathbf{S}_x \end{aligned} \tag{8.77}$$

Combining (8.76) and (8.77) with (8.72) gives

$$\eta = |\mathbf{S}_x^{\mathbf{H}}\hat{\mathbf{I}}^{-1}\mathbf{S}_x|^2/(\mathbf{S}_x^{\mathbf{H}}\mathbf{S}_x)(\mathbf{S}_x^{\mathbf{H}}\hat{\mathbf{I}}^{-2}\mathbf{S}_x) \tag{8.78}$$

Note that $\mathbf{S}_x$ depends on the unitary matrix, $\mathbf{U}$. Define the first column of $\mathbf{U}$ as the unit-length vector colinear with $\mathbf{R}_\mathbf{Z}^{-1/2}\mathbf{S}$. Because all other columns of $\mathbf{U}$ are orthogonal to the first, only the first entry of $\mathbf{S}_x$ is nonzero. Let us denote $\hat{\mathbf{I}}^{-1}$ in partitioned form as

$$\hat{\mathbf{I}}^{-1} = \begin{pmatrix} c^{11} & \mathbf{C}^{12} \\ \mathbf{C}^{21} & \mathbf{C}^{22} \end{pmatrix} \tag{8.79}$$

where c^{11} is a scalar, $\mathbf{C}^{21}$ is an N-dimensional vector, $\mathbf{C}^{12}$ is the Hermitian of $\mathbf{C}^{21}$, and $\mathbf{C}^{22}$ is a square matrix of order N. Then, without loss of generality, we denote the first component of $\mathbf{S}_x$ as unity and the numerator of (8.78) is

$$(1, \phi^\mathbf{H}) \begin{pmatrix} c^{11} & \mathbf{C}^{12} \\ \mathbf{C}^{21} & \mathbf{C}^{22} \end{pmatrix} \begin{pmatrix} 1 \\ \phi \end{pmatrix} \tag{8.80}$$

where ϕ denotes the N-dimension zero vector. The denominator is

$$(1, \phi^\mathbf{H}) \begin{pmatrix} c^{11} & \mathbf{C}^{12} \\ \mathbf{C}^{21} & \mathbf{C}^{22} \end{pmatrix} \begin{pmatrix} \mathbf{c}^{11} & \mathbf{C}^{12} \\ \mathbf{C}^{21} & \mathbf{C}^{22} \end{pmatrix} \begin{pmatrix} 1 \\ \phi \end{pmatrix} \tag{8.81}$$

Expanding (8.80) and (8.81) results in

$$\eta = (\mathbf{c}^{11})^2/[(\mathbf{c}^{11})^2 + \mathbf{C}^{12}\mathbf{C}^{21}] \tag{8.82}$$

Thus, the loss depends only on the entries of $\hat{\mathbf{I}}^{-1}$. As $\mathbf{I}$ depends only on the set of vectors $\{\mathbf{X}_k\}$ and the covariance matrix of $\{\mathbf{X}_k\}$ equals the identity matrix, the properties of η are independent of the covariance matrix, $\mathbf{R}_\mathbf{Z}$.

8.8.7.2 Density Function of the Random Loss

Statistical properties of the estimated covariance matrix have been developed by Goodman [1]. Thus, the first step in determining the statistical properties of the loss is to relate the entries of the inverse matrix and the estimated matrix.

Denote $\hat{\mathbf{I}}$ in partitioned form as

$$\hat{\mathbf{I}} = \begin{pmatrix} c_{11} & \mathbf{C}_{21}^\mathbf{H} \\ \mathbf{C}_{21} & \mathbf{C}_{22} \end{pmatrix} \tag{8.83}$$

where c_{11} is a scalar, $\mathbf{C}_{21}$ is an N-dimensional vector, and $\mathbf{C}_{22}$ is a square matrix of order N. Using (8.79) and (8.83) gives

$$\begin{pmatrix} c_{11} & \mathbf{C}_{12} \\ \mathbf{c}_{21} & \mathbf{C}_{22} \end{pmatrix} \begin{pmatrix} c^{11} & \mathbf{C}^{12} \\ \mathbf{C}^{21} & \mathbf{C}^{22} \end{pmatrix} = \begin{pmatrix} 1 & \phi^{\mathbf{H}} \\ \phi & \mathbf{I} \end{pmatrix} \tag{8.84}$$

so that

$$\begin{aligned} c_{11}c^{11} + \mathbf{C}_{12}\mathbf{C}^{21} &= 1 \\ \mathbf{c}_{21}\mathbf{c}^{11} + \mathbf{C}_{22}\mathbf{C}^{21} &= \phi \end{aligned} \tag{8.85}$$

From the second equation of (8.85), we have

$$\begin{aligned} \mathbf{C}^{21} &= -\mathbf{C}_{22}^{-1}\mathbf{C}_{21}\mathbf{c}^{11} \\ \mathbf{C}^{12} &= -c^{11}\mathbf{C}_{12}\mathbf{C}_{22}^{-1} \end{aligned} \tag{8.86}$$

Combining this with (8.82) gives

$$\eta = 1/[1 + (\mathbf{C}_{12}\mathbf{C}_{22}^{-1})(\mathbf{C}_{22}^{-1}\mathbf{C}_{21})] \tag{8.87}$$

This last equation expresses the loss in terms of the entries of the sample covariance matrix.

The components of the identity matrix, $\hat{\mathbf{I}}$, are random and its statistical properties are given by the central-complex Wishart distribution. Centrality denotes that the true covariance matrix and its estimate both have zero mean. In addition, both matrices are Hermitian. The form of the density for a true covariance matrix, **R**, and its estimate, **A**, formed from K statistically independent data vectors, is

$$\begin{aligned} p(\mathbf{A}) &= [\det(\mathbf{A})]^{K-N} \exp[-\mathrm{tr}(\mathbf{R}^{-1}\mathbf{A})]/\alpha \\ \alpha &= \pi^{K(K-1)/2}\Gamma(N)\Gamma(N-1)\ldots\Gamma(N-K+1)[\det(\mathbf{R})]^{N} \end{aligned}$$

where **R** is $N \times N$. The number of measurement vectors must be at least equal to N so that **A** is nonsingular. Using (8.87) and the properties of the Wishart distribution, after considerable manipulation, we can show that η has a β probability density function. Thus, the density function of the loss, $p(\eta)$ is

$$p(\eta) = [K!/(N-2)!\,(K+1-N)!]\,(1-\eta)^{N-2}\eta^{K+1-N} \tag{8.88}$$

As K tends toward infinity, the probability density function tends toward an impulse at $\eta = 1$.

By using the well known properties of the β distribution, Reed, Mallet, and Brennan [2] have shown that the average loss is

$$E(\eta) = (K + 2 - N)/(K + 1) \tag{8.89}$$

For $N \gg 1$, the loss is approximately 3 dB when $K = 2N$, and this can be thought of as a minimum requirement on the number of vector estimates used to estimate the covariance matrix. More typically, a loss budget may allow for a 1 dB loss so that K is more usually approximately equal to $5N$.

REFERENCES

[1] N.R. Goodman, "Statistical Analysis Based on a Certain Multivariate Complex Gaussian Distribution," *Ann. Math. Stat.*, Vol. 34, March 1963, pp. 152–177.

[2] I.S. Reed, J.D. Mallett, and L.E. Brennan, "Rapid Convergence Rate in Adaptive Arrays," *IEEE Trans. on Aerospace and Electronic Systems*, Vol. AES-10, No. 6, Nov. 1974, pp. 853–863.

Chapter 9
Constant False-Alarm Rate (CFAR) Processors

9.1 INTRODUCTION

The conceptual basic radar system shown in Figure 1.7 makes target-present or target-absent decisions in the various range-angle-doppler frequency resolution cells. As discussed in Section 1.5, in a thermal-noise-only environment, either an operator or automatic decision device makes a target-present decision when the input waveform envelope voltage is greater than a prespecified threshold. The threshold is computed from (1.12) and depends on the thermal-noise power and desired false-alarm probability.

However, when the interference environment comprises the combination of thermal noise and noise jamming with clutter, the output interference power is greater than that due to thermal noise alone. This is true, even if adaptive processing is used to "cancel" the jamming and clutter. As an example, the use of adaptive antennas can achieve significant jammer-power cancellation. However, even for the narrowband optimum weight implementation, the output residue power is higher than that which occurs for the noise-only environment. In addition, the residue power is usually still larger, due to the effects of nonzero bandwidth or channel-to-channel mismatch. Thus, if a constant prespecified threshold is used in a thermal-noise-plus-jammer environment, the false-alarm rate increases due to the residue power increase, and saturation of the data processor will result. To prevent the saturation, a method of preventing the false-alarm rate increase is mandatory. Similar observations hold for the clutter-plus-thermal-noise environment. Again, the output residue power is greater than that due solely to noise, and a method of preventing the false-alarm rate increase is mandatory.

The false-alarm rate increase often is unacceptable, even when the residue power is only slightly different than that due to thermal noise. As a quantitative indication of the severity of the problem, for a prespecified threshold, T, note that (1.12) indicates that the attained false-alarm probability, p_{FA}, is related to the design false alarm probability, p_{FD}, as

$$\begin{aligned} p_{FA} &= \exp(-T/p_{res}) = \exp[-(T/p_n)(p_n/p_{res})] \\ &= p_{FD}{}^{(p_n/p_{res})} \end{aligned} \tag{9.1}$$

Figure 9.1 indicates the attained false-alarm probability for typical false-alarm probability design values of 1×10^{-6}, 1×10^{-8}, and 1×10^{-10} as a function of the ratio of the design interference power (usually thermal-noise power) to the actual residue power. We can see from the figure or (9.1) that a +3 dB power ratio causes the attained false-alarm probability to increase to the square root of the desired value, whereas a −3 dB power ratio causes the attained false-alarm probability to decrease to the square of the desired value. Specifically, when $p_{FD} = 1 \times 10^{-8}$, for the ±3 dB power ratio, the values of p_{FA} are 1×10^{-4} and 1×10^{-16}. Note that, although decrease of the false-alarm rate does not saturate the data processor, it indicates that the threshold is too high. Therefore, target detectability is significantly reduced, which is also unsatisfactory. Clearly, relatively small changes of residue power can cause intolerably large changes of performance. A form of adjustable or adaptive threshold is required in order to implement a CFAR processor action.

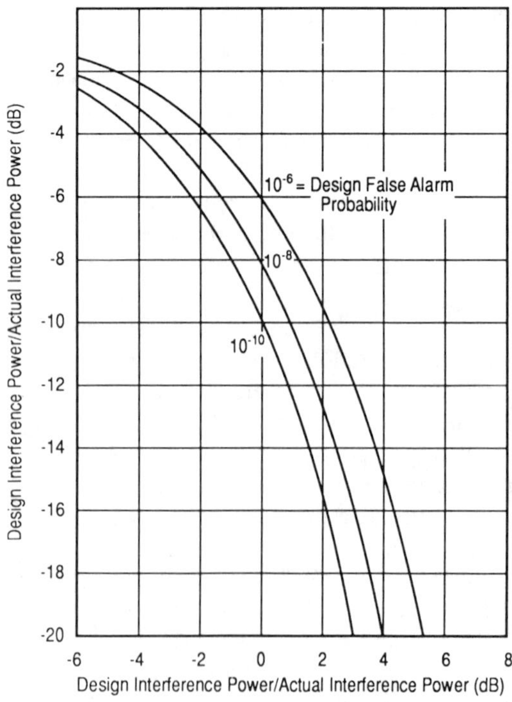

Figure 9.1 Variation of false-alarm probability with interference power.

9.2 CFAR CONCEPT

The basic concept of a CFAR technique is that the voltage of a test cell is compared to that of a set of reference cells. If the test cell voltage is "similar" to those of the reference cells, a target-absent decision is made. If the test cell voltage is not similar, a target-present decision is made. Given this strategy, several questions must be answered to obtain a satisfactory implementation. One is that there is a tacit assumption that it is possible to obtain a set of reference cells with statistical descriptors that are identical to those of the test cell for the target-absent condition. How is this done? A second question is how the dichotomy of "similar" or "not similar" is made.

There are several methods of obtaining a set of reference cells. The proper method depends on the type of interference environment. One common technique, shown in Figure 9.2, uses a range interval bracketing the test cell as the reference cells. Use of this set implies that the characteristics of the interference do not change over the range interval of the reference and test cells. This assumption is true when the interference environment consists of the sum of thermal noise and constant-power jamming noise. The assumption is not true for environments that include ground clutter as the clutter cross section tends to change significantly (and randomly) in relatively small range increments.

A second technique uses an azimuth interval bracketing the test cell's azimuth with all cells at the same range. Clearly, this technique works well only when the interference characteristics are unchanging over the azimuth interval. This approach would not work well for a sidelobe noise-jamming environment as the received jamming power changes with azimuth due to the changing sidelobe gain of the antenna pattern. The approach could work well when the interference is the sum of thermal noise plus main-lobe weather clutter as the weather clutter characteristics are often uniform over azimuth.

A third technique combines the two domains to implement a two-dimensional reference cell configuration. Some radar systems could implement a multidimensional reference set by including doppler frequency, elevation angle, polarization, *et cetera*.

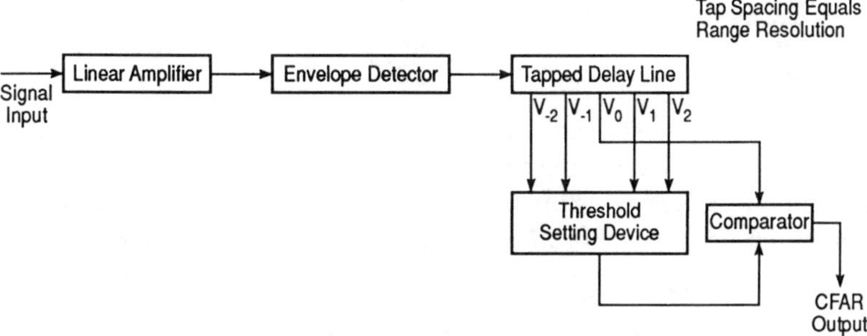

Figure 9.2 CFAR processor using range window for reference cells.
Source: From R. Nitzberg, *Constant-False-Alarm-Rate Signal Processors for Several Types of Interference*. Trans. IEEE, AES-8, No. 1, Jan 1972.

A technique that is appropriate for ground clutter obtains the reference cells from previous scans of the resolution cell. This technique is called *clutter mapping*. In theory, the data from prior scans is used to estimate the clutter plus thermal-noise power of every resolution cell. When a target such as an aircraft enters the cell, the increase in reflected power allows target detection in the presence of the clutter. Often, the clutter map is used with a doppler filter bank and a different map is determined for each filter of the bank.

The following sections discuss several different CFAR processors. The fact that there are so many is indicative of the substantial difficulty in obtaining a single technique that is satisfactory for all interference environments. In general, there is not yet a totally satisfactory solution.

9.3 CELL-AVERAGING CFAR (CACFAR)

Equation (1.12) relates the threshold value, T, the thermal-noise power, p_n, and the design false-alarm probability, p_{FD}, when using a square-law envelope detector. The equation is also correct when the noise power is the sum of thermal noise and jamming noise or the residue power after cancellation. Let us denote the power as p_I for either case. By taking the natural logarithm of both sides of this equation, we derive that the threshold equals the total interference noise power multiplied by a constant, k_1, which depends only on the design false-alarm probability. We denote the square-law envelope detector for the test cell as q_0. Thus, a target-present decision is made when

$$q_0 > \mathrm{T} = k_1 p_I \tag{9.2}$$

When the total interference power is unknown, estimating the interference power as $\hat{p}_I$ and making target-present decision when

$$q_0 > \mathrm{T}_{\mathrm{av}} = k_2 \hat{p}_I \tag{9.3}$$

where T_{av} is an adaptive threshold seems reasonable. In (9.3), k_2 is a new multiplier constant that is usually not equal to k_1.

A common method to estimate the interference power is by averaging the outputs of the reference cells. For M reference cells, the estimate is

$$\hat{p}_I = \sum_{m=1}^{M} q_m / M \tag{9.4}$$

where q_m denotes the square-law detector output of the mth reference cell.

An alternative but similar procedure is to use an envelope detector with an output that is the magnitude of the envelope, rather than a square-law envelope detector with an output that is the magnitude squared of the envelope. The former is often substantially simpler to implement. Then, denoting the envelope of the test and reference cells as e_0 through e_M, the equivalent of (9.3) and (9.4) is to make a target-present decision when

$$e_0 > \mathrm{T}'_{\mathrm{av}} = k'_2 \sum_{m=1}^{M} e_m / M \tag{9.5}$$

Further, target-present or target-absent decisions could be based on any monotonic function of the magnitude of the envelope. Thus, denote the cell voltages for an arbitrary monotonic function as f_0 through f_M. The decision is then based on

$$f_0 > k''_2 \sum_{m=1}^{M} f_m / M \tag{9.6}$$

The question of which is the statistically best procedure requires detailed consideration of the statistics of the cell voltages.

9.4 PERFORMANCE ANALYSIS OF CELL-AVERAGING CFAR (CACFAR) PROCESSOR

The test cell and M reference cells constitute an $M + 1$-dimensional vector. Prior to an envelope detector, each component of the vector is distributed as a complex-valued Gaussian random variable. The probability density function for a general multidimensional complex-valued Gaussian distributed random vector is given by (8.28). For the CFAR problem, there are several important symmetries that simplify the form of the density. One concerns the variance of each cell sample (the diagonal entries of $\mathbf{R}$). Specifically, the averaging used in (9.4) through (9.6) assumes that in the absence of target each cell has the same statistical description. Thus, the diagonal entries of $\mathbf{R}$ are identical for the design environment. This is termed the design environment because later sections will consider the performance degradation when the actual environment differs from the design environment. A second simplification is the assumption that the cell voltages are statistically independent. When the reference cells are contiguous range cells, this is a reasonable approximation when the A/D sampling frequency is approximately equal to the radar bandwidth. For independent cells, the density function of (8.28) factors into the product of $M + 1$ one-dimensional density functions. A third simplification is that for the target-absent condition, the mean of each cell prior to the envelope detectors equals zero and a target can only occur in the test cell. The initial analyses will emphasize these conditions. The later analyses will consider the effects of departures from these conditions.

The zero-mean, one-dimensional, complex-valued Gaussian variate z is defined as

$$z = x + jy \tag{9.7}$$

where x and y are independent, zero-mean, real-valued random variables. The density function of x (or y) is given by

$$p(x) = \exp(-x^2/2v_x)/(2\pi v_x)^{1/2} \tag{9.8}$$

where v_x denotes the variance of x. The joint density of x and y is given by

$$p(x,y) = \exp[-(x^2 + y^2)/2v_x]/(2\pi v_x) \tag{9.9}$$

and note that x and y have the same variance. Note that $|z|^2$ equals $x^2 + y^2$. Thus, the variance of z is

$$v_z = E|z|^2 = E(x^2 + y^2) = 2v_x \tag{9.10}$$

Thus, from (9.8) through (9.10), the density function of z is

$$p(z) = \exp(-|z|^2/v_z)/(\pi v_z) \tag{9.11}$$

The density function can also be expressed in terms of the magnitude and phase of z because

$$\begin{aligned} x &= r\cos\phi;\ y = r\sin\phi \\ z &= r\exp(j\phi) \\ p(r,\phi) &= r\exp(-r^2/v_z)/(\pi v_z) \end{aligned} \tag{9.12}$$

and the factor of r multiplying the exponential is due to the Jacobian of the transformation. The marginal density function for the magnitude can be obtained by integrating over ϕ. This results in

$$p_r(r) = 2r\exp(-r^2/v_z)/v_z \tag{9.13}$$

The first CFAR technique considered uses the outputs of the cells after square-law detection. As the square-law-detected output, q, is a monotonic single-valued function of r, the density function for q is obtained via

$$q = r^2$$
$$p_q(q) = p_r(r = \sqrt{q})/|\mathrm{d}q/\mathrm{d}r| \tag{9.14}$$
$$p_q(q) = \exp(-q/v_z)/v_z$$

From (9.14) and the condition that cell outputs be statistically independent, the joint density of the test cell and reference cells for the target-absent case, when all cells have the common predetector power of v_z (defined as a homogeneous environment), thus is

$$p(q_0, q_1, \ldots, q_M) = \prod_{m=0}^{M} [\exp(-q_m/v_z)/v_z] \tag{9.15}$$

Based on (9.15), we can show that, for any specified false-alarm probability, the highest probability of detection can be obtained by using (9.3) and (9.4). This condition is defined as the optimum CFAR processor for a homogeneous environment.

The first task in evaluating the probability of detection of the optimum CFAR processor is determining the false-alarm characteristics. To use some well known statistical results, it is useful to consider (9.3) in more detail. Combining (9.3) and (9.4) indicates that target present decision is made by the optimum CFAR processor when

$$q_0 > k_q \Sigma q_m \tag{9.16}$$

The probability of false alarm depends on the value of the constant k_q, as the false-alarm probability depends on k_1 in (9.2) for the known interference power condition.

Alternative implementations of the CFAR processor indicated by (9.16) are sometimes desired. As one example, dividing both sides of (9.16) by the summation gives

$$q_0/\Sigma q_m > k_q \tag{9.17}$$

Note that (9.16) and (9.17) are statistically identical. Thus, for any given observation vector, the target-present or target-absent decision is made identically by using (9.16) or (9.17). This CFAR processor is often termed a "normalizer" because of the explicit division of (9.17). The form of (9.17) gives a convenient way of demonstrating the properties of this CFAR processor. Let us assume that "nature" has available a set of independent, identically distributed, unit-variance Gaussian distributed random variables $\{x_m\}$. The set $\{z_m\}$ is formed by multiplying each member of $\{x_m\}$ by the constant $v_z^{1/2}$. Note that this model produces $\{z_m\}$ with the density of (9.11). Then, $q_m = |z_m|^2$. Substituting this into (9.17) shows that the ratio is independent of the value of v_z. Thus, the false-alarm probability is independent of the unknown level, v_z, which is the desired CFAR property.

As discussed in standard probability texts, when the observation vector has the joint density of (9.15), the decision random variable $q_0/\Sigma q_m$ is a well known random variable, denoted as f, called an f variate,and its density function is

$$p_f(f) = M/(1 + f)^{M+1}; \quad 0 \leq f < \infty \tag{9.18}$$

The design probability of false alarm is given by

$$p_{\mathrm{FD}} = \int_{k_q}^{\infty} M\,\mathrm{d}f/(1 + f)^{M+1} \tag{9.19}$$

This relation can be integrated to give

$$p_{\mathrm{FD}} = (1 + k_q)^{-M} \tag{9.20}$$

and solving for the adaptive-threshold multiplier gives

$$k_q = p_{\mathrm{FD}}^{-1/M} - 1 \tag{9.21}$$

Another alternative implementation of this CFAR processor can be obtained from (9.16). Adding $k_q q_0$ to both sides of the inequality and rearranging gives

$$q_0/(q_0 + \Sigma q_m) > k_q/(1 + k_q) = k_\beta \tag{9.22}$$

From (9.20), note that an alternative relation for the false-alarm probability is

$$p_{\mathrm{FD}} = (1 - k_\beta)^M \tag{9.23}$$

Again, note that, in theory, exactly the same target-present or target-absent decisions are made, whether the processor is implemented by using (9.3), (9.16), or (9.22). Although, the decisions of these versions are theoretically identical, the decisions of a practical implementation can differ because of arithmetic precision effects. Specifically, note that the dynamic ranges of the decision random variables of (9.16) and (9.22) are totally different. In (9.16), the range of the decision random variable is from zero to infinity. However, that of (9.22) is from zero to unity. Also, in addition to finite-bit arithmetic differences, as discussed in detail later, there are sometimes significant performance differences that occur for a multiple-carrier-frequency waveform.

The decision random variable of (9.22) is also discussed in standard probability texts, and is called a β variable. Its density function is

$$f_\beta = M(1 - \beta)^{M-1}; \quad 0 \le \beta \le 1 \tag{9.24}$$

Analyses based on this version of the decision random variable sometimes simplify the process. Also, note that (9.23) is easily obtained by integrating (9.24). Computation of the probability of target detection can be done by additional manipulation of (9.7).

For the test cell, the pre-envelope-detector voltage z_0 is

$$z_0 = (x + s_x) + j(y + s_y) \tag{9.25}$$

where x and y denote the thermal noise (or interference residue after adaptive antenna or adaptive doppler processors) portions of z_0 and s_x and s_y denote the target reflection portions of z_0. Target detection depends on the ratio of target reflection energy, $s^2 = s_x^2 + s_y^2$, to the noise (or residue) power. This ratio is denoted as the SINR, which means the ratio of signal energy to the total residue (interference plus thermal noise) power. For nonfluctuating targets, the value of s^2 is a constant; for fluctuating targets, it is a random variable. Different target fluctuation models imply different random properties of s_x and s_y. The later results show that the need to implement an adaptive-threshold processor decreases the target detection probability as compared to that attained in a thermal-noise environment. The decrease of detection probability is termed a *CFAR loss*. Detailed consideration of many target fluctuation models indicates that the CFAR loss is very nearly independent of the actual target model. As the prior results for probability of target detection in a thermal-noise environment was done for a Swerling 1 target fluctuation model, Section 2.8.4.1, this model will also be used to compute the CFAR loss.

The Swerling 1 target model assumes that s_x and s_y are each zero-mean, independent, and identically distributed Gaussian random variables of variance v_s. The target energy, q_s, $s_x^2 + s_y^2$, is a random variable. By using the development leading to (9.14), its probability density function thus is

$$p(q_s) = \exp(-q_s/v_s)/v_s$$

Therefore, the ratio of the target energy to thermal noise (i.e., SNR) is also a random variable. Its density function is of exactly the same form as the previous equation. Denoting the SNR by the variable γ, the density function:

for γ is

$$p_\gamma(\gamma) = \exp(-\gamma/\gamma_a)/\gamma_a \tag{9.26}$$

where γ_a is the average SNR value. This last equation is identical to that introduced in Subsection 2.8.4.1 as (2.9). Equation (9.26) is applicable for the general interference environment, where γ is interpreted as the value of SINR.

Returning to (9.25), for a Swerling 1 target, the probability density function of the test cell prior to the envelope detector now is zero-mean complex Gaussian for either the case of target present or absent. For the target-absent condition, the variance equals the residue power previously denoted as v_z. For the target-present condition, the variance equals the sum of the residue power plus the average target power. This sum equals $v_z(1 + \gamma_a)$. Thus, the density function for the test cell after square-law envelope detection is

$$p_0(q_0) = \exp[-q_0/v_z(1 + \gamma_a)]/v_z(1 + \gamma_a) \tag{9.27}$$

For the target-absent condition $\gamma_a = 0$. For the target-present condition, γ_a is a positive value corresponding to the test cell's SINR.

Then, the viewpoint previously used that the set $\{z_n\}$ is related to a set $\{x_m\}$ for the target-absent condition can be extended to the target-present condition. This is done, as previously, by considering that the set $\{z_m : m = 1, M\}$ is obtained from the set $\{x_m : m = 1, M\}$ by multiplying by the common constant $v_z^{1/2}$. The value of z_0 is obtained by multiplying x_0 by $(1 + \gamma_a)^{1/2}$. Using this relation indicates that the probability density function of the decision random variable for the target-present condition modifies (9.18) to

$$p_f(f) = M/[1 + f/(1 + \gamma_a)]^{M+1} \tag{9.28}$$

The probability of detection is obtained by integrating (9.28) as was done in (9.19). Thus, the probability of detection equals

$$p_D = [(1 + \gamma_a)/(1 + \gamma_a + k_q)]^M$$

Figure 9.3 shows the probability of detection for a false-alarm probability of 1×10^{-6} for the Swerling 1 target parametric with the number of reference cell normalizing samples. Section 9.5 shows that the probability of detection approaches that of the known power-level case as M approaches infinity. The curve labeled "infinite" is computed on the basis of (1.14) for the known power-level case. We can see that, as the number of reference cells increases, the curves approach the known power-level curve. Thus, the CFAR loss approaches 0 dB as the number of reference cells increases. The next section develops equations for the CFAR loss.

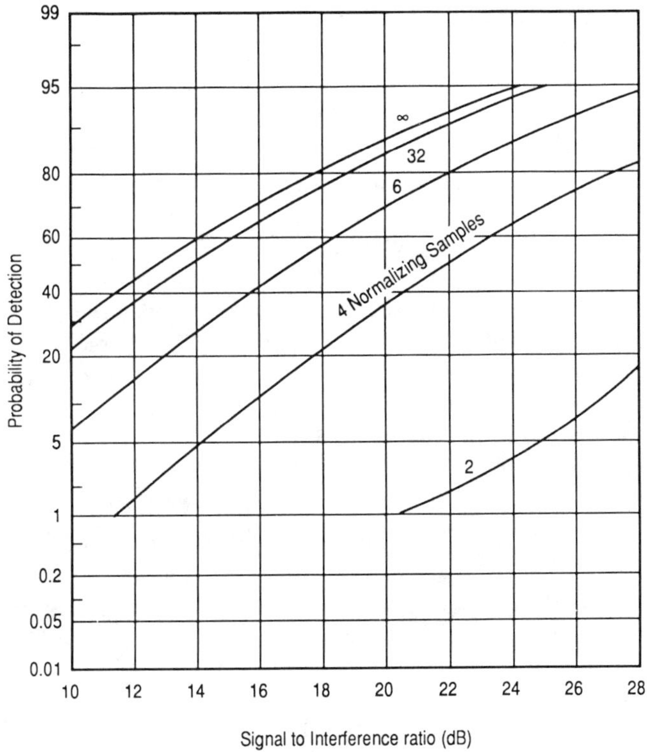

Figure 9.3 Performance curves for optimum CACFAR processor in homogeneous environments.

We observe that the CFAR loss approaches 0 dB for a homogeneous environment as the number of reference cells approaches infinity. Sections to follow consider the performance in nonhomogeneous environments.

9.5 CFAR LOSS OF THE CACFAR TECHNIQUE IN HOMOGENEOUS ENVIRONMENTS

The value of SINR required for a specified p_D and p_{FD} can be obtained by combining (9.20) and (9.28). Thus,

$$\gamma_a(M) = [(p_D/p_{\mathrm{FD}})^{1/M} - 1]/(1 - p_D^{1/M}) \tag{9.29}$$

where the notation $\gamma_a(M)$ is introduced to emphasize the dependence upon M. The limiting value of $\gamma_a(M)$ as M approaches infinity can be determined by L'Hospital's rule. We can show the limit to be

$$\gamma_a(\infty) = \ln(p_{\text{FD}}/p_D)/\ln(p_D) \tag{9.30}$$

Comparison of (9.30) with (1.14) indicates that the limiting value is identical to that for the known level environment. Thus, the CFAR loss tends toward zero as the number of reference cells tends toward infinity.

The CFAR loss equals the ratio of (9.29) and (9.30). This loss is graphed in Figure 9.4 and is approximately independent of the probability of detection. Note that both scales are logarithmic and the CFAR loss, in dB, plots as approximately straight lines. Thus, as an example, the decibel loss is reduced from 1 dB to $\frac{1}{2}$ dB by doubling the number of reference cells. An approximate analysis that develops the straight-line relationship is given below.

First, note the identity:

$$(p_D/p_{\text{FD}})^{1/M} - 1 = \exp[(1/M)\ln(p_D/p_{\text{FD}})] - 1 \tag{9.31}$$

For large M, the argument of the exponential is small and well approximated by the first few terms of a power series. Thus,

$$(p_D/p_{\text{FD}})^{1/M} \approx [\ln(p_D/p_{\text{FD}})/M]\left[1 + \frac{1}{2}\ln(p_D/p_{\text{FD}})\Big/ M\right] \tag{9.32}$$

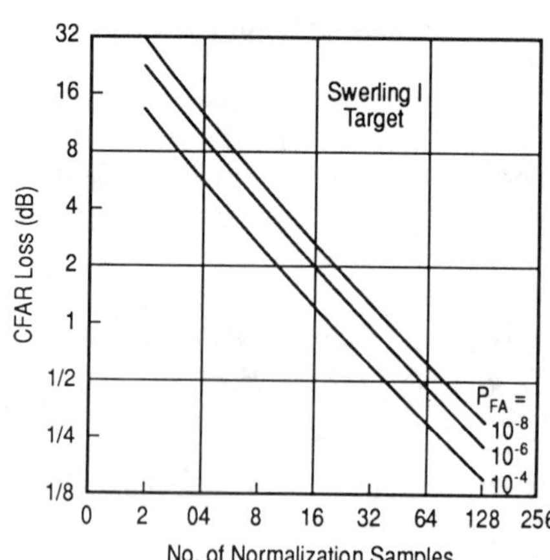

Figure 9.4 Decibel CFAR loss for CACFAR.
Source: From R. Nitzberg, *Analysis of the Arithmetic Mean CFAR Normalizer for Fluctuating Targets.* Trans. IEEE, AES-14, No. 1, Jan 1978.

Similarly,

$$1 - p_D^{1/M} = 1 - \exp[(1/M)\ln(p_D)] \tag{9.33}$$

and combining (9.32), (9.33), and (9.30) results in

$$\gamma_a(M)/\gamma_a(\infty) \approx [1 + \ln(p_D/p_{\mathrm{FD}})/2M]/[1 + \ln(p_D)/2M] \tag{9.34}$$

Because, for small x,

$$1/(1 + x) \approx 1 - x$$

we can further approximate (9.34) as

$$\gamma_a(M)/\gamma_a(\infty) \approx 1 - \ln(p_{\mathrm{FD}})/2M \tag{9.35}$$

where only terms involving first powers of $1/M$ have been retained. The decibel loss is

$$\begin{aligned} L_{\mathrm{DB}} &= 10\log_{10}[\gamma_a(M)/\gamma_a(\infty)] \\ &= 10\log_e[\gamma_a(M)/\gamma_a(\infty)]/\log_e(10) \end{aligned}$$

Because

$$\begin{aligned} &\log_e(1 + x) \approx x \\ &L_{\mathrm{dB}} \approx -5\log_e(p_{\mathrm{FD}})/[M\log_e(10)] \end{aligned} \tag{9.36}$$

Thus, the decibel loss, to a first-order approximation, does not depend on the probability of detection. Also, we can see that doubling the number of reference cells causes the loss to be halved.

9.6 NONHOMOGENEOUS ENVIRONMENTS

The cell-averaging CFAR computes an adaptive threshold to maintain a constant false-alarm rate. The adaptive threshold equals a gain constant (determined by the design false-alarm probability) multiplied by the sum of the instantaneous voltages of the samples in the range window bracketing the test cell. However, the attained false-alarm probability is only equal to the design value when the interference power (thermal noise plus clutter plus jamming plus additional targets) is constant (homogeneous) over the total range window. The required homogeneity does not always occur. Thus, the attained-false alarm probability often differs considerably from the design value. This section considers several nonhomogeneous environments and alternative CFAR techniques that are better able to cope with these environments.

9.6.1 CACFAR Performance in Mismatched Environments

The synthesis and prior analysis of the CACFAR circuit assume that, for the target-absent condition, the powers of the test cell and reference cells (interference and residue power) are all equal. This assumption constitutes the design environment. The actual environment is usually the same as the design environment when the environment is the sum of thermal and jamming noise. However, in the presence of clutter, the actual environment differs considerably from the design environment.

One consequence of this mismatch is that the actual false-alarm rate is not equal to the design false-alarm rate. A second consequence is that the probability of target detection does not necessarily increase monotonically as the number of reference cells increases.

The next subsection derives equations to quantify the effect of the environment mismatch on the false-alarm probability and detection probability. The subsequent subsections discuss and analyze alternative CFAR techniques that can attain a constant false-alarm rate, although all cells do not have the same power. We shall see that attaining a constant false-alarm rate for these more severe environments can be done, but the CFAR loss increases. We should note that the CFAR loss is a loss *per se* only with respect to performance in a known power-level environment. This is not an avoidable loss, but instead is a cost of operating in a non-thermal-noise environment.

9.6.2 Mismatched Environment Analysis

The analysis of the CACFAR circuit was simplified by recognizing that the decision random variable, the left-hand side of (9.17), is the well known f variate and its density function is given by (9.18). For the mismatched environment, the decision variable is not an f variate and its density function is not that of (9.18).

The density function can be found by noting that a target-present decision is made when

$$\begin{aligned} q_0 &> k_q v_N \\ v_N &= \Sigma q_m \end{aligned} \tag{9.37}$$

where v_N is the normalization voltage. By using (9.27), this probability is given by the double integral:

$$p_D = \int_0^\infty p_{vN}(v_N)\left\{\int_{k_q v_N}^\infty \exp[-q/v_0(1 + \gamma_a)]/[v_0(1 + \gamma_a)]\, dq\right\} dv_N$$

where $p_{vN}(\cdot)$ denotes the density function of the normalization voltage and v_0 denotes the predetector test cell power for the target absent condition. Note that, in general, the power of the reference cells $\{v_m\}$ is neither constant nor equal to that of the test cell, v_0. The integral evaluates to

$$p_D = \int_0^\infty \exp[-k_q v_N/v_0(1 + \gamma_a)]p_{vN}(v_N)\, dv_N \tag{9.38}$$

The above integral can be evaluated by using properties of characteristic functions.

9.6.2.1 *Characteristic Function Review*

The characteristic function of a random variable x, $\Phi_x(u)$, with probability density function, $p_x(x)$, is defined by the Fourier transform pair:

$$\begin{aligned} \Phi_x(u) &= \int_0^\infty p_x(x) \exp(jux)\, dx \\ p_x(x) &= \int_0^\infty \Phi_x(u) \exp(-jux)\, du/2\pi \end{aligned} \tag{9.39}$$

Comparing (9.38) and (9.39) indicates that the probability of detection equals the characteristic function of the normalization voltage, evaluated at the argument $ju = -k_q/v_0(1 + \gamma_a)$, so that

$$p_D = \phi_{vN}[jk_q/v_0/(1 + \gamma_a)] \tag{9.40}$$

An important theorem concerning characteristic functions is that when v_N is the sum of independent random variables, as in (9.37), its characteristic function equals the product of the individual characteristic functions of the set $\{q_N\}$.

$$\Phi_{vN}(u) = \prod_{m=1}^{M} \Phi_{qm}(u) \tag{9.41}$$

Denote the pre-envelope detector's mth reference cell power as v_m. Then, using the probability density function of (9.27), the characteristic function of the mth reference cell random variable is

$$\begin{aligned} \Phi_{qm}(u) &= \int_0^\infty \exp(juq)\exp(-q/v_m)/v_m \, dq \\ &= (1 - juv_m)^{-1} \end{aligned} \tag{9.42}$$

Combining (9.38), (9.40), and (9.42) indicates that the probability of detection for a generalized nonhomogeneous environment is given by

$$p_D = \prod_{m=1}^{M} [(1 + k_q v_m)/v_0(1 + \gamma_a)]^{-1} \tag{9.43}$$

Note that for the design environment, $v_m = v_0$, for all values of m, and the last equation reduces to (9.28).

9.6.3 Mismatched Environment Example

One example of a nonhomogeneous environment is a step-function environment. A step-function environment is one where the clutter cross section (reflection power) varies as a step function *versus* range, R. This is an approximate model of reality when there is a sharp boundary between two different types of clutter (such as at a coastline) or a sharp boundary between thermal noise only and the sum of thermal noise plus clutter. For the CACFAR, the false-alarm probability is not the design value if the range window encompasses the step transition. When the test cell is in the high-power region, the false-alarm probability is high, as the adaptive threshold is too low because it is partially formed from the low-power interference region. When the test cell is in the low-power region, the false-alarm probability is too low and as a consequence the probability of target detection is low.

As a qualitative indication of the false-alarm probability attained in a steplike nonhomogeneous environment, consider the environment sketched in Figure 9.5(a). This depicts an interference environment in which power is initially at some constant level (possibly the thermal-noise condition) and then increases linearly, when measured in dB, between R_1 and R_2. The power level is then constant until R_3, where it starts decreasing. The decrease terminates at range R_4, where the power level attains its initial value. Note that the total range over which the normalizer obtains samples (indicated as the *range window*) is less than either the plateau extent or ramp extent. Let us consider the false-alarm probability that will be obtained for the CACFAR normalizer when operating in this environment. First, note that the attained

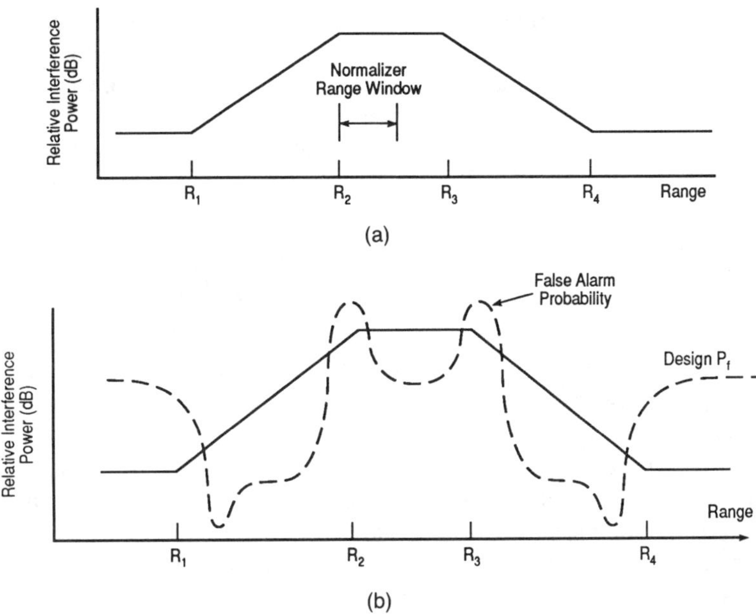

Figure 9.5 Mismatched environmental effect.

false-alarm probability of the circuit equals the design value when the range window is contained completely within a region where the interference power is constant. Thus, at the extreme left of the sketch, the attained false-alarm probability equals the design false-alarm probability. As the range window shifts to the right, a portion of it overlaps the ramp. Note that the average normalization voltage increases as the region of the ramp is entered. Because the normalization voltage is increasing, the false-alarm probability clearly decreases from its design value. Next, consider the case where the range window has moved completely off the ramp and is on the plateau. Again, the attained false-alarm probability equals the design value. Conceptually shifting the range window to the left until its left-hand portion enters the ramp region, we can see that the average normalization voltage decreases as compared to the value attained when completely in the plateau region. This lower value of normalization voltage causes the attained false-alarm probability to be higher than the design value.

Another region of interest is when the normalizer window is completely on the ramp. In this region, the normalization voltage will tend to be high so that the false-alarm probability will be low. To prove that the normalization voltage is high, note that any CFAR technique is essentially estimating the value of the interference power of the test cell. Consider a cell pair symmetrically located about the test cell. Thus, if the test cell were identified as the jth cell, a symmetrically located cell pair of interest would be $j + i$ and $j - i$ for any value of i. Then, in the ramp region, denoting the interference at the center tap as v_j, the interference powers of the other taps would be v_j/b^i and $b^i v_j$. The constant b is related to the slope (in dB) of the interference power *versus* range. The CACFAR normalizer averages these two power values to estimate the value of the test cell's interference power. The difference between the average and actual value is

$$\delta = v_j\left[\frac{1}{2}(b^i + b^{-i}) - 1\right] = \frac{1}{2}v_j(b^{1/2i} - b^{-1/2i})^2 \tag{9.44}$$

We can see that δ is always positive. Thus, the normalization voltage is always high and the attained false alarm probability is always low. The low false-alarm probability occurs whether the interference power variation ramp has positive or negative slope. Thus, the false-alarm probability will have the same value on the up-ramp as on the down-ramp. In addition the values are the same at the corners of the plateau.

A sketch of the expected form of variation of the false-alarm probability as a function of range for the CACFAR normalizer is shown in Figure 9.5(b). We can see that the environment symmetry causes similar symmetry in the range variation of the attained false-alarm probability.

A more quantitative indication of the attained false-alarm probability for a ramp environment when using the CACFAR technique is given by the curves of Figure 9.6. The attained false-alarm probability, as computed by using (9.43), is shown for normalizing with 4, 8, or 16 cells. The design false alarm probability is 1×10^{-6}. As previously indicated, the attained false-alarm probability equals the design value only when the normalization window is completely within a region of constant power level, which occurs at both low and high range-cell numbers. As the test cell increases from low range-cell numbers toward high range-cell numbers, the normalization window encounters a region where the interference power is higher than that of the test cell. This effect tends to increase the normalization voltage and the attained false-alarm probability decreases from the design value. Note that the departure from the design value occurs earlier as the number of normalization cells increases. This occurs because the increased width of the normalization window encounters the discontinuity earlier. We can see that the departure from the design false-alarm probability increases as the number of normalization cells increases. As the normalization voltage will be high in the ramp region, the probability of target detection is low. Thus, although performance increases monotonically with the number of normalization samples in a homogeneous environment, this relationship is not true for all environments.

The difficulty of detecting targets in a ramp environment when using the CACFAR circuit is illustrated by the curves of Figure 9.7. The SINR required for

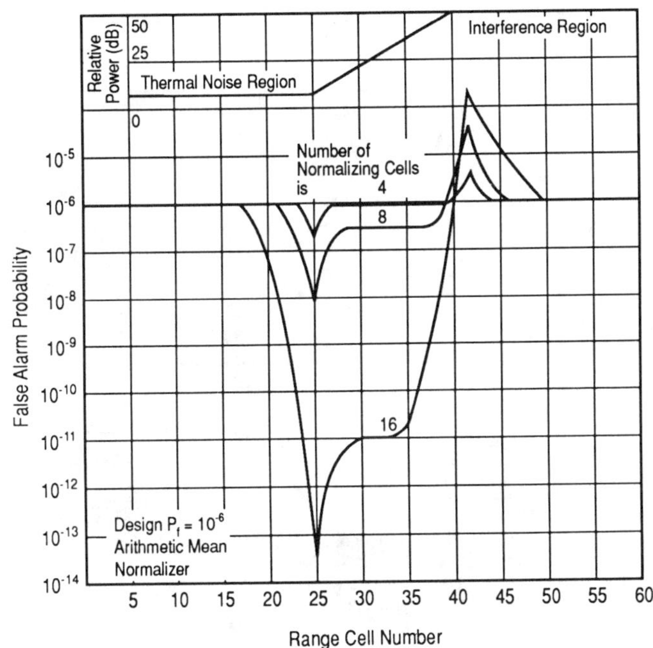

Figure 9.6 False-alarm probability attained for ramp environment.

target detection is shown for four different slopes of a ramp environment. The 0 dB slope is a constant-power (homogeneous) environment. As previously discussed, and as shown by the curve, the required SINR decreases monotonically with the number of normalization cells. For nonzero slopes, the required SINR does not decrease monotonically with the number of normalization cells. Instead, there is an optimum number of cells that minimizes the required SINR. The optimum number decreases as the slope of the ramp increases. As an example, for a slope of 0.5 dB per cell, the optimum number of normalization cells for this environment when using the CACFAR technique is approximately 20. The minimum CFAR loss for this processor thus is 2.4 dB. As another example, for a 2.0 dB slope, the minimum CFAR loss for the CACFAR technique is 7.2 dB.

9.7 RAMP-ENVIRONMENT CFAR TECHNIQUES

A nonhomogeneous environment can be described in a manner similar to that used for the homogeneous environment. For the homogeneous environment, the "nature"

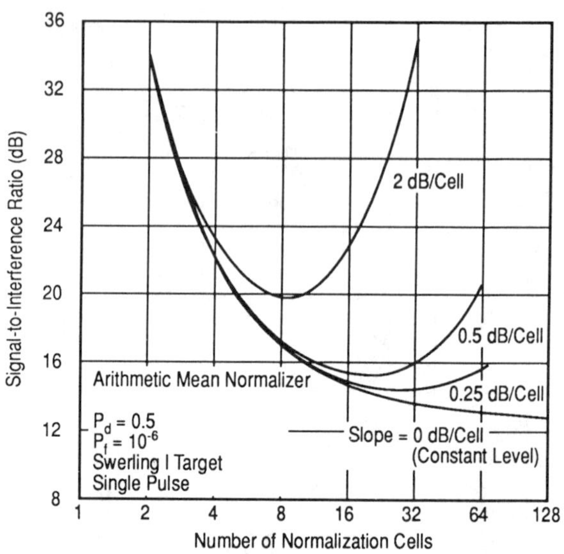

Figure 9.7 SINR required for target detection for ramp environments.

generation model multiplied each member of the set $\{x_m\}$ by the common value $v_0^{1/2}$ to generate the set $\{z_m\}$. For a general environment, each member of $\{x_m\}$ is multiplied by $v_m^{1/2}$, and these values are all different.

A convenient approach is to reindex the cells such that zero corresponds to the test cell and the M reference cells are indexed from $-M/2$ to $M/2$. For convenience, M is assumed to be even. Then, for a ramp environment, we have

$$v_m = v_0 \beta^m \tag{9.45}$$

Note that the power of v_m, when expressed in dB, is a linear function of the index m. The CFAR properties of the CACFAR circuit for the ramp environment can be investigated by combining (9.17) and (9.45). We can see that the test cell's power, v_0, is common to the numerator and denominator of (9.17) and thus cancels. However, when the slope parameter, β, is not equal to unity (the 0 dB slope condition), the denominator depends on the value of β, but the numerator does not. Thus, in general, the attained false-alarm probability depends on the ramp slope.

There are alternative normalizer types of CFAR circuits that can attain constant false-alarm probability in a ramp environment of arbitrary slope. In addition, such processors attain a false-alarm probability that is independent of the test cell's power level in a homogeneous environment. One of these processors is denoted as an *arithmetic-geometric mean normalizer*. It multiplies the voltages from cells that symmetrically bracket the test cell, and then forms the normalization voltage as the sum of the products. Thus, if the test cell is denoted as cell 0, this normalizer computes the product of the voltages from cells 1 and -1 and averages this product with the product of the voltages from cells 2 and -2 and so on.

Another normalizer, the geometric mean normalizer, forms the normalization voltage as the geometric mean of all of the voltages of the cells within the normalization window. Specifically, the CACFAR normalizer and these two additional normalizers form decision random variables $d1$, $d2$, and $d3$ defined as

$$\begin{aligned} d1 &= q_0 \Big/ \sum_{m=1}^{M/2} (q_m + q_{-m}) \\ d2 &= q_0 \Big/ \sum_{m=1}^{M/2} (q_m q_{-m})^{1/2} \\ d3 &= q_0 \Big/ \left[\prod_{m=1}^{M/2} q_m q_{-m} \right]^{1/M} \end{aligned} \tag{9.46}$$

where the first normalizer of (9.46) is identical to (9.17) with slightly modified notation.

The equivalent of (9.45) and the concept of "nature" forming random variables is

$$q_m = |x_m|^2 v_0 \beta^m \tag{9.47}$$

Combining (9.47) and (9.46) shows that the values of each of the decision random variables is independent of the value of v_0. However, the value of $d1$ depends on the value of β, whereas the values of $d2$ and $d3$ are both independent of the value of β. Thus, both processors of $d2$ and $d3$ attain a constant false-alarm probability in ramp environments of arbitrary level and slope.

The normalizer implemented as $d3$ is of greater interest than $d2$ because it can be implemented relatively easily by using a logarithmic envelope detector. To see this, note that any monotonic function of $d3$ is statistically equivalent to $d3$. Thus, if a target-present or target-absent is made by $d3$, the same decision is made using a monotonic function of $d3$ (when the threshold multipliers are properly related by the monotonic relationship). Thus, consider the decision random variable $d4$:

$$d4 = \log(d3) = \log(q_0) - (1/M) \sum_{m=1}^{M/2} [\log(q_m) + \log(q_{-m})] \tag{9.48}$$

We can see that this can be implemented with the circuitry of Figure 9.2 by modifying the detector from square law to logarithmic.

9.7.1 Performance Analysis of Geometric Mean Normalizer

The normalization voltage for the geometric-mean normalizer is given by

$$v_N = \prod_{m=1}^{M/2} (q_m q_{-m})^{1/M} \tag{9.49}$$

The probability density function for the normalization voltage is conveniently expressed in a Laquerre series. Thus,

$$p_{vN}(v) = \sum_{j=0}^{\infty} a_j \exp(-v) L_j(v) \tag{9.50}$$

where the Laquerre polynomial is defined by

$$L_j(v) = (1/j!) \exp(v)\, \mathrm{d}^j/\mathrm{d}v^j [v^j \exp(-v)] \tag{9.51}$$

and the orthogonality relation for the set of polynomials is

$$\begin{aligned} \int_0^{\infty} \exp(-v) L_j(v) L_k(v) &= 1; \quad j = k \\ &= 0; \quad j \neq k \end{aligned} \tag{9.52}$$

The expansion coefficients are given by

$$a_j = \int_0^{\infty} L_j(v) p_{vN}(v)\, \mathrm{d}v \tag{9.53}$$

Substitution of (9.50) into (9.38) gives

$$p_D = \sum_{j=0}^{\infty} a_j \int_0^{\infty} L_j(v) \exp[-v(1 + x)]\, \mathrm{d}v$$

$$x = T/[v_0(1 + \gamma_a)]$$

The integral can be evaluated by (Gradshteyn and Ryzhik [1]):

$$p_D = \sum_{j=0}^{\infty} a_j x^j / [(1 + x)^{1+j}] \tag{9.54}$$

The coefficients, a_j, can be evaluated by using the Laquerre polynomial definition given by Abramowitz and Stegun [2]. Thus,

$$L_j(v) = \sum_{m=0}^{j} (-1)^m {}_jC_{j-m} v^m / m!$$
$$_jC_{j-m} = j!/[(j-m)!m!] \qquad (9.55)$$

where $_jC_{j-m}$ is the binomial coefficient. Substituting (9.55) into (9.53) indicates that

$$a_j = \sum_{m=0}^{j} (-1)^m {}_jC_{j-m} E(v_N^m)/m! \qquad (9.56)$$

where $E(\cdot)$ denotes the statistical expectation. Because v_N, as given by (9.49), is the product of independent random variable:

$$E(v_N^m) = \prod_{m=1}^{M/2} E(q_i^{m/M}) E(q_{-i}^{m/M})$$

Then, using (9.14), for arbitrary exponent value k, we have

$$E(q_i^k) = \int_0^\infty q_i^k \exp(-q_i/v_i)/v_i \, dq_i$$

and this evaluates to

$$E(q_i^k) = v_i^k \Gamma(1 + k)$$

so that

$$E(v_N^m) = \prod_{i=1}^{M/2} (v_i v_{-i})^{m/M} \Gamma^2(1 + m/M) \qquad (9.57)$$

Then, the probability of target detection is obtained by combining (9.54), (9.56), and (9.57).

The resulting equation is used to compute the detection probability for the geometric-mean normalizer for the ramp environments in Figure 9.7. The results for

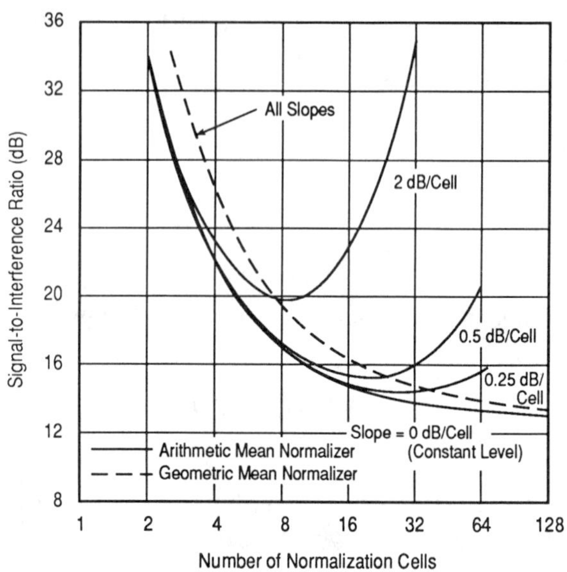

Figure 9.8 SINR required for ramp-environment target detection for arithmetic and geometric mean techniques.

the arithmetic-mean and geometric-mean normalizers are shown in Figure 9.8. Note that the CFAR loss of the geometric-mean normalizer exceeds that of the arithmetic normalizer when the environment is homogeneous. This is expected, as the CACFAR is optimum for an homogeneous environment. Also, the increased loss is indicative of a general property of CFAR circuitry. Thus, when a circuit is derived such that its false-alarm probability tends to remain constant over more general classes of environment, the circuit's ability to detect targets also becomes more difficult and the CFAR loss increases.

9.8 STEP FUNCTION ENVIRONMENT CFAR TECHNIQUES

The split-window *greatest-of-CFAR* (GOCFAR), shown in Figure 9.9, is a circuit modification that is often used. The circuit modification is to split the range window into one-half leading the test cell and one-half lagging the test cell. The voltages of the reference cells of each portion are summed, and the larger sum is multiplied by a gain constant to form the adaptive threshold. This circuit will decrease the high false-alarm rate that occurs for the step environment, but does not alleviate the low false-alarm probability and low detection probability condition. The modification

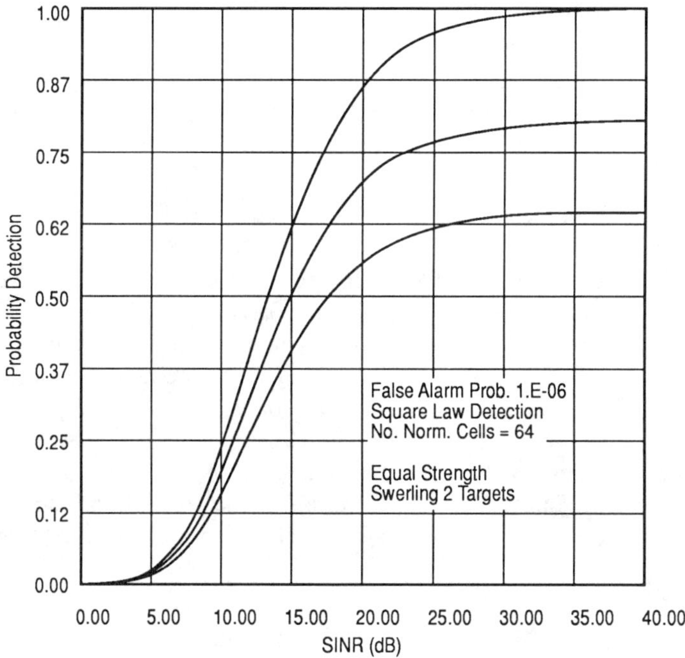

Figure 9.9 CACFAR target detection suppression by additional targets.

slightly increases the CFAR loss in homogeneous environment relative to that incurred by the CACFAR.

9.8.1 GOCFAR Performance Analysis for Homogeneous Environments

Denote the prenormalization voltages obtained by summing the leading and lagging reference cells as y_1 and y_2, respectively. For homogeneous environment, y_1 and y_2 are independent, identically distributed, random variables with probability density and cumulative probability functions, $p_y(\cdot)$ and P_y, respectively. The cumulative probability function of their maximum, the normalization voltage v_N, is the probability that both y_1 and y_2 are less than v_N. Thus,

$$P_{vN}(v_N) = P_y^2(v_N) \tag{9.58}$$

The density function, obtained by differentiating the probability function is

$$p_{vN} = 2p_y(v_N)P_y(v_N) \tag{9.59}$$

The density function of y_1 (or y_2) can be obtained by noting that y_1 is the sum of $M/2$ independent, identically distributed, random variables, each with density function given by (9.14). The density function of (9.14) is termed *exponential,* and is a member of a more general density function family denoted gamma.

The density function of the gamma-distributed random variable, t, is

$$p_t(t) = \alpha^b \exp(-\alpha t)t^{b-1}/\Gamma(b); \quad \alpha > 0, b > 0 \tag{9.60}$$

where α and b are parameters of the density. The density function given by (9.14) is the special case of (9.60) for b equals unity and α equals the reciprocal of v_z.

A pertinent theorem for independent gamma-distributed random variables is that when $t_1, t_2, \ldots, t_N$ are gamma-distributed with a common parameter value, α, and second parameter values, $b_1, b_2, \ldots, b_N$, the random variable, t, defined as the sum of t_1 through t_N, has a gamma density distribution with parameter values of α and the sum of b_1 through b_N. Thus, the density function for y_1 is given by (9.60) with α equal to the reciprocal of v_z and $b = M/2$. The probability function is obtained by integrating (9.60). When $M/2$ is an integer, the probability function is

$$P(y_1) = 1 - \exp(-y_1/v_0) \sum_{n=0}^{M/2-1} (y_1/v_0)^n/n! \tag{9.61}$$

so that

$$p_{vN}(v_N) = \left\{2(v_N/v_0)^{M/2-1} \exp(-v_N/v_0) - 2\exp(-2v_N/v_0) \sum_{n=0}^{M/2-1} (v_N/v_0)^{n+M/2-1}/n!\right\} \Big/ [v_0\Gamma(M/2)] \tag{9.62}$$

The above equation can be combined with (9.38), yielding

$$p_D = 2\left\{[1 + k_g/(1+\gamma_a)]^{-M/2} - \sum_{n=0}^{M/2-1} [2 + k_g/(1+\gamma_a)]^{-(n+M/2)}\Gamma(n+M/2)/[n!\,\Gamma(M/2)]\right\} \tag{9.63}$$

where k_g denotes the adaptive-threshold multiplier for the split-window greatest-of-CFAR. The value of k_g for any specified false-alarm probability can be found by iteratively evaluating (9.63) for $\gamma_a = 0$.

Because, as discussed earlier, the CACFAR is the minimum loss CFAR technique for a homogeneous environment, even without evaluating (9.63), the CFAR loss of the GOCFAR is known to exceed that of the CACFAR. Detailed evaluation of the performance equations indicates that the additional loss is a function of the false-alarm probability and the total number of reference cells. Some typical additional loss values, relative to the CACFAR technique, are given in Table 9.1 for a detection probability of 0.9. We can see from Table 9.1 that the additional loss is on the order of a few tenths of a dB and decreases as the number of reference cells increases.

Table 9.1
Additional CFAR Loss: GOCFAR Relative to CACFAR

	Number of Reference Cells			
False-Alarm Probability	*10*	*20*	*40*	*100*
1×10^{-8}	0.30	0.24	0.16	0.09
1×10^{-6}	0.28	0.21	0.14	0.07
1×10^{-4}	0.24	0.16	0.10	0.05

9.9 MULTIPLE-TARGET CFAR TECHNIQUES

All of the previously discussed techniques explicitly assume that the reference cells do not contain a target. There are many environments where this assumption is erroneous. One example is where some type of cargo is separated from an aircraft in flight. When the large radar cross section aircraft is in the reference cell's window, its presence will substantially increase the value of the adaptive threshold, and detecting the presence of the small radar cross section cargo becomes almost impossible. To quantify the problem, the following section analyzes the multiple-target suppression effect on the CACFAR technique. The succeeding sections discuss and analyze methods of alleviating the problem.

9.9.1 Multiple-Target Effect on CACFAR

The effect of additional targets in the reference cells can be obtained by characteristic functions, as in (9.40). Equation (9.42) derived the characteristic function for a reference cell in the absence of target. For an additional target of average SINR equal to γ_T, replacing the target-absent density function used in (9.42) by the target-present density function of (9.27) gives the characteristic function as

$$\begin{aligned}\phi_{qm}(u) &= \int_0^\infty \{\exp(juq)\exp[-q/v_m(1+\gamma_T)]/v_m(1+\gamma_T)\}\,dq \\ &= [1 - juv_m(1+\gamma_T)]^{-1}\end{aligned} \quad (9.64)$$

Thus, the equivalent of (9.43) for a homogeneous environment and K additional, statistically independent targets is

$$p_D = [1 + k_q/(1+\gamma_a)]^{-(M-K)}[1 + k_q(1+\gamma_T)/(1+\gamma_a)]^{-K} \quad (9.65)$$

The probability of detection is shown in Figure 9.9 for a 64-reference-cell CACFAR. The threshold multiplier is chosen for a false-alarm probability of 1×10^{-6} in a homogeneous environment. The upper curve is the probability of detecting a target in the test cell when the normalization cells contain only homogeneous interference. The middle curve is for the case when one of the reference cells also contains a target. The lowest curve is when two of the reference cells contain targets. The curves show that, when there are additional targets in the reference cells, the probability of target detection asymptotically approaches a value of less than unity as the SNR increases. Thus, the additional CFAR loss due to additional targets depends on the specified detection probability. For a detection probability close to unity, the loss is essentially infinite.

There are several alternative CFAR techniques that substantially reduce the multiple-target additional CFAR loss. One technique is to use the geometric mean of the reference cells, rather than the arithmetic mean used in the CACFAR technique. Another is the censoring CACFAR with or without the split-window greatest-of-CFAR modification. The censoring technique deletes the largest cell voltage (or two largest, *et cetera*) from the window and sums the remaining cells to form the adaptive threshold. For the split-window version, the technique deletes the largest (or two largest, *et cetera*) from each half-window, sums the remaining cells of each half-window, and uses the larger sum as the normalization voltage. One penalty of using the censoring technique is that it slightly increases the CFAR loss in a homogeneous environment. Also, the technique will fail if the reference window contains more additional targets than the design assumes.

A third technique that has relatively small CFAR loss in a homogeneous environment, but can cope with many additional targets in the reference window, is based on using the median of the reference cells to form the adaptive threshold. This technique works well because, although the mean is significantly increased when a few of the reference cells contain additional targets, the median is essentially unaffected. Note that this technique can also be combined with the split-window technique to cope with step-function environments. More generally, instead of using the median, the adapted threshold can be chosen as a multiple of an arbitrarily ordered statistic of the reference cells. As an example, the order statistic can be chosen as the 75% rank or the 50% rank (the median), *et cetera*. The CFAR loss in a homogeneous environment depends on the order statistic used to form the adaptive threshold. We show in a following section that the loss is minimized when the chosen order statistic is in the range of the 75th to 80th percentile.

To aid in comparing the various techniques, the loss for several CFAR processors for 32 reference cells in a homogeneous environment and a false-alarm probability 1×10^{-6} is computed in Table 9.2. The losses are approximately independent of the detection probability, and the analyses are shown in the next section.

Table 9.2
CFAR Loss (dB) in a Homogeneous Environment

CACFAR	0.97
GOCFAR	1.13
75% Rank	1.45
GO 75% Rank	1.66
CACFAR; one-cell censoring	1.01
CACFAR; two-cell censoring	1.06

9.9.2 Censored CACFAR Performance Analysis in Homogeneous Environments*

The analytic procedure used to derive the performance of the censored CACFAR technique is based on the characteristic function of the normalization voltage given in (9.40). Thus, we need to derive the characteristic function of the censored normalization voltage. This is done by first deriving the probability density function for the order statistic. This is a convenient method, as the order statistic's density function is also used in the next section to derive the performance of the order statistic CFAR.

For homogeneous environments, the set of M detected random variables of the reference window cells, $\{q_n\}$, are independent and identically distributed with density function denoted as $p_q(\cdot)$ and probability function denoted by $P_q(\cdot)$. The initial results are derived in terms of generalized probability density functions. When these results

*This section closely follows the paper of J.T. Rickard and G.M. Dillard, "Adaptive Detection Algorithms for Multiple-Target Situations," *IEEE Trans. on Aerospace and Electronic Systems,* Vol. AES-13, No. 4, July 1977, pp. 338–343.

are obtained, the specific exponential density function for square-law detection, as given by (9.14), is inserted.

The probability that the value of a specific cell is less than a specified value, denoted as α, equals $P_q(\alpha)$. The probability that it is greater is $1 - P_q(\alpha)$. The probability that exactly r of the M variates is less than α, and the other $N - r$ are larger than α, denoted as $P_r(\alpha)$, is

$$P_r(\alpha) = {}_rC_M P_q(\alpha)^r[1 - P_q(\alpha)]^{M-r}$$

where ${}_rC_M$ denotes the binomial coefficient. The kth-order statistic is less than α if the number of observations less than α is greater than or equal to k. This probability, $G_k(\alpha)$, is

$$G_k(\alpha) = \sum_{r=k}^{M} {}_rC_M P_q(\alpha)^r[1 - P_q(\alpha)]^{M-r}$$

The derivative of this equals the probability density function of the kth-order statistic. Denoting the density function as $g_k(\cdot)$:

$$g_k(\alpha) = Dp_q(\alpha)P_q(\alpha)^{k-1}[1 - P_q(\alpha)]^{M-k}$$
$$D = M!/[(k - 1)!(M - k)!] \tag{9.66}$$

Let us denote the set of random variables after censoring as $\{c_n\}$. For the homogeneous environment, and censoring of a single value, a particular c_k is equally likely to have any rank from 1 to $M - 1$. Its density function is obtained by averaging (9.66) over the ranks of 1 through $M - 1$ . . . Thus, denoting the density function of the censored random variable as $h_c(\cdot)$:

$$h_c(y) = \sum_{k=1}^{M-1} g_k(y)/(M - 1)$$

Also, because

$$p_q(y) = M^{-1} \sum_{k=1}^{M} g_k(y) = [(M - 1)/M]h_c(y) + M^{-1}g_M(y)$$

From (9.66),

$$g_M(y) = Mp_q(\alpha)P_q(\alpha)^{M-1}$$

so that

$$p_q(y) = [(M - 1)/M]h_c(y) + p_q(y)[P_q(y)]^{M-1} \tag{9.67}$$

Solving (9.67) for $h_c(y)$ gives

$$h_c(y) = [M/(M - 1)]p_q(y)[1 - P_q(y)]^{M-1} \tag{9.68}$$

For the square-law magnitude detectors, the probability density function of q is exponential, as given by (9.14). Without loss of generality, assuming v_z of (9.14) equals unity gives

$$h_c(y) = [M/(M - 1)] \exp(-y)\{1 - [1 - \exp(-y)]^{M-1}\} \tag{9.69}$$

Expanding this last expression in a binomial series and integrating to evaluate the characteristic function gives

$$\Phi(u) = [M/(M - 1)] \sum_{k=1}^{M-1} {}_kC_{M-1}(-1)^{k+1}(1 + k - ju)^{-1}$$

By using (9.40), the probability of detection is the value of the characteristic function at the argument $ju = -k_q/(1 + \alpha_a)$.

The result can be generalized by similar analysis. We can show that when the K largest cells are censored and the remaining $M - K$ cells are summed to form the normalization voltage, the probability of detection in an homogeneous environment is

$$p_D = \{1 + [(M + 1 - K - k)/(M + 1 - k)][k_q/(1 + \gamma_a)]\}^{-1} \tag{9.70}$$

9.9.3 Performance Evaluation of the Order Statistic CFAR in Homogeneous Environments

The density function of the order statistic is given in general form by (9.66). The specific density function for this case is obtained by combining (9.14) and (9.66). The resulting equation is of the same form as that previously derived for (9.69). Again, a binomial expansion can be used to derive the characteristic function. The result for the probability of detection is

$$p_D = \prod_{j=0}^{k-1} (M - j)/[M - j + k_q/(1 + \alpha_a)] \tag{9.71}$$

9.10 CLUTTER-MAP CFAR

Ground clutter is extremely nonhomogeneous and is often modeled by a random cross section *versus* range. The prior CFAR techniques are not applicable, as they explicitly assume a functional variation over range. However, the radar cross section of each cell of the ground clutter tends to change very slowly with time. Thus, the estimate of the cell's interference power can be done by scan-to-scan averaging. Note that, although this technique is quite applicable to ground clutter, it is less applicable to weather, sea, or chaff clutter. This is because, for the latter clutter types, the radar cross section of each resolution cell changes with time due to the wind-induced motion of the basic scatterers.

The previous CFAR techniques are implemented by using a moving-window integrator. Specifically, the power estimate for the kth cell, $p_{\text{est}}(k)$ is derived as

$$p_{\text{est}}(k) = \sum_{m=1}^{M/2} [q(k + m + d) + q(k - m - d)]/M$$

where $q(\cdot)$ denotes the detector output and d denotes the number of cells on each side of the test cell that are omitted from the estimate. Cells are often omitted because of the potential problem of strong sidelobes of pulse compression waveforms. The estimate for the $k + 1$th cell is updated as

$$p_{\text{est}}(k + 1) = p_{\text{est}}(k) + [q(k + 1 + M/2 + d) - q(k + 1 - M/2 - d)]/M$$

Note that this technique requires long-term storage of the voltages of the M cells used to compute the estimate. Similarly, the clutter-map CFAR technique may use a moving-window estimator over several prior scans for each resolution cell. As there can be many resolution cells, the total memory required for storing the data from M scans will be excessive. To reduce this storage requirement, the CFAR map technique often uses a background estimate derived by exponential smoothing of each resolution cell. Thus, for the nth scan and the kth resolution cell, the background estimate is formed from previous scans and the current measurement, $q_n(k)$, as

$$p_{\text{est}_n}(k) = (1 - w)p_{\text{est}_{n-1}}(k) + wq_n(k) \tag{9.72}$$

The performance of the technique can be determined by an iterative analysis. Thus, substituting (9.72) into itself gives

$$p_{\text{est}_n}(k) = wq_n(k) + w(1 - w)q_{n-1}(k) + (1 - w)^2 p_{\text{est}_{n-2}}(k)$$

This then generalizes to

$$p_{\text{est}_n}(k) = w \prod_{m=0}^{\infty} (1 - w)^m q_{n-m}k \tag{9.73}$$

To avoid target self-cancellation, not including the detector output in the background estimate is preferable. Thus, a target-present or -absent decision is made by using $p_{\text{est}_{n-1}}(k)$. If a target-absent decision is made, $q_n(k)$ is used to update the background estimate. If a target-present decision is made, the power estimate is not updated.

The result given by (9.73) is correct regardless of any correlation of the scan-to-scan cell voltages. The remainder of the analysis is based upon the assumption that successive scan outputs from the resolution cell are statistically independent. This occurs when the radar carrier frequency is changed from scan to scan (without duplication) with the spacing between carrier frequencies at least equal to the radar's bandwidth. For constant carrier frequency transmission, an alternate cause of statistical independence is that wind changes the relative positions of the cell's scatterers. Independence occurs when the time period between scans is long.

For the independence assumption $q_n(k)$ and $p_{\text{est}_{n-1}}(k)$ are statistically independent. A target present decision is made when $q_n(k)$ exceeds the threshold multiplier, k_q, times the power estimate, $p_{\text{est}_{n-1}}(k)$. This probability, conditional on the power estimate, equals $\exp[-(k_q p_{\text{est}})/(1 + \gamma_a)]$ and the dependence upon k and n is implicit. Since p_{est} is random, the probability of declaring a target is obtained by integration with respect to the density function of the estimated cell's power. This integral can be recognized as equal to the characteristic function of the estimate. Thus, the probability of detection is computed using the characteristic function technique of (9.40). Since $q_n(k)$ is statistically independent of prior values, the characteristic function of $p_{\text{est}}(k)$ is obtained via the product of the characteristic functions of prior, weighted, values of q so that

$$p_D = \prod_{r=0}^{\infty} [1 + k_q w(1 - w)^r/(1 + \gamma_a)]^{-1} \tag{9.74}$$

For a specified false-alarm probability, the required value of the threshold multiplier, k_q can be determined by iteratively evaluating (9.74) with $\gamma_a = 0$. The probability of detection is then determined by evaluating (9.74) for each specific value of γ_a with the now known value of k_q. Curves of the probability of detection *versus* SINR for several values of the weight, w, are shown in Figure 9.10 for a false-alarm probability of 1×10^{-6}. The weight value equal to zero is equivalent to the case where the total interference power is known exactly *a priori* so that CFAR is not required. The difference between the zero-weight curve and the other curves is the CFAR loss. We can see that the loss decreases monotonically as w decreases. This occurs because decreasing w corresponds to using longer data windows to estimate the resoluton cell power. However, the curves are computed for the tacit assumption that the interference power is constant over time. In a real environment, this tacit assumption is false, and the value of w cannot be too small, as it must be chosen commensurate with the rate of change of the environmental power.

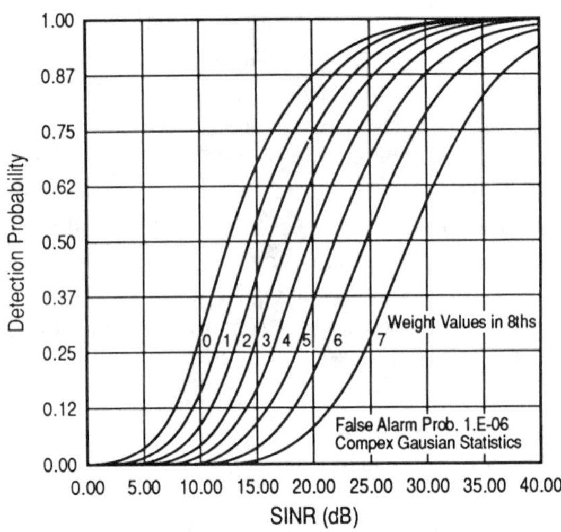

Figure 9.10 CFAR map detection probability curves.
Source: From R. Nitzberg, *Clutter Map CFAR Analysis*. Trans. IEEE, AES-22, No. 4, July 1986.

The CFAR loss is graphed in Figure 9.11 for a detection probability equal to 0.9. Figure 9.11(a) shows the CFAR loss as the weight value ranges over zero to unity. Figure 9.11(b) expands the scale with the range of weight values from 0 to 1/8. For the latter figure, the loss is plotted for false-alarm probabilities of 1×10^{-6} and 1×10^{-8}. We can see that the loss increases as the false-alarm probability decreases. This is the same direction of change as is obtained for the moving-window CFAR techniques.

9.11 SEA AND WEATHER CLUTTER CFAR TECHNIQUES

Some types of clutter, such as sea, weather, or chaff, are neither homogeneous in range nor of constant cross section in a resolution cell from scan to scan. Therefore, none of the previously discussed CFAR techniques are appropriate. The synthesis of an appropriate CFAR technique must be based on the clutter's nonhomogeneous characteristics.

Two different statistical models have been developed to describe these heterogeneous clutter types. Both of these models were developed to be consistent with sets of measured data. Two classes of CFAR techniques, each appropriate for one of the clutter models, have been developed and are discussed below.

The previously assumed clutter statistics are based on the decomposition indicated by (9.7). The statistical description of x and y is that each is a zero-mean Gaussian random variable. This statistical description is consistent with the physical

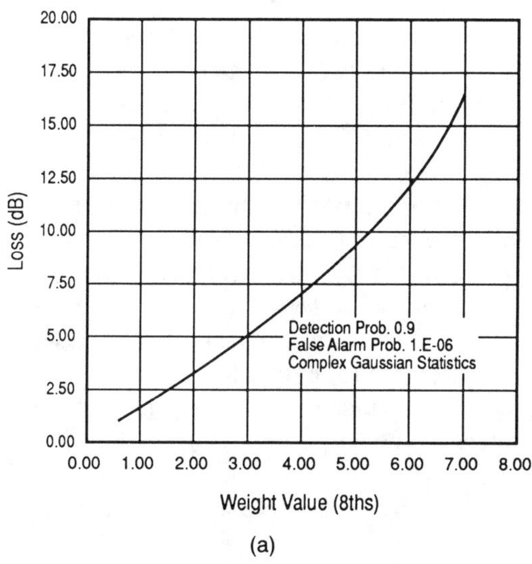

(a)

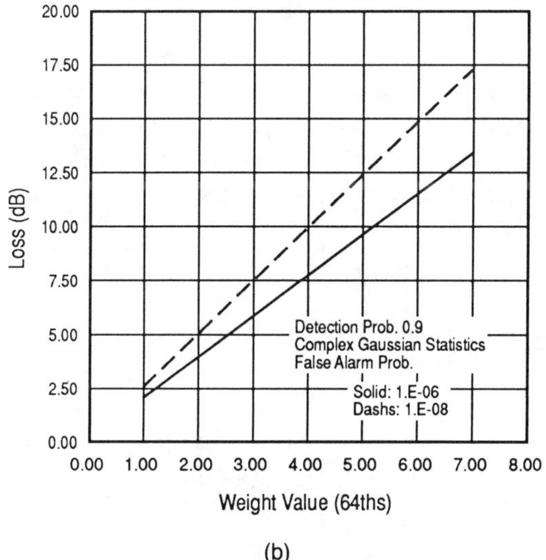

(b)

Figure 9.11 CFAR map-loss, unexpanded and expanded scale.
Source: From R. Nitzberg, *Clutter Map CFAR Analysis*. Trans. IEEE, AES-22, No. 4, July 1986.

model that the clutter return is due to the superposition of the returns of many small reflectors, none of which contributes a significant portion of the total return. For this model, the probability density function of the envelope is Rayleigh. However, many measurements indicate that the tails of the measured probability density function of the envelope are much higher than is consistent with the Rayleigh assumption. Thus, the physical model of many small reflectors is not always accurate. When the CAC-FAR or similar techniques are used for these types of clutter, the large tails cause much higher false-alarm probability than the design value. The measurements indicate that the envelope probability density is more like ones called *log-normal* or *Weibull*. For the log-normal density, the logarithm of the envelope has a Gaussian (normal) density. Both the mean and variance are nonzero and differ for different clutter regions. To attain a constant false-alarm probability when the mean and variance are not constant from region to region, the performance of the CFAR technique developed for this environmental model must be independent of these two parameters.

For Weibull clutter, the density function of the envelope voltage, r, is

$$p(r) = (n/\mu)(r/\mu)^{n-1} \exp[-(r/\mu)^n] \tag{9.75}$$

with parameters n and μ. Methods of implementing processors that have CFAR characteristics for these statistical models are discussed in the next section.

9.11.1 Log-Normal Clutter CFAR

As previously stated, after a logarithmic-law envelope detector, the probability density function of the clutter is log-normal of unknown mean and variance. Denote the detector output for the mth cell as t_m. The unknown mean and variance is equivalent to "nature" forming the mth cell random variable t_m from the zero-mean, unit-variance, Gaussian distributed variable x_m by the generation equation:

$$t_m = \alpha x_m + \beta \tag{9.76}$$

where α and β are arbitrary scalars. Note that the mean of t_m is β and the variance equals α^2. One possible CFAR processor is to form the decision random variable, d, defined as

$$d = (t_0 - \bar{t})^2/(t_m - \bar{t})^2$$
$$\bar{t} = (t_m)/M$$

where the test cell output is denoted as t_0 and there are M reference cells. The proof that this technique has a constant false-alarm probability for any log-normal density function is obtained by substituting the nature-generated random variable definition of (9.76) into the last equation. We can see that the decision random variable is independent of the values of the parameters α and β, which proves the desired CFAR property. A target-present or -absent decision is made by comparing the value of the decision random variable to a constant. This constant is termed the *threshold multiplier*. The density function of the decision random variable for the target-absent condition is an *f*-distribution. The value of the constant can be determined from the known properties of the *f*-distribution. As for the previous techniques, the value of the threshold multiplier depends on the desired false-alarm probability. We can show that the CFAR loss for this type of clutter and processor is very large.

We can also show that this technique has a CFAR property for Weibull clutter. However, the adaptive threshold multiplier is a different value for the two classes of clutter.

9.11.2 Composite Clutter Model

Another heterogeneous clutter model developed to be consistent with measurements is the *composite clutter* model. It retains the assumption of Gaussian statistics of x and y in (9.7), but models the resolution cell's clutter power as a random variable. Specifically, for homogeneous clutter, (9.7) can be rewritten for the nth resolution cell as

$$z_n(t) = p_c(n)^{1/2}[x_n(t) + jy_n(t)] \tag{9.77}$$

where $x_n(t)$ and $y_n(t)$ are zero-mean, unit-variance, Gaussian random processes. The time dependence indicates the change in received voltage for the nth resolution cell from scan to scan. The time dependence is not range. Range dependence is indicated by the subscript n. The multiplier is the square root of the reflected clutter power for the nth cell. Thus, in general, the clutter power is range-dependent; for a homogeneous environment, it is range-independent. A probability density function consistent with measured data is such that the reflected clutter power is a gamma-distributed random variable. Thus, the probability density of the clutter power for the nth cell, $p_n(p_c)$ is

$$p_n(p_c) = [kp_c(n)/p_a]^{k-1} \exp[-kp_c(n)/p_a][k/p_a\Gamma(k)]$$

where p_a is the average reflected power for the clutter region and k is a shaping parameter. Specifically, the mean and variance of the random clutter power, where the dependence on n is implicit, are

$$\begin{aligned} E(p_c) &= p_a \\ v(p_c) &= [p_a]^2/k \end{aligned} \tag{9.78}$$

where v is the variance. The density shape is controlled by k in the sense that, as k approaches infinity, the variance approaches zero. As k approaches infinity, the heterogeneous clutter environment's probability density function approaches an impulse located at p_a. Thus, k tending toward infinity corresponds to a homogeneous environment.

Note that these two models differ considerably in concept. The first is a statistical description of the envelope of the reflected clutter waveform. The second is a statistical description of the predetector waveform. However, both of these models result in a probability density function of the envelope that has large tails. Because of these large tails, the adaptive threshold of a CFAR circuit must be much larger than is necessary for a homogeneous environment. The large threshold thus causes a large CFAR loss.

9.12 MULTIPLE-CARRIER-FREQUENCY CFAR TECHNIQUES

As is well known, the detectability of a fluctuating target in noise or clutter (reverberation) depends on the SINR and the particular form of the target fluctuation characteristic. The fluctuation loss, or increased SINR required for detection as compared to that of a nonfluctuating target, is discussed in standard textbooks. As an example, the fluctuation loss for a Swerling 1 target at a false-alarm probability of 1×10^{-6} is approximately 1.5 dB for a detection probability of 0.5, and 8 dB for a detection probability of 0.9.

The use of a multiple-carrier-frequency waveform to reduce the fluctuation loss is well known, and the theoretical improvement has been experimentally verified. This class of waveform is denoted by some authors as *frequency diversity* and by others as *frequency agility*. A conceptual block diagram of a particular diversity waveform and processor is shown in Figure 9.12. As shown by Figure 9.12(a), each pulse of the pulse train is divided into K subpulses, and each subpulse is at a separate carrier frequency f_1 through f_K. The transmitted spectrum of a pulse is shown by Figure 9.12(b). The spectra are disjoint and like parallel radar transmissions. A processor for the waveform is also shown. The processor consists of parallel channels with bandpass filters, pulse compressors, and envelope detectors for each subpulse. For a thermal-noise environment, the outputs of the test cells of the individual channels are summed before target-present or -absent decisions are made. The modifications to the processor required for a non-thermal-noise environment are discussed below.

A qualitative indication that frequency diversity will decrease the fluctuation loss can be obtained by noting that the total return from a typical target consists of the superposition of the reflections from many individual scattering points. The magnitude of the sum depends on the relative phasing of these reflections. For some

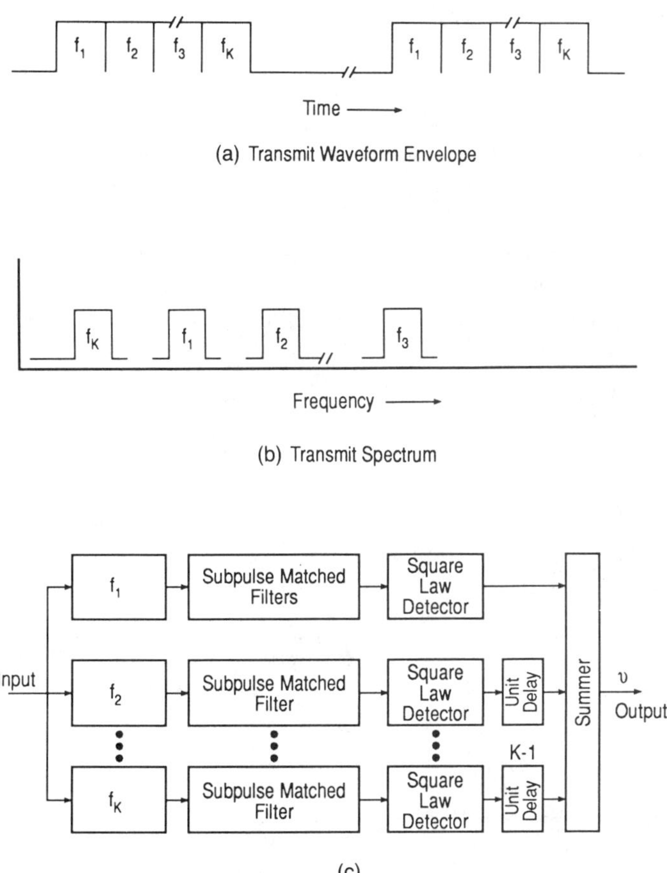

Figure 9.12 Conceptual block diagram of frequency-diversity system.
Source: From R. Nitzberg, *Losses for Frequency Diversity Waveforms Systems*. Trans. IEEE, AES-14, No. 3, May 1978.

phase conditions, the resulting magnitude and the detection probability are small. A change of transmitting frequency changes the relative phasing of the reflections. A small target magnitude occurring on all of several different carrier frequencies is improbable and the fluctuation loss is decreased. In addition to its effectiveness in improving target detection, diversity techniques improve target tracking by reducing the effect of target glint.

The previously discussed CFAR implementations are for a single-frequency-channel waveform and processor. Modification is required for the multiple-frequency waveform. The appropriate modification depends on details of the environmental conditions. Specifically, note that the normalization voltage is an estimate of the test cell's interference power. If the interference is wideband so that the interference power is the same in each channel, an overall normalization voltage ought to be formed by adding the normalization voltages from each channel. The voltages of the test cell of each channel ought to be added and the overall normalizaton voltage is applied to the test cell's sum. However, if the interference power is narrowband so that the interference power is different in each channel, the normalization voltages must not be added. For example, if the external interference is totally contained within a single channel, the normalization voltages for the other channels are to be derived from thermal noise only. The test cell of each channel needs to be normalized separately, and then the normalized outputs are combined. The two configurations are shown in Figure 9.13. As indicated in the following paragraphs, the combination procedure (CFAR algorithm) for a multiple-channel waveform must also be modified.

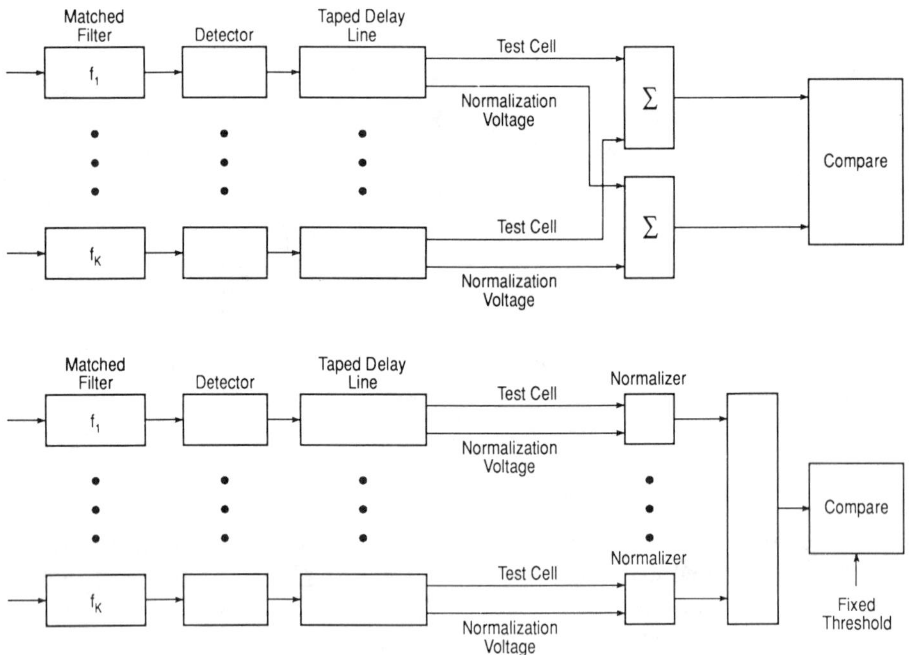

Figure 9.13 Conceptual block diagram of frequency-diversity CFAR techniques.
Source: From R. Nitzberg, *Losses for Frequency Diversity Waveforms Systems*. Trans. IEEE, AES-14, No. 3, May 1978.

Let us consider a diversity channel with normalization voltage derived from only the range cells of that channel. The normalized output, f, is

$$f = q_0/v$$

where v denotes the normalization voltage and q_0 is the test cell voltage. Note that the q_0 notation is consistent with the indicated square-law detectors of the block diagram. As previously discussed, any monotonic function of f can be used for the signal processor target-present or -absent decisional device, and identical target detection performance is obtained. As an example, we have

$$b = (1 + f^{-1})^{-1}$$

which is a monotonic function of f. Combining these last two equations shows that

$$b = q_0/(q_0 + v)$$

Thus, adding the test cell voltage to the normalization voltage gives an alternative processor, which is equivalent in performance. When combining the normalized outputs of several channels, simply summing the f values obtained for each channel is not necessarily best. Summing monotonic functions of the f values for each channel, such as the b values, may be better. In fact, there is no reason to believe that simple summing is sufficient. More complex combining may be superior.

As shown below, optimal combining of the normalized channels generally requires complicated functions of the b variates. The optimum CFAR combiner depends on both the target fluctuation model and the SINR. In the limit, for small SINR values, the optimal processor is to sum the b variates, rather than summing the f variates.

The analysis is first done for a nonfluctuating target and then for a Swerling 2 fluctuating target. For the latter model, the random target fluctuation in each channel is Swerling 1. Thus, the probability density function of the target's cross section is exponential, as given by (9.26). For the Swerling 2 model, the cross section fluctuation is Swerling 1 for each channel, and the channel-to-channel fluctuation is statistically independent.

9.12.1 Nonfluctuating Target Analysis

Let us denote the square-law-detected output of the test cell and the normalization voltage of the jth channel as q_{0j} and v_j, respectively. The normalized output for the jth channel is

$$f_j = q_{0j}/v_j \tag{9.79}$$

For a nonfluctuating target, the probability density function of q_{0j} is noncentral χ^2 with two degrees of freedom. The density function for the target absent case is central χ^2 with two degrees of freedom. Because the normalization voltage is formed from interference alone, for either condition, its density function is χ^2 with $2M$ degrees of freedom. Therefore, the density function of the ratio, f_j, is either the noncentral or central f-distribution with 2 and $2M$ degrees of freedom given by

$$p(f_j,\gamma) = M\exp(-\gamma)\ {}_1F_1[1+M,1,\gamma f_j/(1+f_j)]/(1+f_j)^{1+M}$$

where ${}_1F_1(\cdot)$ denotes the confluent hypergeometric function and γ is the SINR for each channel.

The Neyman-Pearson lemma may be applied to find the optimum method of combining the per-channel normalized voltages. Thus, for K diversity channels, the optimum processor is found as

$$d = \prod_{j=1}^{K} p(f_j,\gamma)/p(f_j,0) \gtrless \mathrm{T}$$

where T is the appropriate threshold value for the desired false-alarm probability. Combining the last two equations gives the optimum processor as

$$d = \prod_{j=1}^{K} {}_1F_1[1+M,\ 1,\ \gamma f_j/(1+f_j)] \gtrless \mathrm{T} \tag{9.80}$$

Note that the optimum processor depends on the value of γ so that there is no single processor that is optimum for all SINR values. Further note that the optimum processor does not directly depend on the variables f_j, but rather on $f_j/(1+f_j)$. In terms of the test cell's voltage and normalizing voltage, we have

$$f_j/(1+f_j) = q_{0j}/(q_{0j}+v_j)$$

Thus, the "natural" normalized variable, when combining diversity channels, includes the test cell voltage in the normalization voltage. The density function of the natural normalized variable is noncentral β. We can obtain the processor that is approximately optimum for small per-channel SINR values by a series expansion of (9.80), resulting in

$$d = \sum_{j=1}^{K} q_{0j}/(q_{0j}+v_j) \gtrless \mathrm{T}' \tag{9.81}$$

Thus, the asymptotic, small-SINR CFAR processor sums the b variates of each channel.

9.12.2 Swerling 2 Target Analysis

The analysis starts from (9.79), as previously done for the nonfluctuating cross section target model. The probability density function of f_j for this target is given by (9.28). Again, applying the Neyman-Pearson lemma gives that the optimum processor is

$$d = \prod_{j=1}^{K} [(1 + f_j)/(1 + \gamma + f_j)]^{M+1} \gtrless \mathrm{T}$$

and this reduces to

$$d = \prod_{j=1}^{K} [1 + \gamma v_j/(q_{0j} + v_j)]^{-(M+1)}$$

Using the identity, $v/(v + q) = 1 - q/(v + q)$, shows that the natural normalized variable for this target model is the same as that found in the nonfluctuating target analysis. We also can show that the small-SINR approximation yields the same processor as previously found.

9.13 HIERARCHAL CFAR TECHNIQUES

One of the drawbacks of using a CFAR processor is that, for a thermal-noise-only environment, an unnecessary CFAR loss is incurred. Although this loss can be reduced by using a large number of reference cells, this is an unsatisfactory solution for the non-thermal-noise environment. Thus, as the number of reference cells increases, the design environment becomes unlikely to match closely the actual environment. An alternative CFAR procedure is to use the adaptive threshold only in a jamming or clutter environment. One method of implementing this concept is to compute a fixed threshold, T, appropriate for the thermal-noise-only environment, and an adaptive threshold, T_a, by a CACFAR circuit or variant. The fixed threshold is used unless the adaptive threshold is much larger. This technique is illustrated by the curve of Figure 9.14.

The technique of using the adaptive threshold only when it is much larger than the fixed threshold is equivalent to using a threshold T′, defined as

$$\mathrm{T}' = \mathrm{T}; \quad \alpha \mathrm{T}_a < \mathrm{T}$$

$$\mathrm{T}' = \mathrm{T}_a; \quad \alpha \mathrm{T}_a > \mathrm{T}$$

where α is a parameter that controls the dichtomy of large and small. Note that when

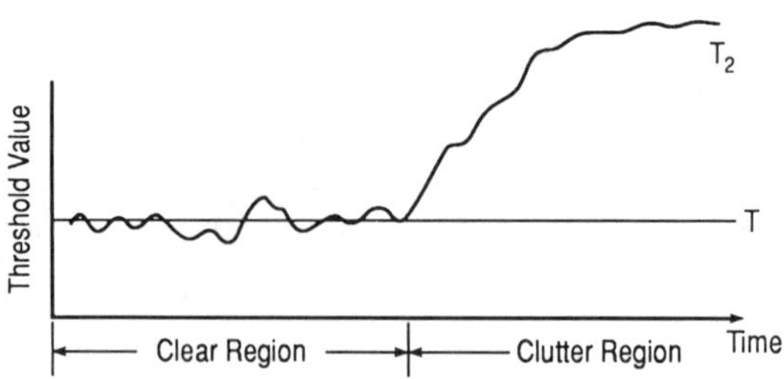

Figure 9.14 Hierarchal CFAR concept.
Source: From R. Nitzberg, *Low-Loss Almost Constant False-Alarm Rate Processors*. Trans. IEEE, AES-15, No. 5, Sept 1979.

$\alpha = 0$, the fixed threshold is always used; when α tends toward infinity, the adaptive threshold is always used. The intermediate values of α control the percentage of time that either threshold is used. Figure 9.15 shows the target detection probability for the thermal-noise environment as a function of SNR with α as a parameter for 8 and 16 statistically independent reference cells. The target fluctuation model is Swerling 1. We can see that the CFAR loss is decreased by decreasing the value of α. Note that, as $\alpha = 0$ corresponds to the fixed threshold case, the CFAR loss can be made as small as desired, even equal to zero. However, the cost of the decreased loss is the nonconstant false-alarm probability in non-thermal-noise environments.

The false-alarm probability, as a function of the environment's interference power level for a homogeneous environment is indicated by the curves of Figure 9.16. The interference level is normalized to the thermal-noise power level. Note that the attained false-alarm probability is less in thermal noise than in clutter. Also, note that the decreased values of α accentuate the change of false-alarm probability between the thermal-noise and clutter environments. Thus, there is a system trade-off betwen small CFAR loss in thermal noise and increased false-alarm probability in clutter.

The previous discussion is in terms of target detection either in an interference environment of known power level (thermal noise) or an additive thermal-noise-plus-clutter environment. The technique is also applicable for a jamming environment of unknown power level or an additive jamming-plus-clutter environment. The required modification is that the fixed threshold, which is appropriate for the thermal-noise environment, must be replaced by an adaptive threshold appropriate for a jamming environment. This can be done by using a very large number of reference cells to develop the adaptive threshold for the jamming environment and a small number of reference cells to develop the adaptive threshold for the combined jamming-plus-clutter environment. The intrinsic assumption is that the jamming environment is statistically stationary so that a very large number of reference cells is satisfactory.

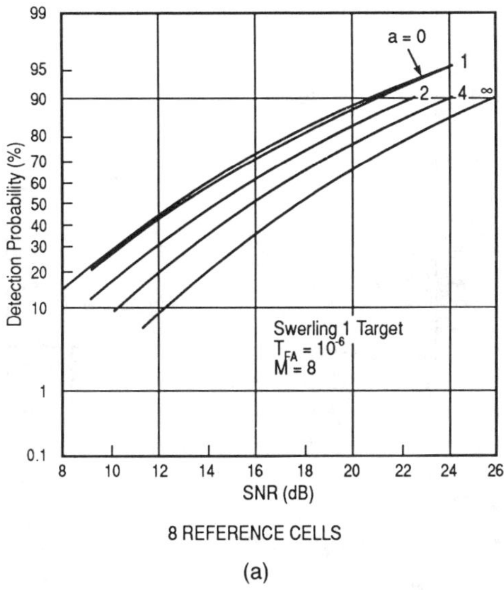

8 REFERENCE CELLS

(a)

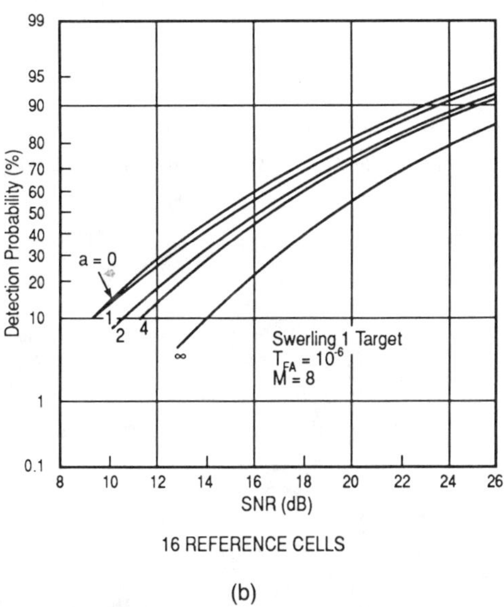

16 REFERENCE CELLS

(b)

Figure 9.15 Detection probability for hierarchal CFAR.
Source: From R. Nitzberg, *Low-Loss Almost Constant False-Alarm Rate Processors*. Trans. IEEE, AES-15, No. 5, Sept 1979.

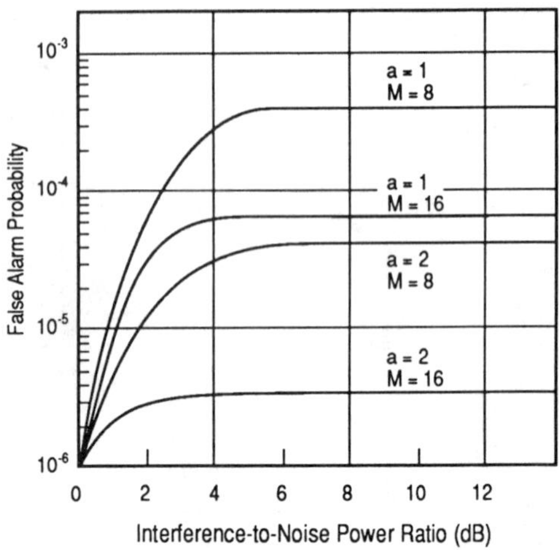

Figure 9.16 Hierarchal CFAR false-alarm probabilities.

9.13.1 Hierarchal CFAR Analysis

A target-present decision is made when the square-law-detected test cell voltage q_0, the thermal-noise power v_n, and the normalization voltage v satisfy the relations:

$$q_0 > kv_n; \quad \alpha v < v_n$$

$$q_0 > kv; \quad \alpha v > v_n$$

where k is a threshold-multiplier constant. Because the conditions of this equation are mutually exclusive, the probability of a target-present decision, p_d, is given by

$$p_d = \text{prob}\{q_0 > k;\ \alpha v < 1\} + \text{prob}\{q_0 > kv;\ \alpha v > 1\}$$

where, without loss of generality, the noise variance has been assumed to equal unity. We can show by a two-dimensional sketch of the regions that the probability of a target-present decision also is

$$p_d = 1 - \text{prob}\{q_0 < k;\ v < 1/\alpha\} - \text{prob}\{q_0 < kv;\ v > 1/\alpha\}$$

Let us define the first of these probabilities as p_1 and the second as p_2. Then, because q_0 and v are statistically independent:

$$p_1 = \text{prob}\{q_0 < k\}\ \text{prob}\{v < 1/\alpha\}$$

For a Swerling 1 target of SINR $= \gamma$ and clutter-plus-noise variance of v this evaluates to

$$p_1 = \{1 - \exp[-k/\text{v}(1 + \gamma)]\}\ P(M, M/\alpha\text{v})$$

where $P(\cdot,\cdot)$ is the incomplete γ function given for this case by

$$P(M, x) = 1 - \exp(-x)[(x^m)/m!]$$

Similarly, denoting the density function of the normalization voltage as $p(v)$, p_2 can be evaluated by

$$p_2 = p(v)\{1 - \exp[-kv/\text{v}(1 + \gamma)]\}$$

Evaluating the integral hence gives the final result:

$$p_d = \exp[-k/\text{v}(1 + \gamma)]P(M, M/\alpha\text{v}) + \{M(1 + \gamma)/[k + M(1 + \gamma)]\}^M[1 - P\{M,(1/\alpha)[k/(1 + \gamma)] + M\}]$$

9.14 IMPLEMENTATION EFFECTS ON CFAR TECHNIQUES

The preceding analyses were based on the assumptions that the outputs of an envelope detector are statistically independent, and the set of outputs are the squared magnitude of the envelope. Often, because of a need to simplify the equipment design or to decrease other losses, both of these assumptions are incorrect. Thus, to decrease signal processing losses, the sampling frequency of the A/D conversion usually exceeds the radar bandwidth so that the voltages of adjacent cells of the normalizing window are not statistically independent. Similarly, the detector output is often equal to the magnitude rather than the squared magnitude. This simplifies implementation because the magnitude is easier to process than the squared magnitude, substantially decreasing the equipment's required dynamic range. We show in the following sections that use of a magnitude detector rather than a squared magnitude detector increases the CFAR loss. An approximation for the CFAR loss increase is that, for M reference cells, the effective number of reference cells equals $0.707M$. Similarly, for M correlated reference cells, the effective number of reference cells is decreased. As an example, when the sampling rate equals twice the bandwidth, the effective number of reference cells equals $0.75M$.

9.14.1 Effect of Correlated Reference Cells

A/D sampling frequencies higher than the radar bandwidth cause adjacent reference cell voltages to be statistically correlated. The effect of the correlation is to increase the errors made in estimating the test cell's power, and therefore to increase the CFAR loss. An approximate representation of the effect of the correlation is that it decreases the effective number of reference cells by a factor depending on the ratio of the radar bandwidth to the A/D sampling frequency. The exact loss depends on the form of the system filter's transfer function. However, the loss is not too sensitive to the exact form of the transfer function. The analysis is done for the special case when the transfer function is a $(\sin x)/x$ function. This transfer function is optimum for a simple pulsed waveform, and also is a reasonable approximation of the transfer function used for other radar waveforms.

When the ratio of A/D sampling rate to the radar bandwidth equals K, the time between samples equals $1/K$. For the assumed transfer function, the autocorrelation output for white noise input is a triangular function. Thus, the correlation between adjacent reference cell samples is equal to $1 - 1/K$. Sampling at a rate equal to the bandwidth (K equal to unity) causes all adjacent samples to have zero correlation so that all samples are statistically independent. As another example, sampling at twice the bandwidth causes the correlation between adjacent samples to equal one-half.

The effective number of samples as a function of K is tabulated below. The derivation is given following Table 9.3.

Table 9.3
Effect of Sampling Rate on CACFAR Loss

Relative Sampling Rate	*Effective Number of Samples*
1	M
2	$0.67M$
4	$0.69M$
8	$0.67M$

We can see that the effective number of samples rapidly approaches the asymptotic value of $2M/3$.

9.14 2 Analysis of Correlated Reference Cells

The normalization voltage of the CACFAR, v_N, is computed by using the quadratic form:

$$v_N = |y_m|^2 = \mathbf{Y}^{\mathbf{H}}\mathbf{Y}/M$$

where the M-dimensional vector $\mathbf{Y}$ denotes the predetector voltages of the M reference cells. The mean value of v_N equals the common variance of each component of the vector $\mathbf{Y}$, which equals the test cell power. The CFAR loss depends on the variance of v_N. The variance can be determined by the second moment of v_N. For convenience and without loss of generality, the variance of each component is assumed to equal unity. Therefore, the second moment equals[3]

$$E(v_N)^2 = \text{tr}(\mathbf{R})^2 + 2\,\text{tr}(\mathbf{R}^2) \tag{9.82}$$

where $\mathbf{R}$ denotes the covariance matrix of $\mathbf{Y}$ and tr(·) denotes the trace of the matrix. Then, the variance of v_N is

$$\text{v}(v_N) = [\text{tr}(\mathbf{R})^2 + 2\,\text{tr}(\mathbf{R}^2)]/(M) - 1 \tag{9.83}$$

As $\text{tr}(\mathbf{R}) = M$, this reduces to

$$\text{v}(v_N) = 2\,\text{tr}(\mathbf{R}^2)/(M^2)$$

Note that $\text{tr}(\mathbf{R}^2)$ equals the sum of the squared length of each column of $\mathbf{R}$.

When the A/D sampling frequency equals the radar bandwidth, the adjacent reference cell samples are uncorrelated so that the entries of $\mathbf{R}$ are zero, except on the main diagonal, where they equal unity. Thus, the $\text{tr}(\mathbf{R}^2) = M$ so that the variance of the normalization voltage is $2/M$.

For twice the sampling frequency the entries of the main diagonal of $\mathbf{R}$ again are unity. The entries of the diagonals above and below the main diagonal change from zero to the value of one-half. All other entries still are equal to zero. Thus, there are M entries equal to unity and $2(M - 1)$ entries equal to one-half. Squaring and summing each entry to compute the trace of the squared matrix gives the variance as approximately equal to $3/M$. Thus, the variance increases from the value achieved for the uncorrelated case so that the effective number of reference cells decreases. As the variance of the normalization voltage increased from the value of $2/M$ for the uncorrelated case to $3/M$, the effective number of reference cells decreases to $2/3M$. Similar computations give the other entries of the previous loss table.

9.14.3 Additional CFAR Loss Due to Magnitude Detection

For magnitude envelope detection, rather than square-law detection as in (9.82), the normalization voltage is given by

$$v_N = \beta[|y_m|]^2$$

The constant β is included so that the expected value of the normalization voltage equals the power of the test cell. As each y_m is distributed as a zero-mean, complex Gaussian random variable of unit variance, the moments of the magnitude are well known to be given by

$$E(|y_m|^p) = (2/\pi)^{1/2} 2^{1/2(p-1)} \Gamma[(p+1)/2]$$

The variance of the normalization voltage can be computed and compared to the variance obtained for the same number of reference cells when using square-law detection. As in the previous section, the ratio of variances defines the effective number of reference cells and equals $0.707M$.

9.15 CFAR TECHNIQUE SUMMARY

As discussed, for a constant-power noise-jamming environment, the CACFAR is statistically optimum. As it also is relatively simple to implement, it is the preferred technique. However, there is no uniformly preferred technique for a clutter environment. The previous material has considered many different CFAR techniques. There also are many other proposed techniques that have not been discussed because the clutter environment is extremely nonhomogeneous. The preferred technique is usually a compromise that is somewhat appropriate for the anticipated classes of clutter environments.

REFERENCES

[1] I.S. Gradshteyn and I.M. Ryzhik, *Tables of Integrals, Series and Products,* New York, Academic Press, 1965.

[2] M. Abramowitz and I.A. Stegun, *Handbook of Mathematical Functions,* New York, Dover, 1965.

[3] C.W. Helstrom, "Statistical Theory of Signal Detection," Pergaman Press, 1960, p. 69.

Chapter 10
Target Detection in Clutter-plus-Noise Environments

10.1 INTRODUCTION

Detection of targets often is significantly affected by the presence of clutter. As discussed in Section 1.8, the detectability of targets in clutter can be enhanced by exploiting the differential doppler frequencies due to different velocities of the target and clutter. Section 1.8 discussed the use of delay-line cancellers and doppler filter banks. The efficacy of these techniques depends on the spectral characteristics of the clutter. The spectrum of ground clutter is relatively narrow and centered at zero doppler. Because delay-line cancellers can be easily configured to have a null at zero doppler, ground clutter is "easily" cancelled by using delay-line cancellers. However, clutter due to reflections from sea, weather, birds, insects, ground traffic, *et cetera* do not have zero-mean doppler, and they are not well cancelled by delay-line cancellers. A doppler filter bank often is a significantly better processor.

Doppler filters can be implemented by using either deterministic weights or adaptive weights. This is similar to the use of deterministic or adaptive weights for antennas to aid in rejecting sidelobe jamming. In either case, the adaptive weight implementation is more expensive, but the performance is significantly improved. This chapter will analyze the relative and absolute performance attained by adaptive and nonadaptive doppler filters. The data can then be combined with cost data to select the preferred architecture. Note that cost data depend on component availability. Thus, if adaptive processing is deemed too expensive in a particular application for a specified time period, this decision may change for the next implementation generation.

The first part of this chapter gives an overview of clutter characteristics and alternative methods of detecting targets in clutter environments. The next part discusses several aspects of delay-line cancellers. These circuits are also called *moving target indicator* (MTI). The latter part is a detailed analysis of the performance of adaptive and nonadaptive doppler filter bank implementations.

10.2 CLUTTER DISCRIMINANTS

The detection capability of a radar system is limited by the phenomenon that the radiation received by the radar often comes from scatterers that are not of interest; this received radiation is termed a *clutter return*. The radiation received from a scatterer of interest is termed the *target return*. These two statements, while fairly trivial and obvious, are intended to emphasize that to a large degree the differences between clutter and target are largely in the mind of the observer. Thus, the procedures used to detect the presence of targets, while ignoring the presence of clutter, are based on the exploitation of any known physical differences between these types of scatterers. Several differences between the scattering properties of targets and clutter exist, and thus can be exploited. Examples of properties that may be useful are below.

Target Elevation

Under some conditions, the scatter that is the target is above the level of the ground and the sources of clutter are at ground level. Techniques that discriminate on the basis of scatterer altitude are thus useful. There are, of course, situations where this is not a useful discriminant.

Target Velocity

This is currently the most common technique used to discriminate between targets and clutter, and it is useful when a significant velocity difference exists.

Transmission-Polarization Sensitivity

The energy backscattered from an object is a function of the polarization of the illuminating energy and different objects usually have differing polarization properties. This property may be exploited by transmitting a polarization for which the backscattered radiation from the undesired objects is relatively low at the receiving polarization. This technique has been used to reduce the weather clutter problem. Transmitting on several different polarizations and discriminating on the resulting observation vector is theoretically conceivable.

Carrier-Frequency Sensitivity

Under some circumstances, the ratio of the energy backscattered from the desired and undesired scatterers is a function of carrier frequency and we make a choice of carrier frequency that maximizes the ratio, consistent with other system requirements. Also, transmitting on several different frequencies and discriminating on the resulting observation vector can be used.

Azimuth-Angle Sensitivity

When the clutter is a collection of point scatterers, the amplitude of the phasor sum is relatively sensitive to the azimuth angle to the radar. The amplitude return from a point target is relatively insensitive to this angle. Thus, comparing the amplitude of the received signal from two antennas, such as a monopulse antenna configuration, is usable as a discriminant technique. This technique can be coupled with carrier-frequency sensitivity for improved performance.

Signal Bandwidth

If the undesired scatterer is essentially a homogeneous sheet of scatterers so that the total undesired backscattered radiation is a linear function of the radar resolution cell and simultaneously the desired target is essentially a point scatterer, the ratio of desired received energy to undesired received energy increases linearly with increased signal bandwidth. Thus, using larger bandwidths is a clutter discriminant.

The choice of which of the above discriminant techniques, singly or in combination, and an evaluation of the relative merits of the techniques clearly require a knowledge of the backscattering properties of different types of reflectors. The backscattering coefficient varies enormously as a function of the specific class of clutter. The consistency of the coefficient is highest for weather, lowest for land; the sea return data are intermediate. This consistency seems to be intimately associated with the fact that there are only a few degrees of rain (possibly five between drizzle and cloudburst), about twice this number of cases for sea variability, and vastly more cases for ground variability. As an example, clutter backscatter for ground return varies with classifications such as rural, urban, or smooth asphalt road, rough asphalt road, wet snow on one-inch grass, *et cetera*. Some general statements can be made. One fairly consistent qualitative rule on backscatter properties, which coincides with physical reasoning, is that smooth surfaces backscatter less radiation than rough surfaces for angles sufficiently far from the vertical so that specular reflection is no longer significant. Thus, as an example, when specular return is not significant, sea return will tend to increase as the sea state increases (a rougher surface) and as the carrier frequency increases (a rougher surface). In addition, the backscattered radiation for a rough surface is less dependent on incidence angle than is that from a smooth surface. This latter condition is qualitatively indicated in Figure 10.1.

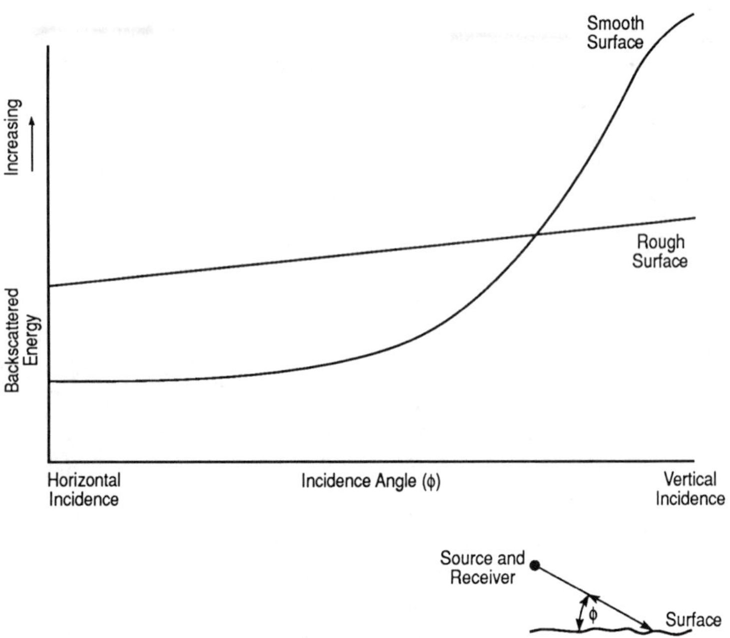

Figure 10.1 Relative backscattering energy for smooth and rough surfaces.

10.3 DEFINITION OF PARAMETERS FOR GROUND AND SEA CLUTTER

The standard definition for the backscattered energy for a point target uses the target cross section σ_0, and this parameter is assumed to be independent of the antenna beamwidths. The standard definition for the backscattered radiation from the earth's surface assumes that the clutter cross section is a linear function of the illuminated area, and this implies an inherent assumption that the surface is homogeneous. With these assumptions, the flat-earth geometry of Figure 10.2 is assumed, and the surface radar cross section is usually taken as

$$\sigma_s = \sigma_0 A_L$$

where A_L is the illuminated area and σ_0 is the normalized surface cross section. An alternative definition is that*

*The relationship between A_i and A_L assumes that the surface resolution cell is determined by the signal bandwidth. This is the common condition for main-lobe clutter and corresponds to the condition that the grazing angle not be too close to the vertical. More precisely, $\tan\phi < 2\Delta\phi R/c$, where $\Delta\phi$ is the elevation beamwidth.

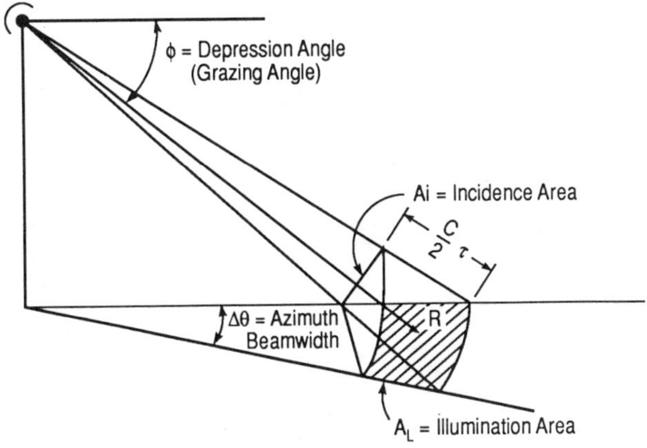

Figure 10.2 Surface echo area geometry.

$$\sigma_s = \gamma A_i = \gamma \tan\phi(Rc\tau/2)$$

$$\sigma_0 = \gamma \sin\phi$$

and some reports present results in terms of the normalized parameter σ_0, while others use the normalized parameter γ. An additional note of caution is that the symbology used is not consistent among all investigators. That chosen for this chapter represents the majority opinion. The major agreement is on the use of the symbol σ_0, and the major source of possible confusion is that γ is sometimes used for depression angle.

The major parameters usually considered to influence the magnitude of the scattering cross section are:

1. Type of terrain (sea, ground, *et cetera*);
2. Additional natural influences (sea state, relative wind direction, wet ground, dry ground, *et cetera*);
3. Carrier frequency;
4. Transmitting polarization;
5. Receiving polarization;
6. Depression angle;
7. Homogeneity of surface.

These parameters are considered in the following sections.

10.3.1 Sea Clutter

The dependence of σ_0 *versus* incidence angle is of the general form indicated by the plot of Figure 10.3; the scale of the depression angle is intended to be logarithmic.

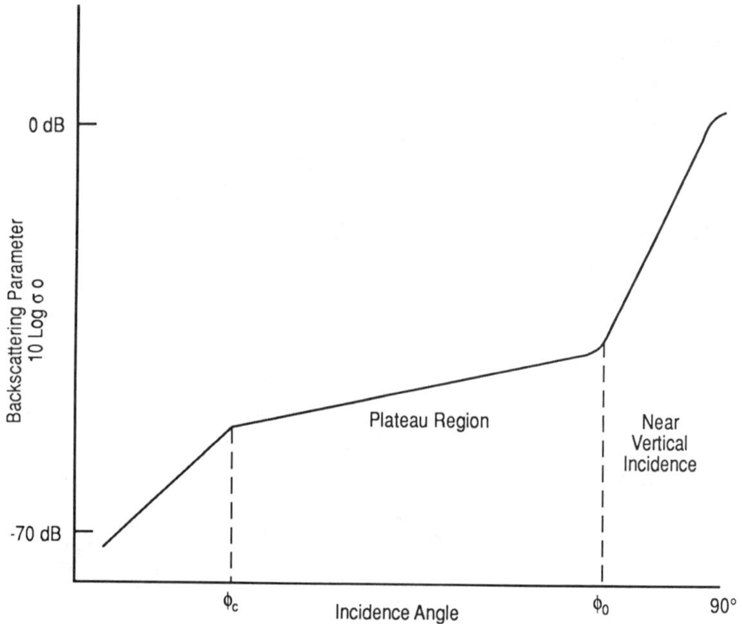

Figure 10.3 General dependence of σ_0 on depression angle.

The measured values of σ_0 range from about $+10$ dB at vertical incidence to as low as -70 dB at low grazing angles. The plateau region is bounded by the depression angles of ϕ_c, the *critical angle,* and ϕ_0. These angles are on the order of one degree and tens of degrees (e.g., 50°), respectively, and their values are dependent on frequency and polarization. For angles less than ϕ_c, the log of σ_0 is approximately proportional to the log of ϕ, as indicated in Figure 10.3, so that

$$\sigma_0 = g_1 \phi^{n_1}$$

This is also approximately true for the plateau region so that

$$\sigma_0 = g_2 \phi^{n_2}$$

The trends as indicated by the values of n_1, n_2, and ϕ_c are important. The following conditions are typical.

As the Sea State Increases

1. σ_0 increases for all depression angles;
2. The value of ϕ_c decreases with increased sea state and typical values are 4° for the calmest sea and 1° for the roughest sea;

3. The values of n_1 and n_2 remain constant at values of about 3.0 and 0.6, respectively, so that denoting the value of σ_0 at the critical angle as σ_c, for angles less than ϕ_c:

$$\sigma_0(\phi) = \sigma_c(\phi/\phi_c)^3$$

and for the plateau region:

$$\sigma_0(\phi) = \sigma_c(\phi/\phi_c)^{0.6}$$

As the Carrier Frequency Increases

1. σ_0 increases as approximately the third or fourth power for calm seas and approximately as the zeroth or first power for rough seas;
2. The value of ϕ_c decreases with frequency;
3. The values of n_1 and n_2 are independent of frequency.

The reader should note that statements 2 and 3 for the sea state variation and the frequency variation are identical in form, and they corroborate the previously stated dependence of σ_c on surface roughness.

Polarization Dependence

The dependence of sea backscatter upon polarization will be summarized by considering the values of σ_0(VV) transmitting and receiving vertical polarization; σ_0(HH) transmitting and receiving horizontal polarizations; and σ_0(VH), which is the cross-polarization component and, for a reciprocal transmission medium, this is equal to σ_0(HV).

The ratio of $\sigma_0(VV)/\sigma_0(HH)$ is essentially independent of depression angle, but is a function of carrier frequency and sea state. The general functional relationship for smooth surfaces (calm seas and the lower portion of the microwave band) is the ratio of $\sigma_0(VV)/\sigma_0(HH)$, which is quite large, on the order of 20 to 30 dB. As the surface roughness increases, the ratio approaches 0 dB and sometimes even is a few dB negative. Information concerning the cross-polarization backscattering cross sections is quite meager.

Another parameter that affects the sea backscatter is the speed and relative direction of the wind. The relative direction is significant because of the differences between the value of σ_0 for upwind, downwind, and crosswind. These differences are on the order of 5 dB and the relative ordering is

$$\sigma_{0u} > \sigma_{0d} > \sigma_{0c} \quad \text{(vertical polarization)}$$
$$\sigma_{0u} > \sigma_{0c} > \sigma_{0d} \quad \text{(horizontal polarization)}$$

where the subscripts u, d, and c refer to upwind, downwind, and crosswind, respectively. In addition to the directional effect, the magnitude of σ_0 is stated to be a function of wind velocity such that σ_0 in dB is a linear function of the logarithm of the wind speed.

10.3.2 Ground Clutter

The general trends concerning the variation of the magnitude of the clutter return for each particular class of ground scatterer seem to have the same form of functional variation as does the energy backscattered from the sea. Thus, the form of the variation is dependent on surface roughness and the operating frequency. The backscattering coefficients tend to follow the functional variation given by Figure 10.1.

10.3.3 Weather Clutter

That the magnitude of the clutter return from precipitation is predictable to a high degree of accuracy is clear from the fact that radar data are often used by meteorologists for determining the precipitation rates and general weather characteristics.

The total magnitude of the clutter return is dependent on the volume of precipitation intercepted by the beam and the backscattering characteristics of the particular type of precipitation. Thus, for an antenna beamwidth of β steradians and a range resolution of L, the volume intercepted at range R, assuming the clutter fills the volume, is given by

$$V = \beta R^2 L = 4\pi R^2 L/G$$

where G is the antenna gain. Under the assumption that the individual scatterers are spherical and small compared to the wavelength, the scattering cross section per unit volume may be theoretically derived as

$$\sigma_V = (\pi^5/\lambda^4)|K|^2 \Sigma d_i^6$$

where d_i is the diameter of the ith drop, and the summation is over all of the drops in the volume. The constant K has been determined empirically, and it is stated to be equal to 0.93 for rain and equal to 0.2 for ice crystals and snow. The variation of σ_V as the inverse fourth power of frequency occurs because of the assumed validity of the Rayleigh approximation for the scattering from spheres that are much smaller than the wavelength. This approximation begins to lose validity for wavelengths smaller than 3 cm. The summation can be related to rainfall rate as

$$Z = \Sigma d_i^6 = \begin{cases} 200 \quad p^{1.6} \text{ (rain)} \\ 2000 \quad p^2 \text{ (snow)} \end{cases}$$

where p is the precipitation rate in mm/hr (the equivalent after melting, for snow). A convenient plot of these relations is shown in Figure 10.4.

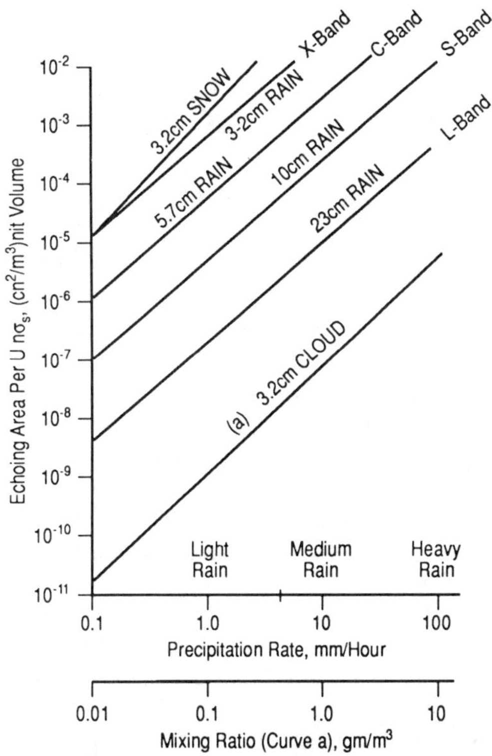

Figure 10.4 Echoing area of precipitation and cloud.

In addition to the standard doppler techniques used for rejection of clutter return, rejection of weather clutter is often implemented by use of polarization-sensitive techniques. The classical implementation has been to transmit and receive on one of the circular polarizations. We can show that if a scatterer is a sphere, the polarization of the scattered energy is orthogonal to the polarization of the receiving antenna, and therefore the system will not detect the presence of spherical targets such as rain drops. Because the polarization scattering properties of targets are dependent on shape, other scatterers such as aircraft will return an electromagnetic wave of a polarization that is not orthogonal to the receiving antenna and the signal-to-clutter ratio will be increased by this technique. In addition, polarization-sensitive techniques are implemented at RF so that these and doppler techniques are compatible. The combination of techniques might be used to produce larger signal-to-clutter ratios than either could produce alone.

There are two limitations for the degree of cancellation that can be achieved by the polarization cancellation technique. One limitation is that raindrops are not precisely spherical, and this degrades the performance of a system that is optimized for spherical shapes. This limitation becomes more severe during rain of heavy intensity (when the clutter is largest) because the departure from spherical shape apparently increases with drop size. The second limitation is caused by the difficulty of designing an antenna with a constant polarization over the full beamwidth.

10.4 DOPPLER FILTERING TECHNIQUES

Doppler processing improves the detection of targets in clutter by exploiting the doppler frequency difference between a target and clutter by transmitting and coherently processing a sequence of N pulses. The period between pulse transmissions is usually constant, denoted as T, and termed the pulse repetition interval. The processor is of the form indicated in Chapter 1, where Figure 1.4 shows a four-pulse processor. Conceptually, more delay segments are used to process greater than four pulses. At any time instant, the output of each delay segment is the voltage corresponding to the return from some range cell for a particular transmitted pulse. Thus, the tapped delay line is simply a mechanism to obtain time alignment of the returns due to the different transmitted pulses. This block diagram is appropriate for either an MTI circuit or a doppler filter. The difference between the two circuits is the design rationale used to choose the values of the weights. The weights are chosen to produce a particular gain *versus* doppler frequency characteristic. The design criterion for an MTI circuit is that the gain characteristic has a null at a specified doppler frequency. This frequency is usually zero, but can be any other frequency. For the doppler filter, there are two design criteria. One is that the circuit has large response at the target's doppler. The other is that the response far from the target is small.

10.4.1 MTI Delay Line Cancellers

The single-delay-element canceller of Chapter 1, shown in Figure 1.5(c), is called a single-delay canceller (or two-pulse canceller). The technique can be generalized to a multiple-delay-element canceller, and a three-pulse canceller is discussed later.

There are two complementary ways of analyzing the gain characteristics of doppler processors. A convenient method for MTI delay-line cancellers is by use of the phasor diagram depicted in Figure 10.5(a). An algebraic technique is later used for the doppler filter banks.

In Figure 10.5(a), the voltage return from a specified range due to the transmission of pulse 1 is shown as the horizontal line. The phase of this return is depicted as zero without loss of generality, as only the phase difference between pulse returns is important. The phase of the return for pulse 2 is denoted as ϕ. The phase difference

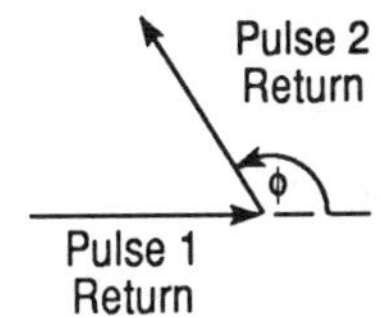

(a) MTI Phasor Diagram

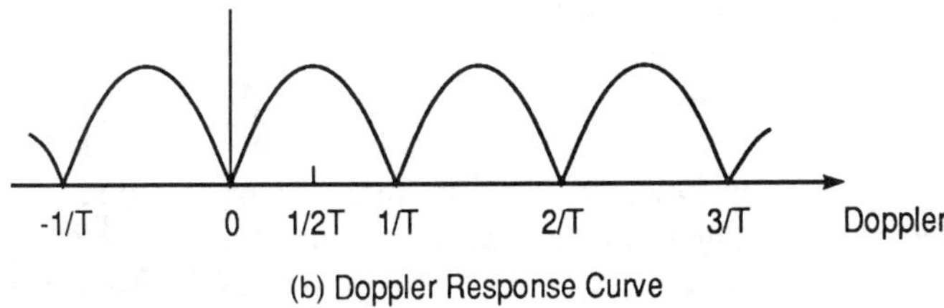

(b) Doppler Response Curve

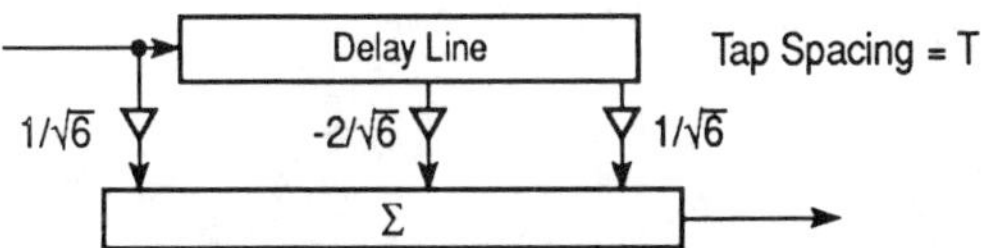

(c) Three Pulse Canceller Block Diagram

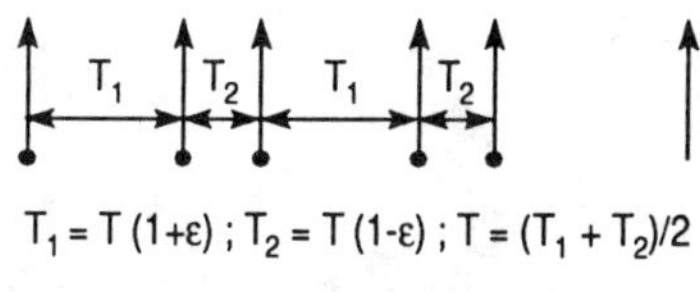

$T_1 = T(1+\varepsilon)$; $T_2 = T(1-\varepsilon)$; $T = (T_1 + T_2)/2$

(d) Stagger Coded Waveform

Figure 10.5 MTI delay-line cancellation.

is due to the range change of the target during the time between pulse transmissions. Thus, for a carrier frequency of f_0, a reflector velocity of v and a velocity of propagation of c,

$$\phi = 2\pi f_0 vT/(c/2) = 2\pi f_D T$$
$$f_D = f_0 v/(c/2)$$

where f_D is the target doppler frequency. Note that at zero target velocity the phase equals zero. When the two successive pulses are subtracted, as in a delay-line canceller, the output voltage equals zero. Similarly, the output equals zero when the phase difference equals 2π, or any integer multiple of 2π. These null conditions are termed *blind speeds*. The doppler frequencies corresponding to the blind or null speeds are obtained by noting from the previous equation that the phase shift consists of integer multiples of 2π when the reflector's doppler frequency comprises integer multiples of $1/T$. This condition is indicated by the transfer function sketched in Figure 10.5(b). Also, note that, because the pulse repetition frequency equals the reciprocal of the pulse repetition interval, the blind doppler frequencies occur at integer multiples of the PRF. The maxima of the transfer function correspond to reflector velocities, termed "optimum" speeds. These correspond to the condition that the phase shift between pulse returns equals odd multiples of π. For this condition, the subtraction of the two returns is equivalent to adding the magnitudes. The transfer function of the single-delay-element MTI canceller is given by (1.20).

To calculate the circuit's performance, note that, for any circuit, the output clutter power, p_{out}, can be calculated by integration as

$$p_{out} = \int_{-\infty}^{\infty} C(f)|H(f)|^2 \, df$$

where $C(f)$ is the clutter power spectrum and $H(f)$ is the circuit's voltage transfer function. For the two-pulse canceller, $H(f)$ is given by (1.20). An alternative computational procedure is to evaluate the output power by using Parsavel's theorem, which states that the following integrals are equal:

$$\int_{-\infty}^{\infty} x(t)y^*(t)dt = \int_{-\infty}^{\infty} X(f)Y^*(f)df$$

where $x(t)$ and $y(t)$ have Fourier transforms $X(f)$ and $Y(f)$, respectively. The applicability of this result is that the Fourier transform of the clutter spectrum equals the autocorrelation function of the clutter. The Fourier transform of the magnitude squared of the transfer function equals the filter's autocorrelation function. This theorem is especially useful for delay-line circuits because the filter's autocorrelation function is given in terms of δ functions for delay line circuits. The resulting integrals are in terms of δ functions and these are simple to evaluate.

The applicable integral relationship for a δ function, $\delta(t)$, and an arbitrary function, $a(t)$, is

$$\int_{-\infty}^{\infty} \delta(t - t_0)a(t)\mathrm{d}t = a(t_0)$$

The impulse response for the single-delay-element canceller, $h(t)$, and autocorrelation function, $R_h(\cdot)$, are

$$h(t) = [\delta(t) - \delta(t - T)]/2^{1/2}$$

$$R_h(\tau) = \int_{-\infty}^{\infty} h(t)h^*(t + \tau)\mathrm{d}t$$

Combining these two equations gives

$$R_h(\tau) = [2\delta(\tau) - \delta(\tau - T) - \delta(\tau + T)]/2$$

Then, applying Parsavel's theorem with the integral relation for δ functions and denoting the clutter's autocorrelation function as $R_C(\tau)$, gives

$$p_{\text{out}} = [2R_C(0) - R_C(T) - R_C(-T)]/2 \tag{10.1a}$$

A special case, usually applicable for ground clutter, is that the clutter spectrum is even and centered at zero. Then, the autocorrelation function is even, and (10.1a) becomes

$$p_{\text{out}} = R_C(0) - R_C(T) \tag{10.1b}$$

A common measure of the performance of an MTI canceller is the ratio of the output clutter power to the input clutter power. However, this measure is not uniquely defined, as the output power depends on the scale of the weight values. The measure can be uniquely defined by constraining the weights, as done in previous sections, to be normalized to unit length. This normalization is used so that, for thermal noise, the output power equals the input power. This constraint can be proved by noting that the spectrum of thermal noise is white. Therefore, its autocorrelation function is zero, except when the lag delay equals zero. At zero lag delay, the value of the autocorrelation function equals the thermal-noise power. Thus, as we can see from (10.1b), for thermal noise, the input power equals the output power.

Therefore, using (10.1b) for the output clutter power shows that the clutter attenuation ratio, CA, is

$$\text{CA} = [R_C(0) - R_C(T)]/R_C(0)$$

A useful example is for the case of a Gaussian power density spectrum. Fourier transformation shows that the autocorrelation function is also Gaussian. Thus, for a clutter power of p_C and bandwidth of B_C, we have

$$C(f) = p_C \exp[-(f/B_C)^2/2]/(2\pi B_C^2)^{1/2}$$

$$R_C(\tau) = p_C \exp[-2(\pi B_C \tau)^2] \qquad (10.2)$$

An approximation valid for the important case of high cancellation is obtained via a series approximation of (10.1b). Because, for small x, $\exp(-x) \approx 1 - x$, for a Gaussian spectrum, the clutter attenuation is approximately

$$\text{CA} \approx 2(\pi B_C T)^2$$

Note that $B_C T$ is equal to the ratio of the clutter bandwidth to the PRF. A typical case might be for this ratio to equal 0.03. Then, evaluating the previous expression shows that the clutter attenuation for the single-delay-element canceller equals −17.5 dB. Often, this is inadequate cancellation and improved MTI circuitry is required.

Improved MTI cancellation can be obtained by using multiple-delay-element configurations. An example is the two-delay-element canceller shown in Figure 10.5(c). It is a three-pulse canceller and can be viewed as a cascade of two-pulse cancellers. The concept is that the cascading ought to improve the rejection of the clutter.

The impulse response of the circuit, when the weights are normalized to unit thermal-noise gain, is

$$h(t) = [\delta(t) - 2\delta(t - T) + \delta(t - 2T)]/(6^{1/2}) \qquad (10.3)$$

By using an analytic procedure that is identical to that used for the single canceller, we can show that the clutter attenuation is given by

$$\text{CA} = \{R_C(0) - [8R_C(T) - 2R_C(2T)]/6\}/R_c(0)$$

By using the same series approximation as previously, this evaluates to

$$\text{CA} = 8(\pi B_C T)^2$$

For the 0.03 bandwidth/PRF ratio previously used, the clutter attenuation is evaluated as −32.0 dB. This is obviously substantially better clutter cancellation than that obtained by the previous circuit. However, the improved cancellation has an associated cost. Specifically, detection of low-speed targets becomes more difficult.

A series expansion of (1.20) shows that the signal power gain for small doppler frequencies approximately equals $2(\pi f T)^2$. For a doppler frequency equal to $0.1/T$, the target gain equals -7.05 dB. The target gain decrease is a significant defect of using an MTI circuit. As an example, when there is no clutter present, the use of an MTI degrades target detectability for a broad range of doppler frequencies.

A comparable evaluation of the target gain for small doppler frequency for the three-pulse canceller can be done, starting from (10.3). The Fourier transform of (10.3) gives the transfer function (voltage gain) of the circuit. A series expansion leads to the signal power gain for small doppler frequencies as approximately equal to $(2\pi f T)^4/6$. For the previous example of the target doppler frequency equal to $0.1/T$, the target gain equals -15.9 dB. Thus, the improved clutter rejection has an additional target detectability cost. Further note that the reduction in target detectability is not only for low-speed targets. Because of doppler ambiguities, the reduction of target detectability occurs for all target velocities that cause a doppler frequency that is close to an integer multiple of the PRF. This is equivalent to stating that higher order canceller circuits increase the difficulty of detecting targets close to the blind speeds.

10.4.1.1 Stagger-Coded MTI Waveforms

The blind speeds of an MTI processor are due to the use of a periodic transmitting waveform. A method of partially alleviating the problems of MTI circuits and blind speeds is to use an aperiodic waveform. An example of an aperiodic waveform is the staggered PRF waveform depicted in Figure 10.5(d). We observe that the waveform's period alternates between two values. The waveform is processed by a two-delay-element circuit, and the delays of the two segments differ to match the different pulse periods. When processing pulses with the relative spacing of the left three pulses of the figure, the circuit's impulse response is

$$h_1(t) = \delta(t + T_1) - 2\delta(t) + \delta(t + T_2)$$

For the other period order, the circuit's impulse response is

$$h_2(t) = \delta(t + T_2) - 2\delta(t) + \delta(t + T_1)$$

For either case, the power transfer function is

$$|H(f)|^2 = 4\{1 + [\cos(2\pi f T)]^2 - 2\cos(2\pi f T\varepsilon)\}$$

The circuit still has a null at zero frequency. However, the blind speed problem has been reduced, as it does not necessarily have nulls at integer multiples of the PRF. Thus, note that, at the first blind speed, the power gain is not equal to zero, but rather is

$$|H(1/T)|^2 = 4[2 - 2\cos(2\pi\varepsilon)] \approx 4(2\pi\varepsilon)^2$$

An example of the doppler response obtained by using stagger coding is shown by the curves of Figure 10.6. The left-hand response is for the case of $\varepsilon = 0.1$. Note that the response is periodic at the doppler frequency of $1/\varepsilon T$. The right-hand response is the no-stagger case. Also, note that, although the problem of the periodic nulls has been reduced by the stagger, the null at zero has widened.

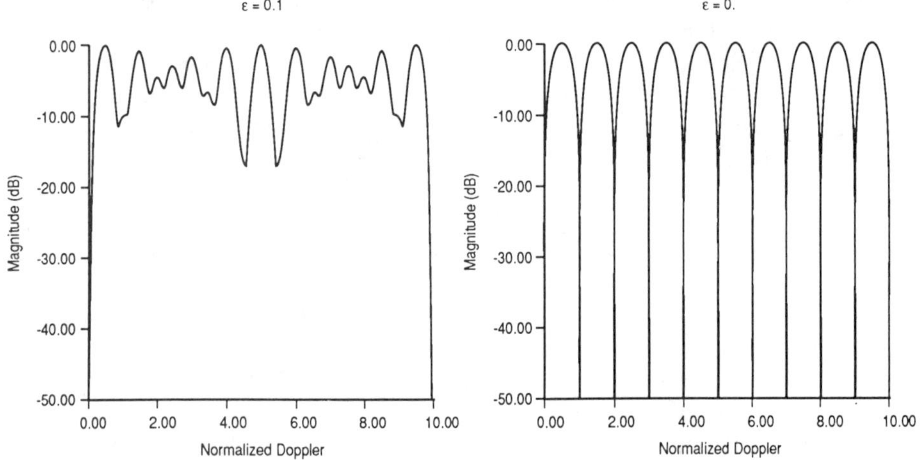

Figure 10.6 Comparative stagger-code MTI performance.

10.4.2 Doppler Filter Banks

The analysis of the doppler processors in Chapter 1 used the real-valued transmitted signal as given by (1.16). Because of the great similarity of the equations obtained by analyzing doppler filters to those previously obtained for adaptive antenna analysis, to continue the analysis using complex valued signals is more convenient. Then, the results of the prior nonadaptive and adaptive antenna analysis can be directly applied to nonadaptive and adaptive doppler processing analysis.

Extrapolating from (1.16), a unit-amplitude N-pulse transmitting pulse train is given by

$$e(t) = \exp(j2\pi f_0 t)\Sigma p(t - nT)$$

where $p(t)$ denotes the pulse modulation waveform. For a target at initial range of R_0 and radial velocity v, the time delay of the reflected pulse due to the two-way travel and returned waveform are

$$t_R = 2(R_0 - vt)/c$$

$$e_R(t) = s_R \exp[j2\pi f_0(t - t_R)]\Sigma p(t - t_R - nT)$$

where s_R denotes the complex amplitude of the reflected waveform. This can be expressed as

$$e_R(t) = s_R \exp[j(2\pi f_0 t - \theta)][\exp(j2\pi f_D t)\Sigma p(t - t_R - nT)]$$

$$\theta = 4\pi f_0 R_0/C; \quad f_D = 2v f_0/c$$

The bracketed term modulates the carrier, and thus is the term of importance. In the remainder, the carrier portion will be deleted.

The delay line delays the n-th pulse by $(N - 1 - n)T$. When the line is filled with returns due to the N-pulse transmission, the voltage at the n-th tap, neglecting the pulse-envelope differential delay due to t_R being time-dependent, is

$$e_n(t) = s_R \exp\{j2\pi f_D[t - (N - 1 - n)T]\}p[t - t_R - (N - 1)T]$$

These voltages are summed by using weights, g_n, so that the output voltage, e_0, and processor voltage gain, $g(f_D)$, are

$$e_0 = s_R p[t - t_R - (N - 1)T] \exp\{j2\pi f_D[t - (N - 1)T]\}g(f_D) \quad (10.4)$$

$$g(f_D) = \Sigma g_n^* \exp(j2\pi f_D nT)$$

The gain response function is like that developed by the phasor diagram of Figure 10.5(a). The pulse envelope is centered at time $t_R + (N - 1)T$. This delay is due to the target range induced delay t_R and the time required for the N-pulse train to fill the delay line.

The processor voltage gain equation is of the exact same form as that developed previously for antennas. Equations (8.3) and (8.20) are the vector equivalents of (10.4). Thus, the effect of weighting for doppler filtering is mathematically identical to weighting for synthesizing antenna patterns.

A typical conceptual implementation of a doppler filter bank is shown in Figure 10.7. Conceptually, each filter of the bank is implemented by using a tapped delay line, the tap spacing of which is equal to the PRI. In digital implementations, the delay indicates the relative timing of storage entries. The filter output is formed as a weighted sum of the tap voltages. The filter response is large over a frequency region of approximately $1/NT$, where N is the number of taps of the delay line and T is the time delay between taps. The response is low over the rest of the frequency band. As each filter has a high response over only a portion of the doppler frequency range, several parallel filters are necessary, each tuned to a different doppler frequency to cover the entire range. A target-present or -absent decision is made for each range cell by incorporating a CFAR technique (normalizer) in each filter. Use of separate CFAR circuits is necessary, as the residue power is usually substantially different in each filter. The residue power tends to be relatively high in filters that encompass or are close to the clutter spectrum and relatively low for the other filters.

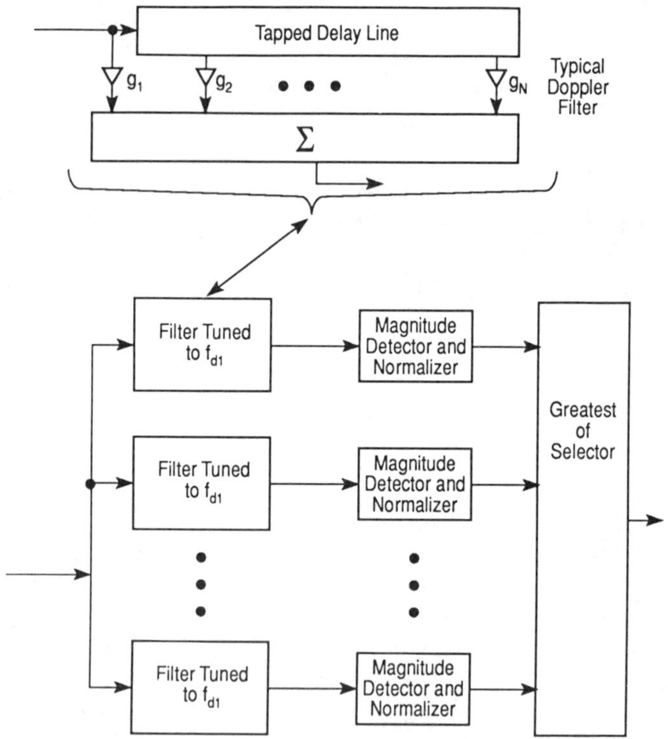

Figure 10.7 Doppler filter bank.
Source: From R. Nitzberg, *Performance Comparison of Adaptive and Tchebychev Weighted Doppler Filtering*. Asilomar Conf. on Circuits and Systems, Pacific Grove, CA, Nov 1983.

The CFAR circuitry can be implemented in two different ways, depending on whether the doppler of a declared target is a required output to the data processor. If it is a required output, each filter output is passed through a CFAR circuit. The CFAR circuit can use an adaptive threshold or a normalized output comparison. Therefore, let us denote the detected output of the test cell of the nth doppler filter as q_n. We denote the normalization voltage for this filter as v_n. Then, the adaptive threshold for each filter equals kv_n. Thus, for each filter, a target-present or -absent decision can be based on the inequality:

$$q_n > T_n = kv_n$$

and if a target-present decision is made, the filter number containing the declared target can be passed to the data processor. Alternately, each test cell voltage can be normalized and compared to the threshold multiplier as

$$q_n/v_n > k$$

Again, if a target-present decision is made, the filter number can be passed to the data processor. Either of the preceding requires N threshold comparison circuits. Often, the target doppler is not required and the implementation can be simplified. This is done by selecting the maximum of the set $\{q_n/v_n\}$ and comparing it to the threshold multiplier. This is the implementation shown in Figure 10.5. The theoretical disadvantage of the simplified implementation is the inability to detect two or more targets at the same range, but at different doppler frequencies.

There are two reasons for considering this circuitry simplification. One is that, for almost all situations, the radar resolution volume is sufficiently small so that never is there more than a single target in the same range-angle resolution cell. A second is that the target doppler often is ambiguous and to deduce target radial velocity from knowledge of the target doppler is very difficult. Specifically, a target doppler equal to f_D cannot be distinguished from one equal to $f_D + m/T$ for any integer value of m. The doppler ambiguities are numerically identical to the blind speeds of delay-line cancellers discussed in Subsection 1.8.1. The example shows that a radar operating at a carrier frequency of 3 GHz with a PRF of 1000 Hz has a velocity ambiguity equal to 97.2 knots. As target velocities are often in excess of 500 knots, we can see that knowledge of the filter number that exceeds the threshold does not imply knowledge of the target's radial velocity.

As previously stated, the response of each filter is high over a relatively small frequency range and low elsewhere. The details of the response in this low region (the sidelobe response) determines the ability of the radar to detect a target. Specifically, for target detection, the filter's output power due to a target with its doppler frequency in the main-lobe region must be large compared to the output power due to the sidelobe region's clutter plus noise. When the tap weights are of equal magnitude, the level of the sidelobes of the filter's frequency response is relatively high. To reduce the sidelobe level and thus increase the ratio of the output signal to clutter plus noise (SCNR), special weight values, such as Chebyschev or Taylor weighting are commonly used. While these weighting techniques can produce substantially higher SCNR values than obtained by using uniform weights, these techniques have disadvantages. One is that weighting increases the main-lobe width and the widening can degrade performance. Another is that choosing a particular sidelobe level is necessary and the design choice will be improper for some conditions. These disadvantages are avoided by the use of adaptive weighting.

The primary advantage of adaptive weighting as compared to the fixed-weighting technique is that the latter reduces the sidelobe level everywhere. The former reduces the sidelobe level only over the clutter's frequency extent. As a consequence, the sidelobe level over the clutter region can be tens of dB lower for adaptive weighting than for fixed weighting without significant main-lobe widening. The cost of this performance improvement is a requirement to estimate the adaptive weights by use of an algorithm.

10.5 OPTIMUM DOPPLER FILTER WEIGHTS

The optimum weights criterion for adaptive arrays is based on the assumption that the target angle is known *a priori*. The known location is equivalent to a known amplitude and phase of the voltage induced by the target at each element of the adaptive array. These voltages are components of the vector **S** used in (8.7) to compute the optimum array weights to maximize output SINR. In (8.7), the entries of the covariance matrix $\mathbf{R}_Z$ are due to the interelement correlations of the interference (the sum of thermal and jamming noise). As the mathematical formulation of a nonadaptive-adaptive array is identical to that of a nonadaptive-adaptive doppler filter, as shown by (10.4). Equation (8.7) also can be used to derive optimum weights for the adaptive doppler filter. Note that the discussion is in terms of a single filter, rather than a bank. Each filter of the bank is designed under the assumption that the target doppler is known *a priori,* but is a different value. This is an appropriate strategy for a search radar and is similar to that used for the adaptive array, where the beam is scanned over time.

One difference between the optimum weight design and performance evaluation procedure for antennas and doppler filters is that, for the former, the interference is essentially a collection of discrete point sources, whereas, for the latter, the interference is a collection of distributed sources. Thus, the jammers are at particular angles and the performance is determined by the depth of the antenna null at these angles. For doppler filters, the equivalent of the jammer angle is the clutter spectrum, and it is not discrete points at particular doppler frequencies but is characterized by the spectrum center frequency, its bandwidth, and shape (rectangular, Gaussian, *et cetera*).

A typical clutter-plus-thermal-noise spectrum is sketched in Figure 10.8. The horizontal axis, normalized frequency, is the ratio of the doppler frequency to the PRF. This normalization is convenient because the doppler frequency ambiguity interval equals the PRF. The mean frequency of the clutter is shown relative to the target doppler frequency.

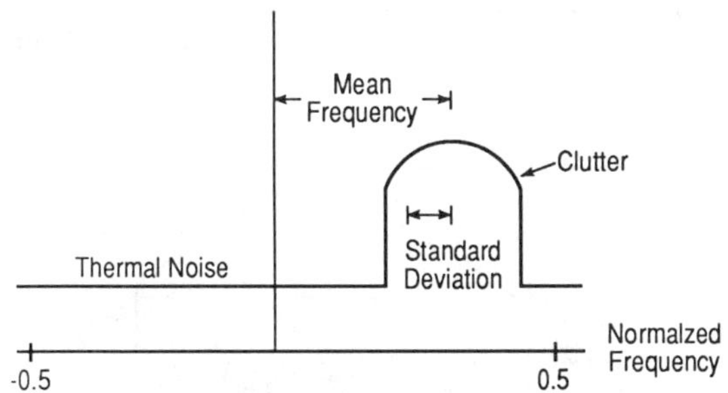

Figure 10.8 Typical clutter-plus-thermal noise spectrum.
Source: From R. Nitzberg, *Performance Comparison of Adaptive and Tchebychev Weighted Doppler Filtering*. Asilomar Conf. on Circuits and Systems, Pacific Grove, CA, Nov 1983.

The concept and performance of nonadaptive and adaptive doppler filtering is illustrated by the responses of Figure 10.9. All responses are for a 16-pulse processor. Figure 10.9(a) shows the response for uniform weights. These are the optimum weights for a thermal-noise environment (interference spectrum is white, constantly over the doppler frequency range), but unsatisfactory when clutter is present. Figure 10.9(b) shows the response for −40 dB Chebyschev weights. The peak sidelobes are 40 dB below the filter's peak response. However, the main lobe is quite broad as compared to that of the uniform weights. This broadening negates the effectiveness of the lower sidelobes as the clutter spectrum becomes close to the target's doppler frequency. Figure 10.9(c) is an example of the filter response for adaptive weights. The region of the clutter spectrum is clear. The filter develops a very low response over this region. Note that the main-lobe width for the optimum weights is much narrower than that of the Chebyschev weights. The width is approximately the same as that for uniform weights. The relative main-lobe widths can be used to generate a first-order approximation of the relative performance of deterministic low-sidelobe weights and adaptive weights.

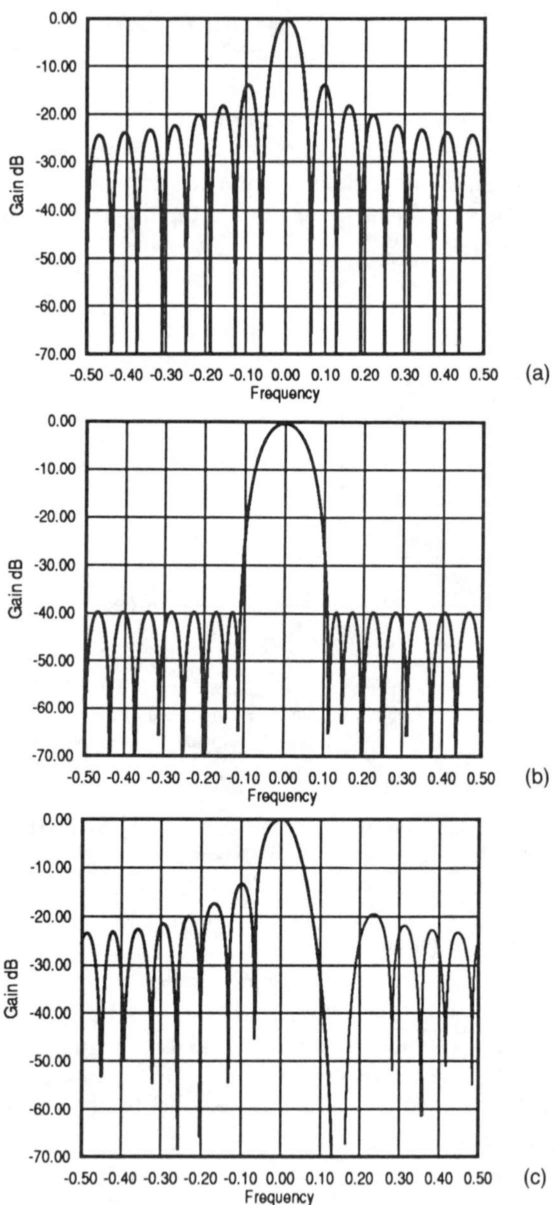

Figure 10.9 Filter's responses.
(a) Uniform weights.
(b) Chebyschev weights.
(c) Optimum weights.

10.5.1 Approximate Performance of Nonadaptive and Adaptive Doppler Weighting

As previously indicated, the target's doppler is highly ambiguous for the PRFs typically used in ground search radars. Because of this, a reasonable approach is to model the doppler frequency difference between the target and the mean of the clutter spectrum as a random variable that is uniformly distributed over the PRF. For high-powered clutter, if the clutter doppler frequency is within the main-lobe region, the probability of target detection is assumed to be zero. If the clutter doppler frequency is outside the main-lobe region, the probability of detection is assumed to equal unity. The main-lobe region for either filter is defined as the region centered at zero and extending either side to the first null. As can be seen from Figure 10.7(b), for 16 pulses, the defined normalized main-lobe width is approximately 0.25. More generally, for N pulses, the width is approximately $4/N$. The adaptive weight's main-lobe width is essentially that of the uniform weight's filter and this is equal to $2/N$ when using adaptive doppler weighting as compared to deterministic doppler weighting. Thus, the probability of clutter being in the main-lobe region of the doppler filter equals $4/N$ for Chebyschev weighting and $2/N$ for adaptive weighting. Therefore, a first-order indication of relative performance of adaptive weighting *versus* Chebychev weighting is equal target detectability is obtained by transmitting one-half the number of pulses. This can have a very large effect on the system, as it implies that search dwell time can be halved. This decreases the data processor update time and improves tracking.

10.5.2 Optimum Doppler Filter Weights—Equation Details

As previously indicated, the optimum weights for adaptive antennas or adaptive doppler filtering are given by (8.7). For adaptive antennas, the typical covariance matrix is given by (5.49) and (5.50). A similar form can be obtained for the adaptive doppler formulation.

For an N-pulse processor, the covariance matrix, $\mathbf{R}$, is an $N \times N$ matrix. The nth row, mth column entry of $\mathbf{R}$ is the cross-correlation of the nth and mth tap voltages. The tap voltages are the sum of clutter plus noise. As the clutter and noise are statistically independent, $\mathbf{R}$ is the sum of a covariance matrix due to noise alone, $\mathbf{R}_n$, and one due to clutter, $\mathbf{R}_c$. Thus,

$$\mathbf{R} = \mathbf{R}_n + \mathbf{R}_c \tag{10.5}$$

Because the noise is statistically independent from tap to tap, we have

$$\mathbf{R}_n = p_{\text{noise}}\mathbf{I} \tag{10.6}$$

where p_{noise} is the noise power and $\mathbf{I}$ is the identity matrix.

The entries of the clutter covariance matrix can be developed as follows. Denote the voltage at the delay line input of Figure 10.7 as $z(t)$. The voltage at the nth delay-line tap equals $z[t - (n - 1)T]$. A particular value of t denotes a particular range cell, and the set of voltages at the delay taps correspond to the set of returns for the particular range cell due to the returns of the N-pulse transmission. Let us denote this set of voltages as $\{z_n\}$. In the absence of a target, the voltage z_n is the sum of thermal noise plus clutter. As the clutter and thermal noise are statistically independent, the correlation of z_n and z_m is the sum of a term solely due to the thermal noise and another solely due to clutter. For noise, the term is given by components of (10.6). For the clutter, this term $r_{nm}(c)$ is

$$r_{nm}(c) = E\{c[t - (n - 1)T]c^*[t - (m - 1)T]\} \tag{10.7}$$

where $c(t)$ denotes the clutter waveform at the delay-line input. In general, the correlation varies with range (the time, t). As a simple example, the cross section of clutter varies with range, which makes the cross-correlation dependent on range. In addition, for weather clutter, the bandwidth varies with range, and there can be other range-dependent effects.

For unambiguous evaluation of (10.7), clarifying the role of the variable t is necessary, as there are two entirely different time variables involved. One is a "delay time," which really means range. The second is the variation of voltage with time at a specified range. Thus, the mathematical model is that, at each range, the return due to clutter is a stationary random process. The power of the random process varies with range because the clutter cross section varies with range. The spectral shape and, in particular, the bandwidth of the random process also can vary with range and an example is shown in Figure 10.8. For some cases, the power density spectrum of weather clutter is well approximated by a rectangular spectrum of total width B_c, centered at the doppler frequency of f_c and total power p_c. Thus, the one-sided power density spectrum, $P(f)$, is

$$P(f) = (p_c/B)\,\mathrm{Rect}[(f - f_c)/B_c] \tag{10.8}$$

The complex-valued autocorrelation function, $r_c(\tau)$, is

$$\begin{aligned} r_c(\tau) &= \int_{-\infty}^{\infty} P(f)\exp(j2\pi f\tau)\,df \\ &= p_c \exp(j2\pi f_c\tau)\sin(\pi B_c\tau)/(\pi B_c\tau) \end{aligned} \tag{10.9}$$

Combining this with (10.7) gives

$$r_{nm}(c) = p_c \exp[j2\pi f_c T(m - n)]\sin[\pi B_c(m - n)T]/[\pi B_c(m - n)T]$$

An alternative assumption is that the clutter power density spectrum and autocorrelation function are Gaussian, as given by (10.2) so that

$$r_{nm}(c) = p_c \exp[j2\pi f_c T(m - n)] \exp\{-2[\pi B_c T(m - n)]^2\}$$

From (10.4), the nth tap voltage solely due to a target equals $s_R[\exp(j2\pi f_D nT)]$. The vector of target returns, similar to (8.20), is defined as

$$\begin{aligned} \mathbf{S} &= s_R \exp(j2\pi f_d T)\mathbf{S}_f \\ \mathbf{S}_f^T &= [1, \exp(j2\pi f_d T), \exp(j4\pi f_d T), \exp(j6\pi f_d T), \ldots] \end{aligned} \tag{10.10}$$

and $\mathbf{S}_f$ is a steering or direction vector.

The resulting filter peaks exactly at the doppler frequency of f_d in the absence of clutter, as shown in Figure 10.9, and approximately at f_d in the presence of clutter. Thus, there is a tendency for the peak to shift away from the clutter. The shift increases as the difference between the target and clutter dopplers decreases. Incorporating these equations into (8.7) allows evaluation of the optimum doppler filter weight.

10.6 RELATIVE PERFORMANCE OF NONADAPTIVE AND ADAPTIVE DOPPLER FILTERS

The target detection performance of a doppler filter depends on the clutter characteristics (total power, center-frequency bandwidth, spectral shape), number of pulses processed, target energy, target fluctuating characteristics, and effect of the CFAR circuitry. To avoid confusion due to intermingling of effects, the performance results are obtained for the assumption that a perfect lossless CFAR is implemented and the false-alarm probability equals 1×10^{-6}. In all cases, the target's doppler frequency is assumed to be known exactly and this knowledge is used in (10.10).

The first series of results are for an eight-pulse processor and a Swerling 1 target with the SNR = 20 dB. The SNR is the total target energy for the eight pulses. Thus, the 20 dB corresponds to a numeric value of 100 so that the per-pulse SNR = 12.5. From (1.14), the probability of detection in a noise environment equals 0.872 for a single pulse with SNR = 20 dB.

Figure 10.10 shows the probability of target detection in a combined clutter and noise environment when the filter weights are of equal magnitude. These filter weights are optimum in a noise-only environment. The optimum processor coherently sums the eight-pulse returns and produces a probability of target detection equal

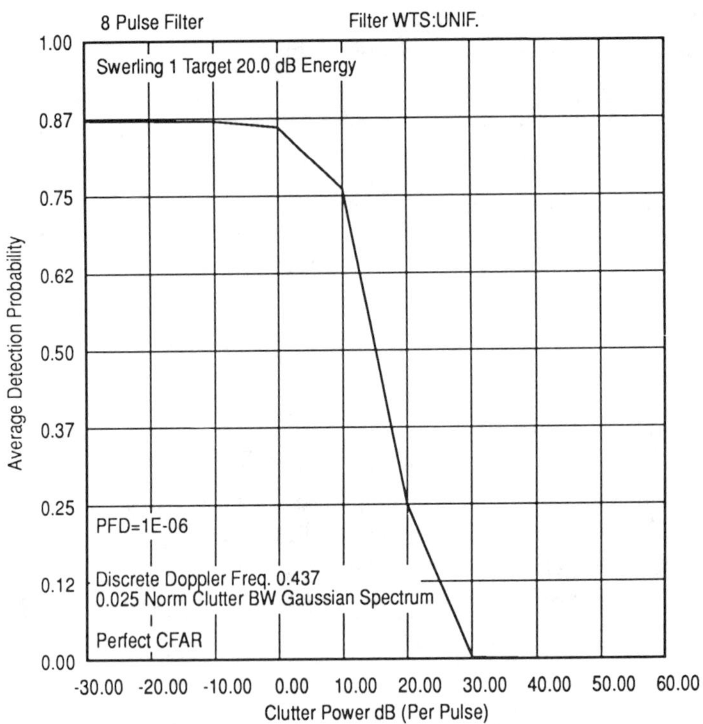

Figure 10.10 Detection probability for doppler filters—Uniform weights f_C = .44.

to that attained by a single-pulse radar when the SNR equals the sum of the per-pulse SNR. When the per-pulse CNR = −30 dB, this is almost a pure noise environment. We can see that, for this clutter level, the probability of detection is approximately equal to the maximum possible value of 0.872. We also can see that the probability of target detection decreases as the clutter power increases.

The detailed computation of the probability of detection is for a Gaussian-shaped clutter spectrum when the ratio of the clutter bandwidth and PRF equals 0.025. For the first set of curves, the center frequency of the clutter equals 0.4375 times the PRF. Thus, the clutter center frequency is approximately at a sidelobe peak and far from the main-lobe region. We can see that the probability of target detection begins decreasing significantly when the CNR is as small as 0 dB.

Figure 10.11 shows the probability of detection for two doppler filters. One is a duplicate of the uniform magnitude weight's filter of the previous figure. The other is for a −40 dB relative sidelobe Chebyschev weight's filter. Note that, at the −30 dB CNR, the probability of detection for the Chebyschev filter is less than that of the uniform weight's filter. This demonstrates the well known signal processing loss associated with a low-sidelobe design. The Chebyschev weighted filter has a higher

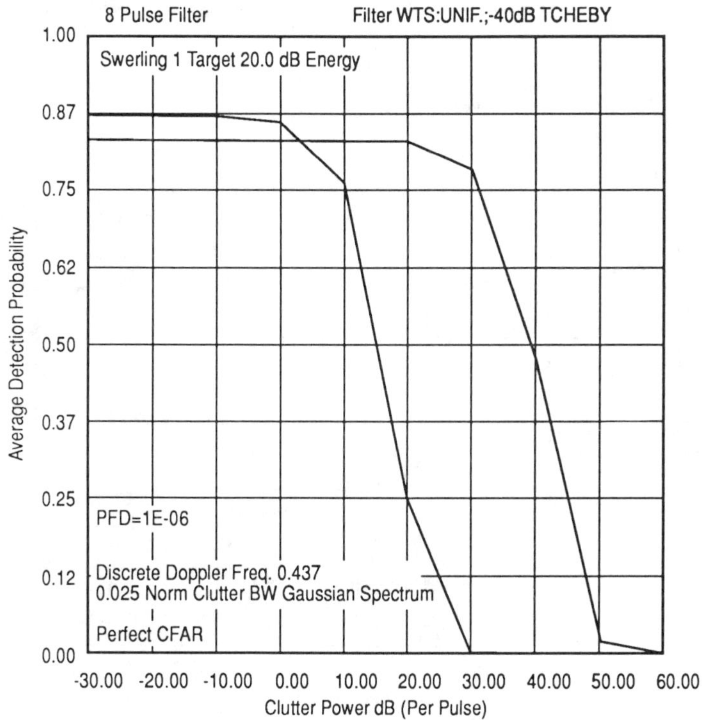

Figure 10.11 Detection probability for doppler filters—Chebyschev weights f_C = .44.

detection probability than the uniform weight's filter when the per-pulse CNR exceeds approximately 0 dB. Figure 10.12 shows the detection probability for the previous two doppler filters plus a third doppler filter, configured by using optimum weights. Note that for the low CNR values, the detection probability of the uniform magnitude weight's filter and the optimum filter are identical. Also note that the performance of the optimum weight's filter is much better than that of the other two filters for high CNR values.

As previously discussed, a defect of Chebyschev weighted filters is that the main-lobe width becomes wider, and this causes difficulty in detecting targets with a doppler frequency which is close to that of the clutter. This defect was not significant for the curves of Figures 10.8 through 10.12, as the clutter's center frequency is not close to the main lobe. Figure 10.13 shows the probability of detection for the three filters when the clutter's center frequency is close to the main lobe. For this figure, the center frequency equals 0.1875 times the PRF. The curves for the three filters are easily distinguishable: that for Chebyschev weighting has the lowest detection probability at the low CNR value; that for optimum weighting has the highest detection probability at the high CNR value. Comparing Figures 10.12 and

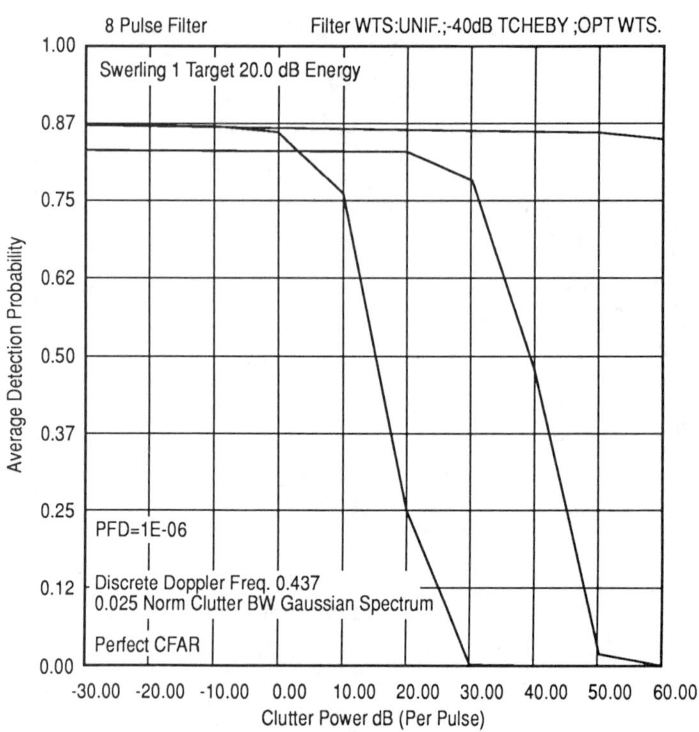

Figure 10.12 Detection probability for doppler filters—Optimum weights $f_C = .44$.

10.13, we can see that the detection probability significantly decreases for all three filters. Also, note that there is almost no advantage in using Chebyschev weighting as compared to uniform weighting.

The final case considered is for the condition that the doppler frequency difference between the target and clutter's center frequency is a random variable uniformly distributed over the unambiguous doppler range. Note that this assumption can cause the clutter spectrum to be within the filter's main-lobe region. As discussed earlier, high-speed targets and the relatively low PRFs for ground-based search radars cause the target's doppler frequency to be highly ambiguous so that the uniformly distributed random-variable model is appropriate. The method used to evaluate the average detection probability for the random-variable model was to set the clutter's center frequency equal to 50 different values, uniformly spaced over the PRF for each CNR value. At each of the 50 values, the detection probability was computed for the two deterministic weight sets by computing the residue power, and thus the filter's output SINR and detection probability. The average of the 50 detection probabilities was computed. For the optimum filter, the optimum weight set was computed at each one of the 50 frequency values for each CNR value. For each optimum

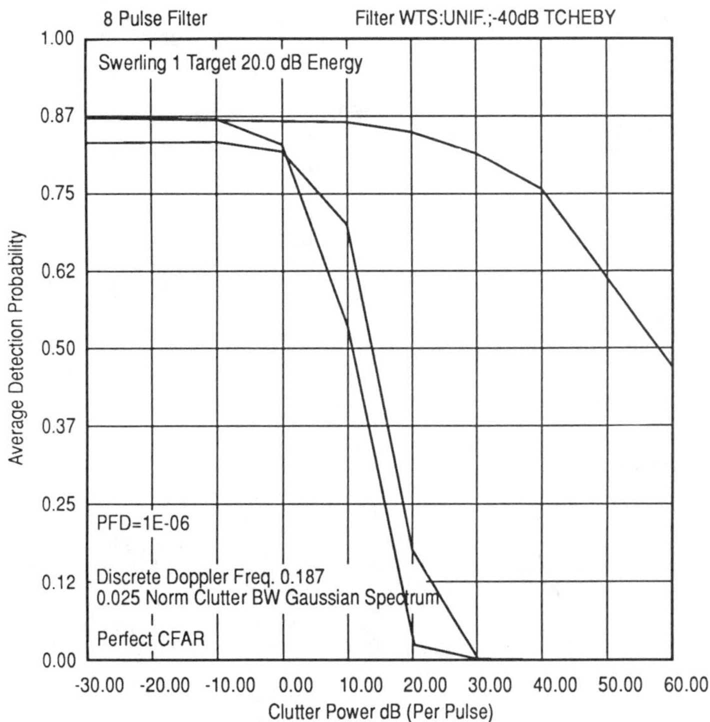

Figure 10.13 Detection probability for doppler filters—Uniform weights $f_C = .19$.

weight set, the resulting output SINR, and thus the detection probability, was computed. Then, the 50 values were averaged. The detection probability of the three filters for this condition is shown by Figure 10.14. Again, optimum weighting is substantially superior to fixed weighting. Figure 10.15 is similar, except that it is done for a 16-pulse filter. We can see that the performance improves significantly when the number of pulses processed is doubled.

10.7 MTI IMPLEMENTATION DETAILS AND MTD

The preceding discussion of MTI delay-line cancellers and doppler filter banks has concentrated on theoretical aspects and neglected several important implementation limitations. The most important limitation is the analog implementation that was used prior to the successful development of digital circuitry. These implementation limitations have had significant effects on the achievable cancellation of MTI circuitry. An example of the limitation is the analog implementation of the delay element of the MTI canceller. Thus, the prior discussion tacitly assumed that the delay-line output was a delayed version of the IF waveform, whether the delay is implemented

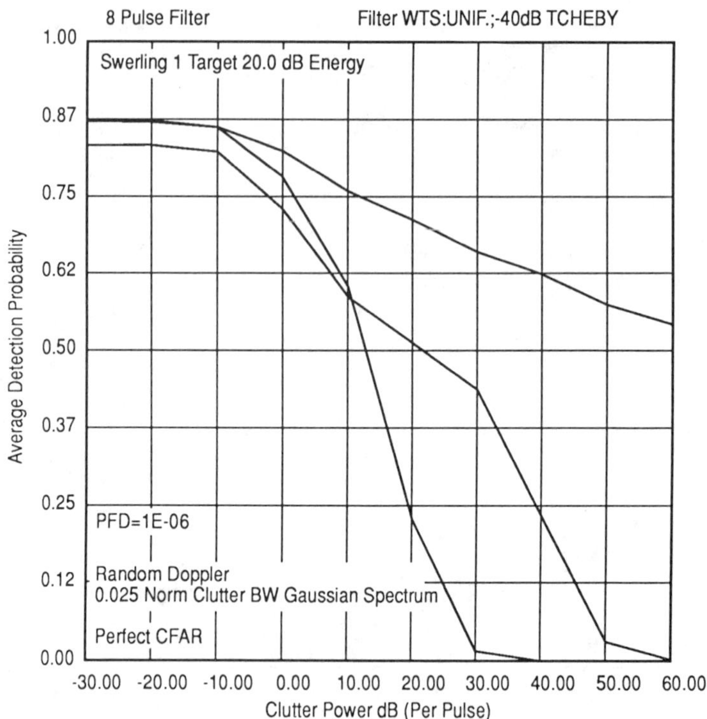

Figure 10.14 Detection probability for doppler filters—Uniform weights random relative clutter center frequency.

by analog or digital circuitry. Specifically, an arbitrary waveform, denoted as $z(t)$, is given as

$$z(t) = x(t)\cos(2\pi f_0 t) - y(t)\sin(2\pi f_0 t) \tag{10.11a}$$

$$z(t) = r(t)\cos[2\pi f_0 t + \theta(t)] \tag{10.11b}$$

$$r(t) = [x(t)^2 + y(t)^2]^{1/2}; \quad \theta = \arctan[y(t)/x(t)]$$

The synchronous demodulator depicted in Chapter 1 (Figure 1.1) forms the outputs $x(t)$ and $y(t)$ by multiplying $z(t)$ by $\cos(2\pi f_0 t)$ and $\sin(2\pi f_0 t)$, respectively, and then filtering. The two outputs are digitized and form the inputs to a digital processor. The two outputs often are considered as a single complex-valued waveform, $z'(t)$. Therefore,

$$z'(t) = x(t) + jy(t)$$

The delay-line implementation is then simply a matter of when data are read out

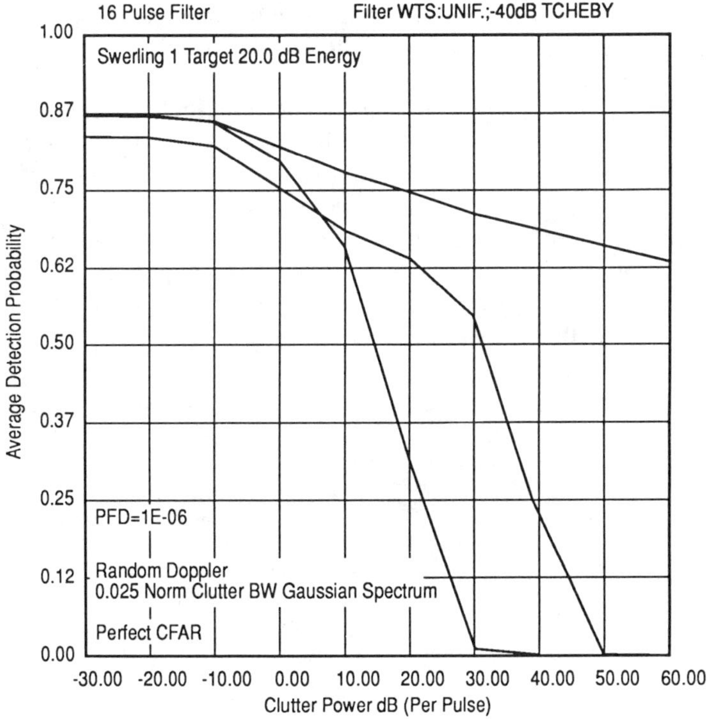

Figure 10.15 Detection probability for doppler filters—Uniform weights random relative clutter center frequency, 16 pulses.

from a memory register. In particular, to ensure that the delays of the two portions of the waveform, $x(t)$ and $y(t)$, are *identical* becomes a simple matter.

The delay of an IF waveform by analog techniques requires a similar technique, but one does not exist. As a consequence, only a single channel can be delayed and the delay circuitry usually implements a version of (10.11b). Often, a limiter is used so that the output is nominally a constant-magnitude, phase-modulated waveform. Limiting was thought to be a desirable operation, as it induced a type of CFAR.

Another problem was that the source of carrier frequency energy was not a gated constant-frequency carrier, but rather a magnetron. This source produced a pulsed carrier with random starting phase. Incorporating phase-locking circuitry was necessary to compensate for the random starting phase. In addition, the long- and short-term stability of the magnetron and various other oscillators was marginal.

In a series of papers (1974), MIT Lincoln Laboratory personnel described the performance shortcomings of the then-current *airport surveillance* radars with regard to detection of targets in clutter. Some of the shortcomings were due to limitations imposed by the current technology when the equipment was designed. Some limitations were due to inadequate analytical techniques used to predict performance during the design phase.

The original radar's design reports predicted a 40 dB clutter improvement factor when using the MTI canceller. However, this value was based on the analysis of a linear canceller, whereas the actual circuitry used a limiter. The results of the Lincoln Laboratory analysis are shown by the curves of Figure 10.16. The curves show that with limiters the improvement factor for the three-pulse canceller equals 23 dB, rather than 40 dB. The reports also discussed that the use of doppler cancellers degrades detection of low-velocity targets and, in particular, prevents the detection of tangential motion targets that are in nonclutter regions.

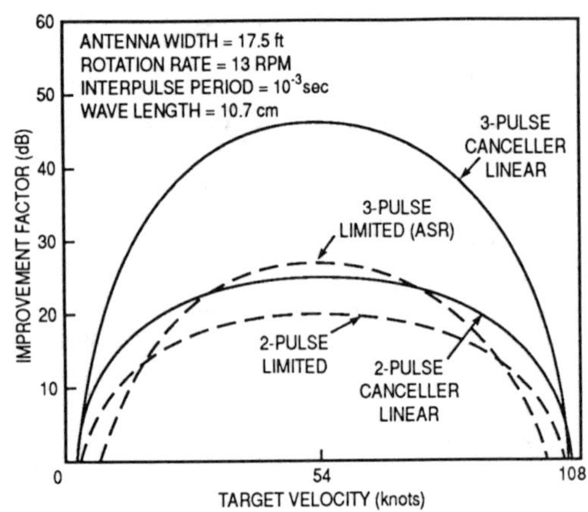

Figure 10.16 MTI canceller improvement factor with and without limiters.
Source: From C.E. Muehe et al, New Techniques Applied to Air-Traffic Control Radars. Proc. IEEE, Vol 62, No. 6, 1974.

Lincoln Laboratory's analysis recommended a multifaceted radar improvement program named, *moving target detection* (MTD). One recommendation was to increase the improvement factor by replacing the nonlinear delay-line canceller MTI with a linear processor. The reports recommended approximate versions of optimally weighted doppler filter banks. Several versions were suggested at different times as a function of the current technology in digital equipment. An example of the expected performance is indicated by the curves of Figure 10.17. The dotted curve is the improvement factor's response curve for one of the filters of the bank. The peak value is a point on the solid curve. The solid curve is the envelope of the peak improvement factor for all possible filters (an infinite number of filters). By comparing the two previous figures, we can see that there is a potential for significantly

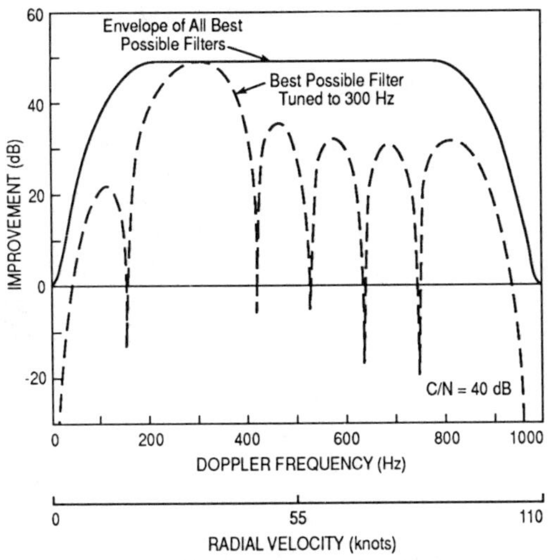

Figure 10.17 Optimum processing improvement factor.
Source: From C.E. Muehe et al, New Techniques Applied to Air-Traffic Control Radars. Proc. IEEE, Vol 62, No. 6, 1974.

superior performance. In particular, note that the improvement factor at the blind speeds equals 0 dB so that there is no target detection degradation. Also, the notch widths have been decreased so that the detection of targets with velocities close to blind speeds has been improved. Other changes were also recommended, including changing the transmitter from a magnetron to a klystron, changing the transmitting waveform to no longer using stagger-coded PRFs and to use modern CFAR processing, rather than the implemented logarithmic FTC.

A block diagram of the earliest MTD system is shown in Figure 10.18. The input is to a "ping-pong" memory. As shown, the top memory is being loaded while the data, stored in the bottom memory during the preceding coherent processing interval, are being processed. The earliest MTD doppler processor was chosen for relatively simple implementation because of the severe constraint of available components. With this constraint, a reasonable approximation to an optimum doppler processor for ground clutter ws obtained by cascading a three-pulse canceller and an eight-pulse doppler filter bank processor, configured by using a FFT. The outputs of these filters are combined in weighted triplets to reduce the sidelobe response. Specifically, the output of each filter is combined with the pair of adjacent filters by using weights of relative values of $-1/4$, 1, and $1/4$. In addition, a separate parallel filter is implemented for zero doppler targets. As indicated, the recommended CFAR technique is the clutter map. Later versions also added a parallel CACFAR so as to produce a version of an hierarchal CFAR.

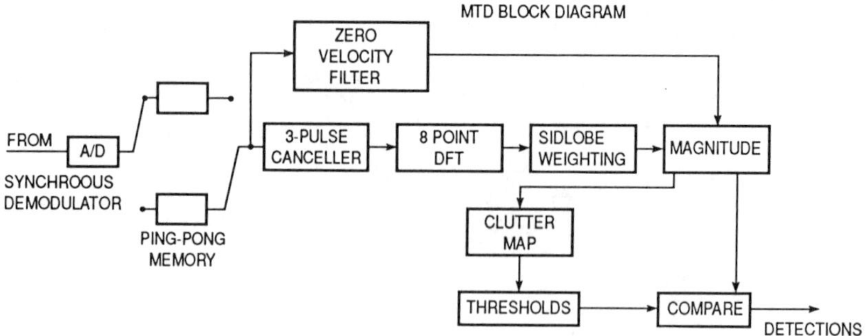

Figure 10.18 Conceptual block diagram of MTD Technique.
Source: From C.E. Muehe et al, New Techniques Applied to Air-Traffic Control Radars. Proc. IEEE, Vol 62, No. 6, 1974.

10.8 COMBINED CFAR-SMI ALGORITHM

Any of the algorithms discussed for adaptive antennas are also applicable to adaptive doppler processing. The SMI algorithm is of special interest because of an inherent CFAR capability by way of a minor modification. The modified SMI algorithm can also be used for adaptive antennas. However, because of its applicability to heterogeneous environments, it is particulary of interest for adaptive doppler processing.

The technique's concept is sketched in Figure 10.19. The estimated optimum doppler filter (one of several of a bank) is cascaded with an adaptive gain. Because, for a heterogeneous clutter environment, the clutter cross section and spectrum may change from one range cell to another, the value of the adaptive gain and the estimated weight vector of the adaptive doppler filter can be recomputed from range cell to range cell. As shown below, when the adaptive gain is properly chosen, the residue power level of the detector input is independent of the power level and spectrum of the adaptive filter input. Thus, the combination of the adaptive filter's weight vector and adaptive gain produces a CFAR action.

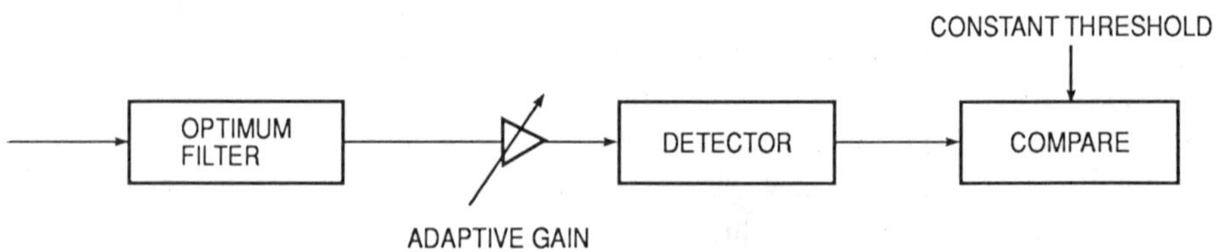

Figure 10.19 Conceptual block diagram of combined SMI and CFAR.

Extrapolating from (8.7), the estimated optimum weight for the SMI algorithm is

$$\mathbf{G}_{\text{est}} = \alpha(\mathbf{R}_{\text{est}})^{-1}\mathbf{S}$$

where $\mathbf{R}_{\text{est}}$ is the covariance matrix estimate, $\mathbf{S}$ is the steering vector for the filter (or antenna beam), $\mathbf{G}_{\text{est}}$ is the estimated optimum weight vector, and α is an arbitrary constant. Note that α is totally arbitrary, and its value can be changed from range cell to range cell. Thus, α can serve as the adaptive gain. Because the residue power when using optimum weights is

$$p_{\text{res}} = \mathbf{S}_{\mathbf{H}}\mathbf{R}^{-1}\mathbf{S}$$

an intuitive choice of α is

$$\alpha^{-2} = \mathbf{S}^{\mathbf{H}}\mathbf{R}_{\text{est}}^{-1}\mathbf{S}$$

so that a possible modification of the previous equation is

$$\mathbf{G}_{\text{est}} = (\mathbf{R}_{\text{est}})^{-1}\mathbf{S}/[\mathbf{S}^{\mathbf{H}}(\mathbf{R}_{\text{est}})^{-1}\mathbf{S}]^{1/2}$$

As proved below, this technique, derived on an intuitive basis, produces a CFAR action. Thus, when the adaptive weights are computed by using this algorithm, the residue power does not depend on the actual covariance matrix.

The proof uses the same transformation technique as in (8.72) through (8.82). Thus, the covariance matrix is estimated from K data vectors $\{Z_k\}$ as

$$\mathbf{R}_{\text{est}} = \Sigma(\mathbf{Z}_k\mathbf{Z}_k^{\mathbf{H}})/K$$

Imagine that "nature" generates the vectors $\{\mathbf{Z}_k\}$ from vectors $\{\mathbf{X}_k\}$ by the transformations:

$$\mathbf{Z}_k = \mathbf{R}^{1/2}\mathbf{U}\mathbf{X}_k; \quad \mathbf{X}_k = \mathbf{U}^{\mathbf{H}}\mathbf{R}^{-1/2}\mathbf{Z}_k$$

$$\mathbf{S}_x = \mathbf{U}^{\mathbf{H}}\mathbf{R}^{-1/2}\mathbf{S}$$

$\mathbf{R}$ is the true covariance matrix. Note that $\mathbf{S}_x$ depends on the unitary matrix $\mathbf{U}$. The matrix $\mathbf{U}$ is defined such that the first column is the unit-length vector colinear with $\mathbf{R}^{-1/2}\mathbf{S}$. Then, because the other columns of $\mathbf{U}$ are orthgonal to the first, only the first entry of $\mathbf{S}_x$ is nonzero. Note that $\mathbf{U}$ and $\mathbf{R}$ are unknown to the observer, who only measures $\{\mathbf{Z}_k\}$, but these matrices are known to nature. As shown by (8.74), the covariance matrix of each $\mathbf{X}_k$ is the identity matrix, and the estimated covariance matrix is given by (8.75):

$$\mathbf{R}_{\text{est}} = (\Sigma \mathbf{Z}_k \mathbf{Z}_k^{\mathbf{H}})/K = \mathbf{R}^{1/2}\mathbf{U}(\Sigma \mathbf{X}_k \mathbf{X}_k^{\mathbf{H}})\mathbf{U}^{\mathbf{H}}\mathbf{R}^{1/2}$$

We denote $(\Sigma \mathbf{X}_k \mathbf{X}_k^{\mathbf{H}})/K$ as $\mathbf{I}_{\text{est}}$. Then, because $\mathbf{U}$ is unitary, $\mathbf{U}^{\mathbf{H}}$ equals $\mathbf{U}^{-1}$. Thus,

$$(\mathbf{R}_{\text{est}})^{-1} = \mathbf{R}^{1/2}\mathbf{U}(\mathbf{I}_{\text{est}})^{-1}\mathbf{U}^{-1}\mathbf{R}^{-1/2}$$

Then, substituting the definition for $\mathbf{S}_x$, the estimated optimum weight vector is

$$\mathbf{G}_{\text{est}} = \mathbf{R}^{-1/2}\mathbf{U}(\mathbf{I}_{\text{est}})^{-1}\mathbf{S}_x/[\mathbf{S}_x^{\mathbf{H}}(\mathbf{I}_{\text{est}})^{-1}\mathbf{S}_x]^{1/2}$$

By noting that $\mathbf{I}_{\text{est}}$, $\mathbf{R}$, and all powers of each are Hermitian, the result is

$$p_{\text{res}} = \mathbf{S}_x^{\mathbf{H}}(\mathbf{I}_{\text{est}})^{-2}\mathbf{S}_x/\mathbf{S}_x^{\mathbf{H}}(\mathbf{I}_{\text{est}})^{-1}\mathbf{S}_x$$

Therefore, because only the first component of $\mathbf{S}_x$ is nonzero:

$$p_{\text{res}} = [(\mathbf{I}_{\text{est}})^{-2}]_{11}/[(\mathbf{I}_{\text{est}})^{-1}]_{11}$$

where the subscripts 11 denote the first row and column entry of the respective matrix.

This result proves that the residue power is independent of the covariance matrix, $\mathbf{R}$, so that the adaptive gain produces a CFAR action. Also, note that this CFAR technique is compatible with the previously discussed techniques. Thus, the techniques can be cascaded for possibly improved performance.

Chapter 11
Application of Modern Spectral Estimation Techniques to Doppler Processing

11.1 INTRODUCTION

The previously discussed SMI algorithms were based on maximum-likelihood estimation of either a Hermitian or persymmetric Hermitian covariance matrix, depending on the inherent symmetry of the problem. For constant PRF waveforms, the covariance matrix for doppler processors has the additional symmetry of Hermitian-Toeplitz. As previously discussed, ML estimation of these matrices is complicated because of the extreme nonlinearity of the resulting equations. However, the use of some type of estimator that exploits this symmetry is desirable. For example, the SMI algorithms that average data from adjacent range cells explicitly assume environment homogeneity over the range window. However, the characteristics of clutter change substantially over small range extents. As demonstrated below, the use of modern spectral estimation techniques allows estimation of the covariance matrix by use of a single-range-cell data vector. Thus, the technique copes with the worst possible clutter nonhomogeneity.

The main property of a Hermitian-Toeplitz matrix is that the first column is obtained from the autocorrelation function of the random process, and the entire matrix is specified from the values of the first column. Spectral estimation techniques are applicable to the estimation of Hermitian-Toeplitz matrices because of the Fourier transform relationship between an autocorrelation function and the power spectral density (psd) of a random process. Therefore, Fourier transformation of the estimated psd is one method of estimating the covariance matrix. As discussed below, there are many different spectral estimation techniques. The relative performance of each has not yet been determined because of the difficulty of analysis. The technique that is emphasized is the *maximum entropy method* (MEM) or all-pole modeling. Experimental and simulated results indicate that the technique works well for doppler processing. In addition, the spectral estimate is relatively simple to implement and can be formed from the data of a single range cell so that it is compatible with heterogeneous clutter. Note that the emphasis is on doppler processing. As discussed in more detail later, adaptive doppler processing can succeed with relatively low-fidelity estimates of the psd. Other applications, such as target identification, may require high-fidelity estimates, and hence other estimation techniques.

11.2 MAXIMUM ENTROPY SPECTRAL ESTIMATION

Modern spectral estimation was initially developed by use of an entropy concept. *Entropy* here refers to the information theoretic concept. Shortly after this initial derivation, the entropy concept was shown to be equivalent to assuming that the observed N-point data sequence was formed by passing white noise through an all-pole filter of $N - 1$ poles. Because of the form of the difference equation describing the input-output relations for an all-pole filter, it is also called an *autoregressive filter*. This observation expands the available estimation procedure to modeling the data as produced by more general circuits.

An example of this concept is shown in Figure 11.1. We should emphasize that, although this filter also uses a tapped delay line, the concept is *not* related to the delay line structure used for doppler filtering. The circuit of Figure 11.1 is an imaginary circuit and only intended as a model of nature generating the observed data.

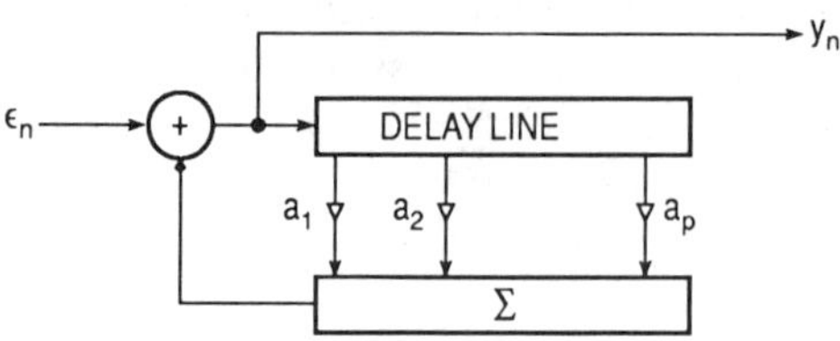

Figure 11.1 All-pole network example.
Source: From R. Nitzberg, *Implementation of an Adaptive Processor by Modern Spectral Estimation Techniques*. Int'l Conf on Communications, Denver, CO, July 1981.

Also possible is to model the observed data as due to white noise exciting an all-zero network or a network with both poles and zeros. The all-pole model has received the major interest because the estimation procedure (estimating the coefficients a_1 through a_P, and from these coefficients computing the power spectrum estimate) is relatively simple. An example of a typical estimated spectrum as compared to a true but unknown spectrum is sketched in Figure 11.2. The dashed curve shows the spectrum estimated when using a four-pole network model. The four peaks of the estimated spectrum are due to the four poles of the modeling network. Because the estimated spectrum is not identical to the true spectrum, the adaptive doppler filter weights computed by the estimation technique are not correct. However, note

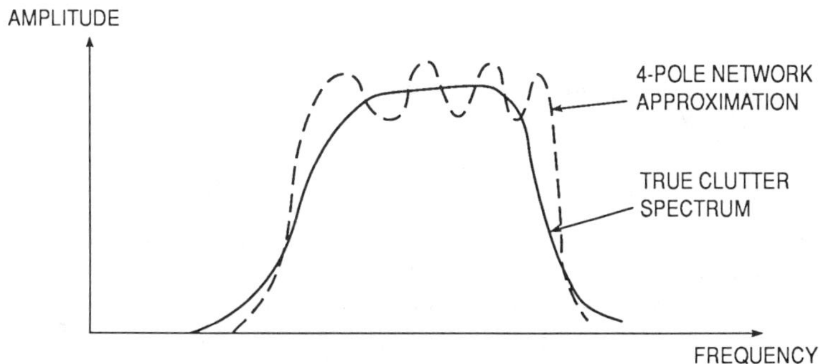

Figure 11.2 Theoretical spectral model mismatch.
Source: From R. Nitzberg, *Implementation of an Adaptive Processor by Modern Spectral Estimation Techniques*. Int'l Conf on Communications, Denver, CO, July 1981.

that the region of low response for the adaptive doppler filter must be in approximately the correct region so that an approximately correct optimum doppler weight will be computed. There will be a loss, termed a *mismatch loss,* because the estimated spectrum is not identical to the true interference spectrum. However, because any spectrum can be approximated with arbitrarily small error as the number of poles increase, we expect that the mismatch loss can be made small. Thus, we expect the output SINR of the adaptive doppler filter to approach that of the optimum doppler filter as the number of poles in the modeling network increases. However, as shown below, increasing the number of poles of the modeling network will increase the complexity of the implementation so that there is a required implementation trade-off.

11.3 THEORETICAL MISMATCH LOSS

To quantify the mismatch loss, we analyze the case where the first $P + 1$ lag values of the autocorrelation function of the observed data sequence is known exactly. Specifically, denoting $R(k)$ as the autocorrelation function with time lag of kT, the values $R(0)$, $R(1)$, $\ldots$, $R(P)$ are assumed to be known exactly. For P less than $N - 1$, where N is the number of pulses processed by the adaptive filter, this information is inadequate to compute the weights of the optimum doppler filter. The unknown but required autocorrelation values to compute the adaptive doppler filter weights are estimated by using the MEM technique.

Given the values $R(0)$ through $R(P)$, there is a unique P-pole autoregressive filter, the output autocorrelation function of which equals the specified values. The extrapolated autocorrelation function of the observed process is then computed as that of the modeling P-coefficient all-pole filter. As the optimum filter theory is applicable when N values of the interference autocorrelation sequence are known, the "optimum" adaptive doppler filter weights can now be computed. However, as the true spectrum is the sum of thermal noise plus clutter, it cannot be generated by an all-pole network, and there must be a difference between the true optimum Doppler weights and those computed by this procedure. As shown by some examples to follow, this loss is often quite small.

A specific example is given by the curves of Figure 11.3. It shows the mismatch loss for a particular weather-clutter spectrum for a tactical radar. The total interference is the sum of clutter with a Gaussian spectrum and white thermal noise. The curves show the mismatch loss *versus* the number of poles of the modeling network for three CNR ratios and a 16-pulse processor. The curves show that the mismatch loss decreases with decreased CNR and decreases monotonically as the number of poles of the circuit model increases. The maximum number of poles is shown to be eight because the complexity of the MEM algorithm increases substantially when the number of poles exceeds $N/2$.

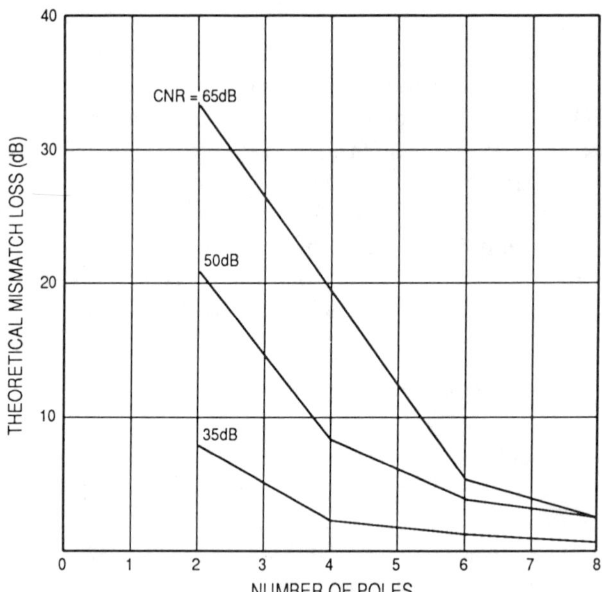

Figure 11.3 Theoretical mismatch loss for volume clutter.
Source: From R. Nitzberg, *Implementation of an Adaptive Processor by Modern Spectral Estimation Techniques*. Int'l Conf on Communications, Denver, CO, July 1981.

An indication of the variation of the mismatch loss with clutter bandwidth is given by Figure 11.4(a) and (b). The horizontal axis is the standard deviation of the

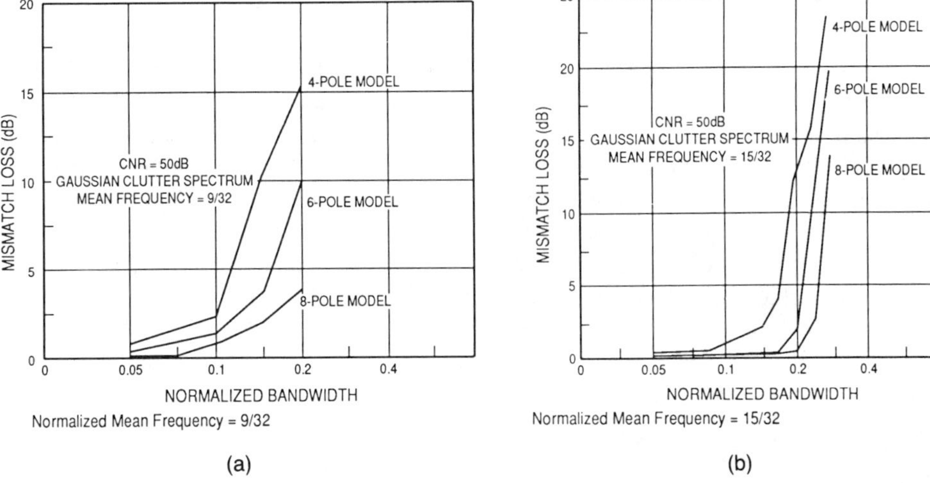

Figure 11.4 Theoretical mismatch loss.
Source: From R. Nitzberg, *Some Design Details of the Application of Modern Spectral Estimation Techniques to Adaptive Processing*. ASSP Workshop on Modern Spectral Estimation, Hamilton, Ontario, August 1981.

Gaussian spectrum normalized to the radar PRF. In Figure 11.4(a), the clutter's center frequency has a normalized mean shift of 9/32, relative to the target doppler frequency. In Figure 11.4(b), the normalized shift equals 15/32. In both cases, CNR = 50 dB. We can see that the mismatch loss increases with bandwidth and decreases with the number of poles in the autoregressive circuit model. Note that the mismatch loss is substantially greater for the 9/32 mean shift. We conjecture that this is due to the fact that the skirts of the autoregressive spectrum decreases at a much lower rate than the actual spectrum. At the smaller mean shift, this causes the skirts of the autoregressive spectrum to have a greater effect in the region of the target doppler than that of the true clutter spectrum.

11.4 ESTIMATION LOSS

The previous mismatch loss curves were based on the operationally unrealistic assumption that the first $P + 1$ values of the autocovariance sequence would be exactly known. For this assumption, the performance loss is only due to the mismatch of the true noise-plus-clutter spectrum and the modeled spectrum. A more realistic case is where the network parameters are estimated from data. Thus, in addition to the modeling loss, there is an estimation loss that depends on the number of data range cells used to form the estimate as well as the degree of homogeneity over range. The curves of Figures 11.5 and 11.6 indicate that the loss is small, even if only one range cell of data is used. The three curves of Figure 11.5 are the theoretical spectrum of the noise plus clutter and the estimated spectrum for two Monte Carlo simulations of a four-pole spectral model. The estimation algorithm is discussed in detail in Secton 11.5. The improvement factor when using optimum weights for the true spectrum is 35.9 dB and for the theoretical four-pole model equals 27.5 dB. The difference, 8.4 dB, is the mismatch loss. For the first Monte Carlo experiment, the experimental improvement factor is 27.2 dB; for the second, it equals 24.3 dB. As both these values are close to the theoretical mismatch loss, we can see that the estimation loss is small.

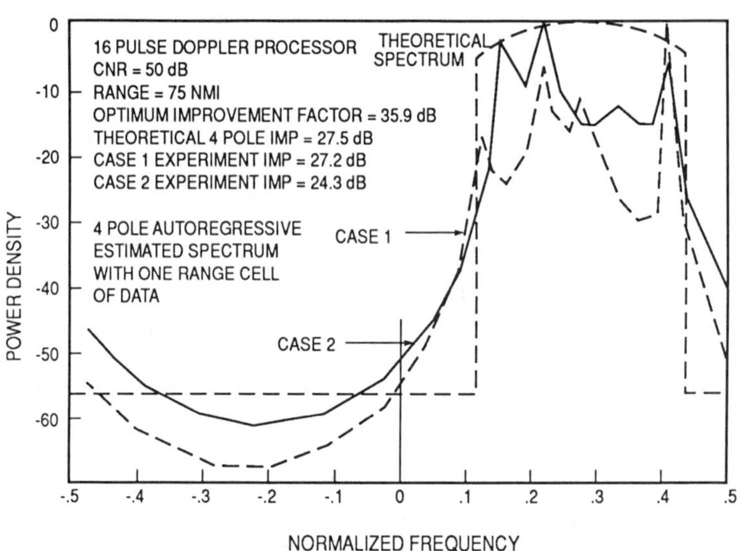

Figure 11.5 Examples of estimated four-pole spectrum.
Source: From R. Nitzberg, *Implementation of an Adaptive Processor by Modern Spectral Estimation Techniques*. Int'l Conf on Communications, Denver, CO, July 1981.

The curves of Figure 11.6 are from a single Monte Carlo experiment for modeling a slightly narrower spectrum by an eight-pole autoregressive spectrum. Note that the estimated spectrum differs considerably from the true spectrum. However, the loss of the adaptive doppler filter, the design of which is based on this estimated spectrum, is only 3.2 dB greater than that of the optimum (interference spectrum

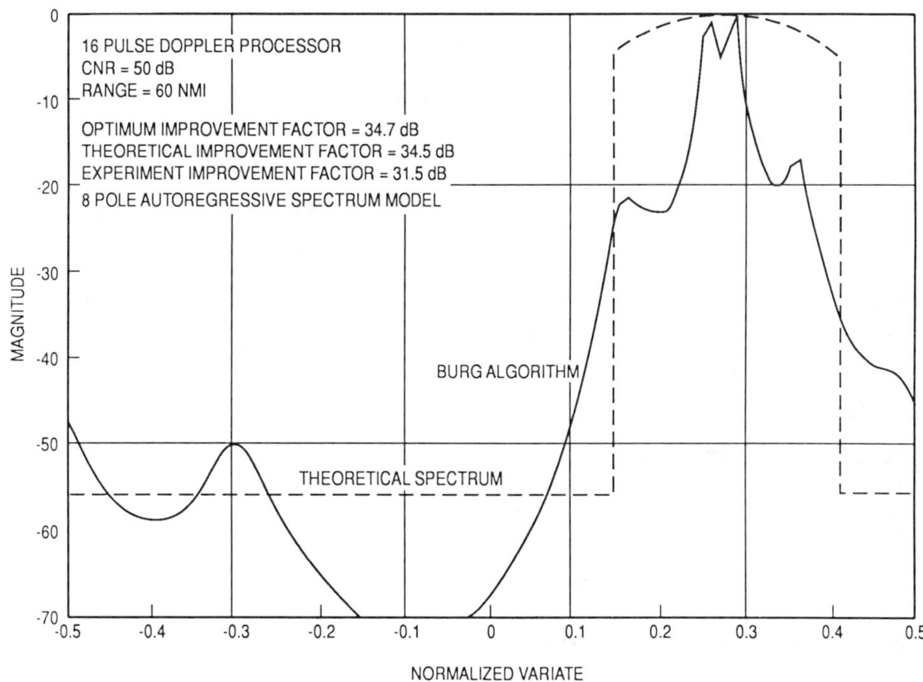

Figure 11.6 Example of estimated eight-pole spectrum.
Source: From R. Nitzberg, *Some Design Details of the Application of Modern Spectral Estimation Techniques to Adaptive Processing*. ASSP Workshop on Modern Spectral Estimation, Hamilton, Ontario, August 1981.

known) doppler filter. Thus, the curves of Figures 11.5 and 11.6 verify that low-fidelity spectral estimates are satisfactory. This occurs because, although the estimated spectrum is poor, the estimate correctly indicates the spectral region of high interference power.

A decomposition of the loss relative to optimum processing is given by the data of Table 11.1. The loss values were obtained from 25 Monte Carlo simulations for each case and computing the average loss. The table shows the losses for three different normalized clutter bandwidths. We can see that the mismatch loss decreases monotonically with increased number of poles for all three bandwidths. However, there is a tendency for the estimation loss to increase with the number of poles used in the autoregressive model. Thus, the combined loss sometimes increases or decreases with the number of poles.

Table 11.1
Losses Relative to Optimum Processing
(CNR = 50 dB)
Mean Clutter Frequency = 9/32

Bandwidth	*Model*	*Mismatch Loss (dB)*	*Estimation Loss (dB)*	*Combined Loss (dB)*
0.12745	4-Pole	8.40	1.41	9.81
	6-Pole	4.10	4.40	8.50
	8-Pole	2.50	4.66	7.16
0.0881	4-Pole	1.80	2.62	4.42
	6-Pole	1.00	2.04	3.04
	8-Pole	0.60	2.37	2.97
0.05	4-Pole	0.80	0.99	1.79
	6-Pole	0.30	1.97	2.27
	8-Pole	0.20	2.72	2.92

11.5 MEM ESTIMATION EQUATIONS

The MEM technique was invented by J.P. Burg* in the early 1960s to overcome the well known disadvantage of "poor" spectral resolution associated with "classical" spectral estimation. The estimation technique has evolved considerably since its introduction. The evolutionary history will not be discussed; only the present form will be emphasized here.

The viewpoint of modern spectral estimation is to assume that all random data sequences can be approximated as being generated by passing a white noise sequence through the filter of Figure 11.7. A difference equation that relates the input and output sequences is

$$x(t) = \sum_{k=1}^{P} a_k x(t - kT) + \sum_{k=0}^{Q} b_k n(t - kT) \tag{11.1}$$

The Fourier transform of (11.1) gives the filter transfer function:

$$H(f) = \sum_{k=0}^{Q} b_k \exp(-j2\pi f kT) \Bigg/ \left[1 - \sum_{k=1}^{P} a_k \exp(-j2\pi f kT)\right]$$

*D.G. Childers, Ed., "Modern Spectrum Analysis," IEEE Press, 1978.

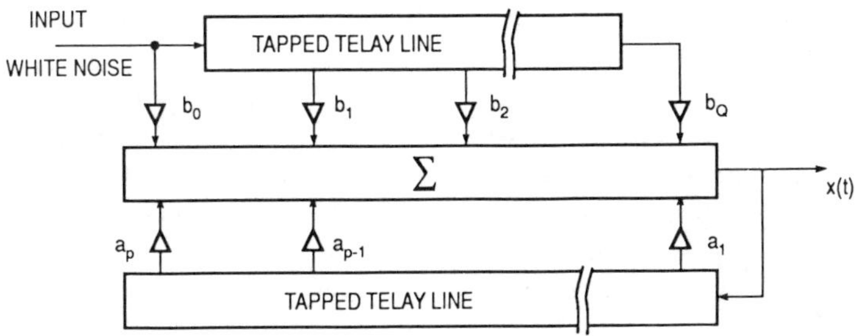

Figure 11.7 General filter model.

This function can be restated as a z-transform by noting that $z = \exp(j2\pi fkT)$, so that

$$H(z) = x^{P-Q} \sum_{k=0}^{Q} b_k z^{Q-k} \Big/ \sum_{k=0}^{P} a_k z^{P-k} \tag{11.2}$$

where a_0 is defined as equal to unity. Because the filter input is white, the z-transform psd of the output, $S(z)$, is given by

$$S(z) = H(z)H^*(1/z^*) \tag{11.3}$$

The transfer function has Q zeros and P poles. The locations of the poles are determined by the values of the feedback coefficients, a_k, and that of the zeros by the values of the feed-forward coefficients, b_k. The spectral estimation problem then consists of using the data to estimate the sets of coefficients $\{a_k\}$ and $\{b_k\}$. After these coefficients are estimated, the estimated spectrum is computed by inserting the coefficient estimates into (11.2) and (11.3).

The Burg algorithm uses the output data at the taps of the doppler filter, as shown in Figure 11.7 to estimate the feedback coefficients of an all-pole network. The input data to the algorithm is denoted as the complex voltages $\{y(n)\}$ at the N taps of the adaptive doppler processor. These voltages correspond to the reflected energy in a particular range cell from the last N transmitted pulses. To develop the form of the algorithm, note from Figure 11.1 that the output of an all-pole network, $y(n)$, is given by the difference equation:

$$y(n) = \varepsilon(n) + \sum_{k=1}^{P} a_{Pk} y(n-k)$$

where $\varepsilon(n)$ is the white noise sequence exciting the network and a_{Pk} denotes the filter coefficients. The double subscript is used because the algorithm employs recursive relations that successively approximates the data by a one-pole network, two-pole network, *et cetera,* up to a P-pole network. The coefficients $\{a_{Pk}\}$ are determined as those most consistent with the data, given appropriate constraints.

The equations for the best set of coefficients is as follows. Define forward and backward fitting sequences at the mth step of the iterative algorithm as $f_m(n)$ and $b_m(n)$. As the first step, these sequences are initialized as

$$f_1(n) = y(n + 1); \quad b_1(n) = y(n); \quad 1 \leqslant n \leqslant N - 1 \tag{11.4}$$

The coefficients with identical subscripts, a_{mm}, are denoted as reflection coefficients. The first reflection coefficient is computed as

$$a_{mm} = \sum_{n=1}^{N-m} 2f_m(n)[b_m(n)]^* \Big/ \sum_{n=1}^{N-m} [|f_m(n)|^2 + |b_m(n)|^2] \tag{11.5}$$

with m equal to unity. A recursive relation for the higher order reflection coefficients is obtained via the update relations

$$\left.\begin{aligned} f_m(n) &= f_{m-1}(n + 1) - a_{m-1,m-1}b_{m-1}(n + 1) \\ b_m(n) &= b_{m-1}(n) - [a_{m-1,m-1}]^* f_{m-1}(n) \end{aligned}\right\}; \quad 1 \leqslant n \leqslant N - m$$

and then reusing (11.5). This is continued until $m = P$, the number of poles in the approximating network. The other filter coefficients are then calculated from the recursive relation:

$$a_{m,k} = a_{m-1,k} - a_{mm}[a_{m-1,m-k}]^*; \quad 1 \leqslant k \leqslant m - 1 \tag{11.6}$$

11.6 OPERATIONS COUNT IMPLEMENTATION CONSIDERATIONS

For a real-time processor, the number of required multiplication operations to implement an algorithm has a major effect on the architecture.

The algorithm computation starts by computing a_{11}. This requires $N - 1$ complex multiplications for the numerator and $2(N - 1)$ complex multiplications for the denominator (ignoring multiplying by the factor of $1/2$). The division is usually done by a reciprocal table look-up and multiply. Thus, the total number of complex multiplications to compute $a_{11} = 3N - 2$. Now, m is taken as equal to 2 and, using (11.4), $f_2(n)$ and $b_2(n)$ are updated for $N - 2$ values. Thus, there are $2 \times (N - 2)$

updates, with each requiring one complex multiplication. In general, the updates require $2 \times (N - m)$ complex multiplications. Computing a_{mm} requires $3 \times (N - m)$ complex multiplications to evaluate the numerator and denominator, and one for the division. Thus, the number of complex multiplications required for computing an a_{mm} value for $m > 1$ entails $5 \times (N - m) + 1$ complex multiplications. The number required to compute a_{11} through a_{PP} is

$$\#_1 = 3N - 2 + \Sigma(5N - 5m + 1)$$

The coefficients a_{P1} through $a_{P,P-1}$ are obtained by successive application of the recursive relationship of (11.6). The required number of complex multiplications for this computation is

$$\#_2 = 3N - 2 + \Sigma(5N - 4m)$$

The estimated optimum weight vector can now be computed by the estimated covariance matrix. Whereas previous discussions indicated that the preferred computational procedure for SMI avoided computing the inverse covariance matrix, it is computed when using this technique because there is a simple and direct relation between the network feedback coefficients $\{a_k\}$ and the inverse covariance matrix. This relation is so simple that there is a preference to use the autoregressive circuit model, even if others produce better estimates.

An example of the direct relation of the inverse covariance matrix of a two-pole autoregressive filter is given below:

$$\mathbf{R}^{-1} = \begin{bmatrix} 1 & -a_1^* & -a_2^* & 0 & 0 & 0 \\ -a_1 & 1 + |a_1|^2 & -a_1^* + a_1 a_2^* & -a_2^* & 0 & 0 \\ -a_2 & -a_1 + a_1^* a_2 & 1 + |a_1|^2 + |a_2|^2 & -a_1^* + a_1 a_2^* & -a_2^* & 0 \\ 0 & -a_2 & -a_1 + a_1^* a_2 & 1 + |a_1|^2 + |a_2|^2 & -a_1^* + a_1 a_2^* & -a_2^* \\ 0 & 0 & -a_2 & -a_1 + a_1^* a_2 & 1 + |a_1|^2 & -a_1^* \\ 0 & 0 & 0 & -a_2 & -a_1 & 1 \end{bmatrix}$$

Note that the matrix has a quasi-Toeplitz structure, as the entries on the diagonals are identical after a *transient*. Let us denote the elements in the ith row and jth column of the inverse matrix as r^{ij}. The general expression for the quasi-Toeplitz portion of the matrix entries is

$$r^{i,i+k} = \Sigma a_n [a_{n+k}]^*; \quad 0 \leq k \leq P, P + 1 - k \leq i \leq N - P$$

For $k > P$, the values of the matrix entries are zero. The general expression for the transient entries is

$$r^{ij} = \Sigma a_n[a_{n+j-i}]^*; \quad N - P + 1 \leq i \leq N, i \leq j \leq N$$

The other matrix entries are obtained by using the Hermitian persymmetric properties

$$r^{N+1-j,N+1-i} = r^{ij}$$

$$r^{ji} = [r^{ij}]^*$$

To use these expressions to compute the inverse matrix, note that the value of a_0 is defined as unity. Therefore, the number of complex multiplications required to compute the quasi-Toeplitz form of each diagonal equals $P - k$. For k diagonals, the number of complex multiplications to compute all quasi-Toeplitz entries equals $\Sigma(P - k)$. For the other terms, note from the previous example that computing r^{33} had produced the r^{22} term so that recomputing this term was not necessary. Similarly, all the other nonquasi-Toeplitz terms have been obtained in the intermediate steps. Thus, the total number of complex multiplications to obtain the inverse matrix is

$$\#_3 = 5NP - 3P(P + 1)/2 - 2(N + 1)$$

The last step is the computation of the adaptive weighting. The Toeplitz nature of the covariance matrix and the special symmetry of the steering vector combine to simplify the weight-vector computation. Specifically, with proper normalizaton of phase, the components of the steering vector are related as

$$s_k = [s_{N-k+1}]^*$$

and the adaptive weight components are therefore similarly related. Thus, it is only necessary to compute the first $N/2$ adaptive weights. There are additional simplifications in computing the weights because the inverse matrix has many entries of zero and one. Multiplying by either of these is trivial and does not count as an operation. Again, using the example inverse covariance matrix, note that computing the first adaptive weight requires P complex multiplications. Computing the second requires $P + 2$ complex multiplications. This count increases by one for each weight until the maximum possible number of complex multiply operations of $2P + 1$ is reached. Thus, we have the number of complex multiplication operations required to compute the weight vector after the inverse covariance matrix is computed:

$$\#_4 = N(2P + 1)/2 - \left(\frac{1}{2}P^2 + \frac{1}{2}P + 1\right)$$

In addition, applying the CFAR normalization discussed in Section 10.8 requires N complex multiplications to compute the inner product of the steering and weight vectors. Combining all these counts gives the required number of complex multiplication operations to update and apply the adaptive weight vector:

$$\# = N + (6NP - N/2 - 2P - 2P^2 + 1)$$

which is approximately equal to $6NP$. Thus, the complexity of the equipment implementation increases linearly with the number of poles of the modeling network and a performance *versus* equipment complexity trade-off is required. Also note that the complexity of this technique is small compared to SMI which requires approximately N^3 multiplications to compute and apply the adaptive weights.

INDEX

The Artech House Radar Library

David K. Barton, *Series Editor*

Active Radar Electronic Countermeasures by Edward J. Chrzanowski

Airborne Pulsed Doppler Radar by Guy V. Morris

AIRCOVER: Airborne Radar Vertical Coverage Calculation Software and User's Manual by William A. Skillman

Analog Automatic Control Loops in Radar and EW by Richard S. Hughes

Aspects of Modern Radar by Eli Brookner, *et al.*

Aspects of Radar Signal Processing by Bernard Lewis, Frank Kretschmer, and Wesley Shelton

Bistatic Radar by Nicholas J. Willis

Detectability of Spread-Spectrum Signals by Robin A. Dillard and George M. Dillard

Electronic Homing Systems by M.V. Maksimov and G.I. Gorgonov

Electronic Intelligence: The Analysis of Radar Signals by Richard G. Wiley

Electronic Intelligence: The Interception of Radar Signals by Richard G. Wiley

EREPS: Engineer's Refractive Effects Prediction System Software and User's Manual, developed by NOSC

Handbook of Radar Measurement by David K. Barton and Harold R. Ward

High Resolution Radar by Donald R. Wehner

High Resolution Radar Cross-Section Imaging by Dean L. Mensa

Interference Suppression Techniques for Microwave Antennas and Transmitters by Ernest R. Freeman

Introduction to Electronic Defence Systems by F. Neri

Introduction to Electronic Warfare by D. Curtis Schleher

Introduction to Sensor Systems by S.A. Hovanessian

Logarithmic Amplification by Richard Smith Hughes

Modern Radar System Analysis by David K. Barton

Modern Radar System Analysis Software and User's Manual by David K. Barton and William F. Barton

Monopulse Principles and Techniques by Samuel M. Sherman

Monopulse Radar by A.I. Leonov and K.I. Fomichev

MTI and Pulsed Doppler Radar by D. Curtis Schleher

Multifunction Array Radar Design by Dale R. Billetter

Multisensor Data Fusion by Edward L. Waltz and James Llinas

Multiple-Target Tracking with Radar Applications by Samuel S. Blackman

Multitarget-Multisensor Tracking: Advanced Applications, Yaakov Bar-Shalom, ed.

Over-The-Horizon Radar by A.A. Kolosov, et al.

Principles and Applications of Millimeter-Wave Radar, Charles E. Brown and Nicholas C. Currie, eds.

Principles of Modern Radar Systems by Michel H. Carpentier

Pulse Train Analysis Using Personal Computers by Richard G. Wiley and Michael B. Szymanski

Radar and the Atmosphere by Alfred J. Bogush, Jr.

Radar Anti-Jamming Techniques by M.V. Maksimov, *et al.*

Radar Cross Section Analysis and Control by A.K. Bhattacharyya and D.L. Sengupta

Radar Cross Section by Eugene F. Knott, *et al.*

Radar Detection by J.V. DiFranco and W.L. Rubin

Radar Electronic Countermeasures System Design by Richard J. Wiegand

Radar Evaluation Handbook by David K. Barton, *et al.*

Radar Evaluation Software by David K. Barton and William F. Barton

Radar Propagation at Low Altitudes by M.L. Meeks

Radar Range-Performance Analysis by Lamont V. Blake

Radar Reflectivity Measurement: Techniques and Applications, Nicholas C. Currie, ed.

Radar Reflectivity of Land and Sea by Maurice W. Long

Radar System Design and Analysis by S.A. Hovanessian

Radar Technology, Eli Brookner, ed.

Receiving Systems Design by Stephen J. Erst

Radar Vulnerability to Jamming by Robert N. Lothes, Michael B. Szymanski, and Richard G. Wiley

RGCALC: Radar Range Detection Software and User's Manual by John E. Fielding and Gary D. Reynolds

SACALC: Signal Analysis Software and User's Guide by William T. Hardy

Secondary Surveillance Radar by Michael C. Stevens

SIGCLUT: Surface and Volumetric Clutter-to-Noise, Jammer and Target Signal-to-Noise Radar Calculation Software and User's Manual by William A. Skillman

Signal Theory and Random Processes by Harry Urkowitz

Solid-State Radar Transmitters by Edward D. Ostroff, *et al.*

Space-Based Radar Handbook, Leopold J. Cantafio, ed.

Spaceborne Weather Radar by Robert M. Meneghini and Toshiaki Kozu

Statistical Theory of Extended Radar Targets by R.V. Ostrovityanov and F.A. Basalov

The Scattering of Electromagnetic Waves from Rough Surfaces by Peter Beckmann and Andre Spizzichino

VCCALC: Vertical Coverage Plotting Software and User's Manual by John E. Fielding and Gary D. Reynolds